MW00814693

Next Generation Intelligent Environments

Stefan Ultes • Florian Nothdurft • Tobias Heinroth
Wolfgang Minker
Editors

Next Generation Intelligent Environments

Ambient Adaptive Systems

Second Edition

Editors
Stefan Ultes
University of Ulm
Ulm, Germany

Florian Nothdurft
University of Ulm
Ulm, Germany

Tobias Heinroth
University of Ulm
Ulm, Germany

Wolfgang Minker
University of Ulm
Ulm, Germany

ISBN 978-3-319-23451-9 ISBN 978-3-319-23452-6 (eBook)
DOI 10.1007/978-3-319-23452-6

Library of Congress Control Number: 2015958520

Springer Cham Heidelberg New York Dordrecht London
© Springer International Publishing Switzerland 2011, 2016
This work is subject to copyright. All rights are reserved by the Publisher, whether the whole or part of the material is concerned, specifically the rights of translation, reprinting, reuse of illustrations, recitation, broadcasting, reproduction on microfilms or in any other physical way, and transmission or information storage and retrieval, electronic adaptation, computer software, or by similar or dissimilar methodology now known or hereafter developed.
The use of general descriptive names, registered names, trademarks, service marks, etc. in this publication does not imply, even in the absence of a specific statement, that such names are exempt from the relevant protective laws and regulations and therefore free for general use.
The publisher, the authors and the editors are safe to assume that the advice and information in this book are believed to be true and accurate at the date of publication. Neither the publisher nor the authors or the editors give a warranty, express or implied, with respect to the material contained herein or for any errors or omissions that may have been made.

Printed on acid-free paper

Springer International Publishing AG Switzerland is part of Springer Science+Business Media (www.springer.com)

Foreword

The technological changes we have all witnessed in our lifetime are breathtaking and raise the intriguing question as to what the future will bring. It is hard to believe that the Internet as the public knows it (e.g. the World Wide Web) has been in our homes barely 25 years, but in that short time has embedded itself inseparably into our daily lives radically changing, for example, the nature of commerce, entertainment and education. Even bigger changes beckon with the Internet of Things promising to move the Internet from managing data to controlling the physical fabric of our lives in the form of buildings, transportation, factories and other physical environments. The sheer complexity of such visions, involving ecosystems comprising tens, hundreds or even thousands of collaborating embedded computers and people, is mind-boggling and represents major challenges to technology designers as to how to manage such systems, in ways that ensure people fully benefit from such technology. While there are various views for solutions to this challenge, this book is rooted in the concept of intelligent environments, which takes it inspiration from aspects of human intelligence and how we deal with the management of complex systems, notably by the application of reasoning, planning and learning.

The notion of people being in control of technology is a central tenant of the intelligent environments paradigm which, in turn, has huge implications for the design of the technology. For instance, it has given rise to the mantra frequently voiced by designers of intelligent environment technology, that "the user is king (or queen)", meaning that technology should, with only a few exceptions (e.g. safety), always do what the user requests. In addition, the drive towards humanising technology has led to a desire for the human interface to be as natural as possible by, for example, using voice dialogue or gesture recognition. This strategy has implications for the artificial intelligence which seeks also to adopt a human-friendly form of reasoning (fuzzy logic). The overall hope is to drive technology in a more human amenable direction.

Towards these goals, this book describes research, funded by the European Community's 7th Framework Programme that aimed to investigate how the vision for intelligent environment ecologies comprising people and technology might be best realised. In general terms, the work sought to create context-aware environments that used sensors to perceive the world and applied ontology (and fuzzy logic) to interpret the context and operate environments. As was emphasised earlier, at the core of intelligent environments are the needs of people who do not think in computational terms and value privacy and trust. As a consequence, the work incorporates trust models, fuzzy decision making mechanisms and speech dialogue which, collectively, function in a more human-like way. A particularly noteworthy aspect of the research was the adoption of a novel conceptualisation, a "bubble", that can be viewed as a virtual container of computational resources belonging to an owner and used to achieve tasks while providing clearly defined boundaries to aid the enforcement of privacy requirements.

Beyond being a valuable record of a key European project, the book is an important compendium of knowledge on key areas of intelligent environments. For instance, the book opens with two highly illuminating chapters that discuss network adaptation and middleware issues based on the use of Service-Oriented Architecture, OSGi and UPnP technology, which are especially suitable for intelligent environments which require highly dynamic adaptation and reconfiguration. Knowledge is a key aspect of any intelligent system, and Chap. 3 presents a particularly interesting scheme for mixing and aligning heterogeneous ontologies. Privacy and trust are critical factors for user acceptance of intelligent environments, an issue which is extensively discussed in Chap. 4. Chapter 5 explains how fuzzy logic can form a bridge between the imprecise and often probabilistic reasoning favoured by people and numeric calculations employed in computers. An especially important contribution is the use of type 2 fuzzy logic to facilitate set adaptation to match perception drift, such as the meaning of warm as seasonal weather changes. Chapter 6 provides an in-depth discussion of human-computer interfaces for intelligent environments, making the case that some kind of adaptation in the user interfaces is a vital factor in improving their usability in intelligent environments and presenting a novel adaptable interface, namely, the Interaction Agent. Speech dialogue provides a particularly natural means of interacting with intelligent environments. Chapter 7 describes an approach based on user-centred adaptation which alters the content, flow and structure of the ongoing dialogue using short-term and long-term strategies. Chapter 8 complements the fuzzy rule-based reasoning by introducing a framework for hybrid planning which implements multiple search strategies. Finally, Chap. 9 overviews the key user experience principles, concluding by describing the project outcomes. It is evident, of course, that the journey on researching intelligent environments is far from complete. As was voiced in the opening sentence of this paragraph, predicting the future is challenging, but for those of you interested in ensuring that the direction of travel leads to a better future, you can perhaps take heart from the 1971 Alan Kay quotation: "The best way to predict the future is to invent it", which might be reworded as, while the future cannot be predicted, we can have a significant hand in shaping it. In that respect, this book represents a major milestone in the advancement of intelligent environments, providing a valuable single point resource for anyone interested in getting a comprehensive insight into the main technologies and issues surrounding the development of intelligent environments and the development of a better future. I applaud the authors for their valuable contribution to the field.

Colchester, UK
April 2015

Vic Callaghan

Preface

This book is based on the work that has been conducted within the ATRACO (Adaptive and TRusted Ambient eCOlogies) project[1] as part of the European Community's 7th Framework Programme (FP7/2007–2013) under grant agreement no 216837. The aim of ATRACO project is to contribute to the realisation of trusted ambient ecologies. Interactive appliances, collaborative devices and context-aware artefacts, as well as models, services, and software components, are parts of ambient ecologies. A context-aware artefact, appliance or device uses sensors to perceive its context of operation and applies an ontology to interpret this context. It also uses internal trust models and fuzzy decision-making mechanisms to adapt its operation to changing context. Finally, it employs adaptive dialogue models to communicate its state and interact with people.

Ambient ecologies form the infrastructure that supports user activities. In ATRACO, each activity is modelled as a "bubble" using finite resources to achieve the goals of its owner and having clearly marked borders, which realise the privacy requirements. The user tasks that compose an activity are supported by an ad hoc orchestration of ubiquitous computing services, which are manifested via an ecology of smart artefacts. The bubble adapts to different contexts by renegotiating its borders, adopting suitable interaction modes and employing resource management models. In ATRACO, adaptation will be researched in terms of artefact operation, ecology composition, network election and man-machine interaction with respect to user context and behaviour.

The book edition consists of nine chapters each covering a detailed look on a specific scientific area within the field of intelligent environments. For the second edition, most chapters have been updated to include recent advances in the respective fields. Furthermore, a chapter dealing with user-centred dialogue has been added as human-machine dialogue is an important pillar of communicating with intelligent environments. Because of this, the whole book has been reorganised to better reflect the dependencies of the scientific issues addressed in each chapter.

[1] http://www.uni-ulm.de/in/atraco.

We are convinced that computer scientists, engineers, and others who work in the area of ambient environments, no matter if in academia or in industry, may find the edition interesting and useful to their own work. Graduate students and PhD students specialising in the area of intelligent environments more generally, or focusing on issues related to the specific chapters in particular, may also use this book to get a concrete idea of how far research is today in the area and of some of the major issues to consider when developing intelligent environments in practice. We would like to express our sincere gratitude to all those who helped us in preparing this book. Especially we would like to thank all reviewers who through their valuable comments and criticism helped improve the quality of the individual chapters as well as the entire book.

Ulm, Germany
June 2015

Stefan Ultes
Florian Nothdurft
Tobias Heinroth
Wolfgang Minker

Acknowledgements

The editors would like to thank Vic Callaghan for his interesting foreword that allows the audience to quickly step into the book. We furthermore want to thank the reviewers for their valuable input and for all the objections and suggestions they had. A list of the reviewers can be found in the Appendix at the end of the book.

The research leading to these results has received funding from the European Community's 7th Framework Programme (FP7/2007–2013) under grant agreement no 216837 and from the Transregional Collaborative Research Centre SFB/TRR 62 "Companion-Technology for Cognitive Technical Systems" funded by the German Research Foundation (DFG).

Contents

List of Contributors

Yacine Bellik National Center for Scientific Research (LIMSI-CNRS) BP 133, Orsay cedex, France

Julien Bidot Institute of Artificial Intelligence, Ulm University, Ulm, Germany

Aysenur Bilgin The Computational Intelligence Centre, School of Computer Science and Electronic Engineering, University of Essex, Colchester, UK

Susanne Biundo Institute of Artificial Intelligence, Ulm University, Ulm, Germany

Dimitris Economou inAccess, Maroussi, Athens, Greece

Christos Goumopoulos The Hellenic Open University, Patras, Hellas

Hani Hagras The Computational Intelligence Centre, School of Computer Science and Electronic Engineering, University of Essex, Colchester, UK

Tobias Heinroth Institute of Communications Engineering, Ulm University, Ulm, Germany

Achilles Kameas The Hellenic Open University and DAISy Research Unit, Patras, Hellas

Bastian Könings Institute of Media Informatics, Ulm University, Ulm, Germany

Ioannis Liverezas inAccess, Maroussi, Athens, Greece

Apostolos Meliones Department of Digital Systems, University of Piraeus, Piraeus, Greece

inAccess Networks S.A., Maroussi, Athens, Greece

Wolfgang Minker Institute of Communications Engineering, Ulm University, Ulm, Germany

Florian Nothdurft Institute of Communications Engineering, Ulm University, Ulm, Germany

Gaëtan Pruvost National Center for Scientific Research (LIMSI-CNRS) BP 133, Orsay cedex, France

Florian Schaub School of Computer Science, Carnegie Mellon University, Pittsburgh, PA, USA

Lambrini Seremeti The Hellenic Open University and DAISy Research Unit, Patras, Hellas

Stefan Ultes Institute of Communications Engineering, Ulm University, Ulm, Germany

Joy van Helvert University of Essex, Colchester, UK

Christian Wagner The Computational Intelligence Centre, School of Computer Science and Electronic Engineering, University of Essex, Colchester, UK

Michael Weber Institute of Media Informatics, Ulm University, Ulm, Germany

Acronyms

AA	Artefact Adaptation
AE	Ambient Ecology
AS	Activity Sphere
DM	Dialogue Manager
FTA	Fuzzy Task Agent
IA	Interaction Agent
IE	Intelligent Environment
LO	Local Ontology
NA	Network Adaptation
OM	Ontology Manager
PA	Planning Agent
PM	Privacy Manager
POMDP	Partially Observable Markov Decision Process
SA	Sphere Adaptation
SM	Sphere Manager
SO	Sphere Ontology
TM	Trust Manager
UBA	User Behaviour Adaptation
UIA	User Interaction Adaptation
UPnP	Universal Plug and Play
UX	User Experience

Chapter 1
A Middleware Architecture for Ambient Adaptive Systems

Christos Goumopoulos

Abstract Ambient adaptive systems have to use mechanisms to regulate themselves and change their structure in order to operate efficiently within dynamic ubiquitous computing environments. First of all we outline a survey on existing middleware solutions for building ambient adaptive systems. After, discussing the limitations of the existing approaches, we present our propositions for a middleware architecture to support dynamic adaptation within ambient environments. Our approach is based on the service-oriented architecture (SOA) paradigm which can be considered as an evolution of the component-based design paradigm. The aim is to use component interfaces for the identification and automated connection of components acting as service providers/consumers. The proposed middleware provides a solution that supports the adaptation of applications at the structural level, where the structure of the application can change through dynamic service composition. We call this adaptation "polymorphism" in analogy with the synonymous term found in the object-oriented programming paradigm. Besides SOA, we use a set of intelligent agents to support adaptive workflow management and task realization based on a dynamically composed ontology of the properties, services, and state of the environment resources. An experimental prototype is provided in order to test the middleware developed.

1.1 Introduction

Intelligent environments (IE), like smart homes, offices, and public spaces, are featured with a large number of devices and services that help users in performing efficiently various kinds of tasks. Combining existing services in pervasive computing environments to create new distributed applications can be facilitated by middleware architectures, but this should accommodate special design considerations, including context awareness, adaptation management, device heterogeneity, and user empowerment [6].

C. Goumopoulos (✉)
Research Academic Computer Technology Institute, Patras, Greece
e-mail: goumop@cti.gr

© Springer International Publishing Switzerland 2016

S. Ultes et al. (eds.), *Next Generation Intelligent Environments*,
DOI 10.1007/978-3-319-23452-6_1

Traditional middleware, such as Remote Procedure Calls [4], OMG CORBA [10], Java Remote Method Invocation (RMI) [44], and Microsoft Distributed Component Object Model (DCOM) [19] facilitate the development of distributed applications and help to resolve problems such as tackling the complexity of programming inter-process communication and the need to support services across heterogeneous platforms. However, traditional middleware is limited in its ability to support adaptation.

Ambient adaptive systems which are a special category of distributed systems operate in a dynamic environment. The dynamicity of the environment may relate with evolving user requirements and varying execution context due to the diversity of available devices, user preferences, and services. Consequently there is a need for both applications and infrastructure to be designed for change. The evolution of user requirements calls for system evolution. The dynamic execution environment calls for dynamic adaptation. In order to allow evolution, the internal structure of the system must be made open in order to support proactive and reactive system reconfiguration.

In this work, we present firstly a survey of the state-of-the-art on existing middleware solutions for building adaptive ambient systems. After, discussing the limitations of the existing approaches, we present our propositions for middleware architecture to support dynamic adaptation within ambient environments. Our approach uses the service-oriented architecture paradigm coupled with agents and ontologies. The aim is to use component interfaces for the identification and automated connection of components acting as service providers/consumers. The proposed middleware provides a solution that supports the adaptation of applications at the structural level, where the structure of the application can change through dynamic service binding. Behavioral adaptation, not examined here, is also possible when the application logic is changed as a result of learning. An experimental prototype is provided in order to test the middleware developed.

1.2 Related Work

Three key paradigms that can be used to build adaptive systems are computational reflection, aspect-oriented programming (AOP), and SOAs. Researchers have also explored the possibility to combine different paradigms such as AOP and reflection in middleware systems to increase support for the development of dynamic distributed systems [18]. In the following we examine how each one of these paradigms can support the development of adaptive systems.

1.2.1 Reflective Middleware

In principle *computational reflection* allows a program to observe and modify its own structure and behavior at runtime, by providing a self-representation that can be accessed and changed by the program [31]. More importantly, these changes must be causally reflected to the actual computations performed by the program. In particular, the architecture of reflective systems follows a kind of "white-box" approach that provides comprehensive access in the internal details of a system allowing dealing with highly dynamic environments, for which runtime adaptation is required. This is conceptually contrary to the encapsulation principle in object-oriented programming followed by traditional middleware that adopts the remote object model. The reflection technique was used initially in the domain of programming languages as a means to help designing more open and extensible languages. The reflection is applied also in other domains including operating systems and distributed systems. Recently, reflection has been also applied in middleware, which needs to adapt its behavior to changing requirements when operating in a dynamic environment [27]. The dynamic modification of the middleware implementation allows for the adaptation of the behavior of distributed applications that are based on this middleware. Typically, reflective middleware provides adaptation of behavior of distributed applications in terms of non-functional requirements such as QoS, security, performance, and fault tolerance.

A reflective system is organized into two levels called *base-level* and *meta-level*. The former represents the basic functionality of the system. The latter models the structural and computational aspects of the base-level in order to observe or modify the behavior of the objects that exist in the base-level, assuming an object-oriented system. The reflective approach supports the inspection and the adaptation of the underlying implementation (base-level) in runtime. A reflective system provides a *meta-object protocol* (MOP) in order to determine the services that are available in meta-level and their relationship to the base-level objects [25]. The meta-level can be accessed via a process called reification. Reification is the disclosure of certain hidden aspects of the internal representation of the system in terms of programming entities that can be manipulated at runtime. The "opening of" implementation offers a simple mechanism in order to interpose some behavior (e.g., add a method in an object, save the state of the object, check security issues, etc.) with a view to watch or alter the internal behavior of the system.

Reflection enables an application to adjust its behavior based on a reflective middleware that allows inspecting and adapting its own behavior according to application's needs. Figure 1.1 shows schematically a simple example of this situation. A meta-object (`mObj`) has been defined at the meta-level and is associated through MOP to a base-level object (`Obj`) belonging to some middleware implementation. An application calls a method of the middleware (`Obj.Method()`) which is reified through MOP and a defined object with a specified reference (`ref`). This object is passed to the associated meta-object that executes a method (`mObj.mMethod(ref)`). This method executes a logic specified by the MOP

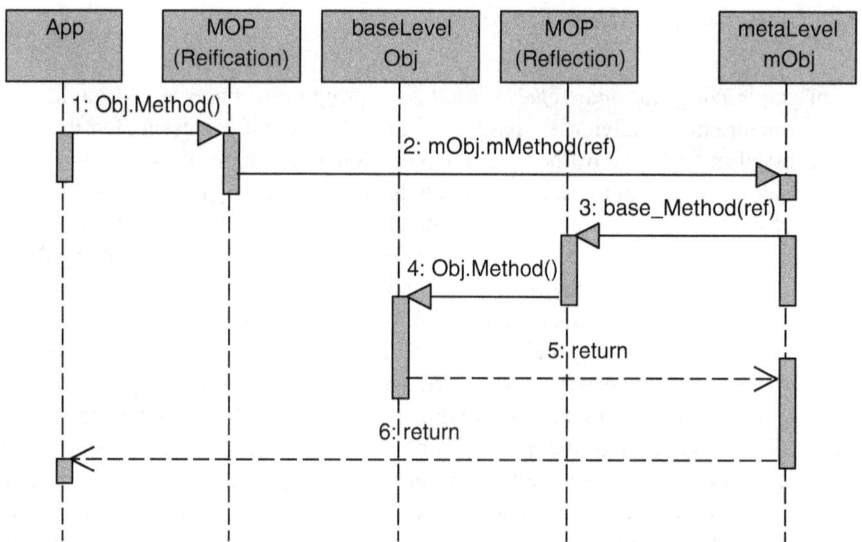

Fig. 1.1 An application calling a method in a reflective middleware

interface and then passes control to the original method through a reflection method (`base_Method(ref)`). The meta-object receives the results, performs any postprocessing specified by the MOP and returns to the calling application.

A category of reflective systems support a higher level reflection in the sense that they can add or remove methods from the objects and classes dynamically and even change the class of an object in runtime. The practical result is to be able to restrict the size of middleware with a minimal total of operations that can run in devices with limited resources. On the contrary, other systems are focused in simpler reflective forms in order to achieve a better performance. Their reflective mechanisms are not part of the normal flow of control and are only called when needed.

Middleware systems that integrate reflection in their architecture have been developed as research prototypes. In the following we cite a few examples of reflective middleware. A number of early systems such as FlexiNet [21], OpenCorba [29], dynamicTAO [26], and OpenORB [5] were based on CORBA and targeted flexibility and dynamic reconfigurability of Object Request Broker (ORB). However, these systems suffered from the heavy computational load imposed by CORBA. Capra et al in [7] discuss CARISMA, which uses reflection to support dynamic adaptation of middleware behavior to changes in context (e.g., adapting a streaming encoder binding in variable QoS conditions) and ReMMoC, which uses reflection to handle heterogeneity requirements imposed by both applications and underlying device platforms. Both approaches target a minimal reflective middleware for mobile devices where pluggable components can be used by developers to specialize the middleware to suit different devices and environments, thus solving heterogeneity issues. QuA middleware explores the principle of mirror-based reflection to design a reflective API according to the programming abstractions defined by a language [13]. In the QuA middleware approach a mirror can be defined to reflect a service, in terms of middleware abstractions like type, interface, service, and binding, without being dependent the running instances.

Even though reflection is a powerful mechanism to construct adaptive systems there are still issues that need to be understood and solved. The performance of reflective middleware is a matter that is open for further research. The majority of reflective systems impose a rather heavy workload that would cause significant performance deterioration in devices with limited resources and there is always a trade-off issue between performance and scope of adaptability. Another issue that needs to be addressed is dynamically tuning the scope of changes when reconfiguring the system based on adaptation semantic information.

1.2.2 Aspect-Oriented Programming

AOP [23] is a software development paradigm that emphasizes decomposition of complex programs in terms of intervened cross-cutting aspects, such as QoS, security, persistence, fault tolerance, logging, and resource utilization. This is different from other programming paradigms which emphasize functional decomposition breaking a problem into units like procedures, objects, and modules. For instance, object-oriented programming uses inheritance hierarchies to abstract commonalities among classes, however, global aspects (affect many classes) are implemented in an ad-hoc manner and become tightly intermixed with classes, which makes changes to the program difficult and error prone. On the other hand, AOP supports the concept of separation of concerns to counter this problem. AOP defines the methods and tools to separate cross-cutting aspects during development time. The source program consists of modules that deal with the different aspects that are described independently. All these modules are integrated during compile (statically) or runtime (dynamically) to form a global application with new behavior using a composition tool called *aspect weaver*.

AOP combines principles of object-oriented programming and computational reflection discussed in the previous section. AOP languages have functionality similar to, but more restricted than MOPs and use a few key concepts: *join points*, *point cuts*, and *advices*. A *join point* is a place, in the source code of the program, where aspect-related code can be inserted. A join point needs to be addressable and understandable by an ordinary programmer to be useful. It should also be stable across typical program changes in order for an aspect to be stable across such changes. Aspect weaving relies on the concept of *point cut*, i.e., the specification of a set of join points according to a given criterion, and *advice*, i.e., the definition of the interaction of the inserted code with the base program. An advice specifies whether the inserted code should be executed before, after, or in replacement for the operations located at the point cuts. Two Java-based composition tools that implement the AOP paradigm are AspectJ [24] and JAC [34].

AOP benefits outlined above are important to adaptive middleware. Such approach enables the separation of middleware cross-cutting concerns (e.g., security, logging) at development time and later at compile or runtime, where these concerns can be selectively woven into application code. Using AOP, tailored versions of middleware can be generated for application-specific domains. In the following we cite a few examples of adaptive middleware developed based on AOP principles. Yang et al in [45] proposes a systematic approach for preparing an existing program for adaptation and defining dynamic adaptations. The approach uses a static AOP weaver at compile time and reflection during runtime. This basic scheme has been followed by others researchers also. Frei et al in [12] present an architecture supporting dynamic AOP that establishes an event infrastructure to extend existing application's behavior at runtime. When the application extension is activated, the dynamic AOP platform inserts an AOP aspect into the AOP platform which intercepts the application's execution and monitors its progress. Whenever the application reaches selected points in the execution, the AOP platform redirects the execution to the appropriate application extension. The memory footprint of the platform is however quite heavy (1MB) to run on resource-constrained devices. Similarly, Maciel da Costa et al in [30] discuss an adaptive middleware architecture, based on aspects, which can be used to develop adaptive mobile applications. A mail server prototype was implemented based on Web Services, Java, and AspectJ technologies to evaluate the architecture regarding operation adaptation depending on resource utilization (e.g., power consumption).

AOP has advantages, such as separation of cross-cutting concerns, but presents also difficulties. Programming in terms of aspects requires much more than just identifying the different aspects of concern. It requires being able to express those aspects of concern in a way that is precise and that makes the relations among the aspects of concern precise. This is what enables the aspect weaver to work, and is also what makes possible reasoning about the code or debugging the code. Another important problem related to AOP is the composition of aspects. For instance, if different pieces of aspect-related code are inserted at the same join point, the order of insertion may be relevant if the corresponding aspects are not independent. Such issues cannot usually be settled by the weaver and call for additional specification. Finally, most AOP approaches do not support adequate point cut descriptions to capture join points based on context data and business-level semantics.

1.2.3 Service-Oriented Architecture

The SOA paradigm has been envisioned as an evolution of the component-based engineering paradigm centered on the concept of service [11]. This can be applied in the design of distributed applications that are seen as a composition of services. In addition, the service concept can be applied recursively, since a system component can provide a service, but simultaneously it can encapsulate a composition of services from its service requestors.

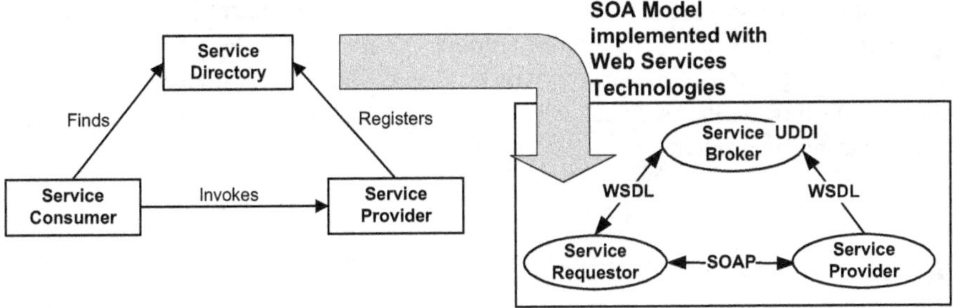

Fig. 1.2 SOA conceptual model

In an SOA environment, resources on a network are made available as independent services that can be accessed without knowledge of their underlying platform implementation. A provided service is usually embodied in a set of interfaces, each of which represents an aspect of the service. In general this set contains the operations that a service supports, and some information on how to access these operations. Service interfaces can be published in registries, which also provide services themselves (publish and discovery services), allowing the potential service requestors to discover and access these services (Fig. 1.2).

The independent deployment of services enables late binding which is essential for adaptive systems. *Late binding* provides the opportunity for dynamic composition of services or for swapping two compatible services at runtime through a well-defined interface. In the SOA paradigm, we can view two abstraction levels of the service concept. *Elementary services* are basic functionalities, usually provided by resources (e.g., devices) in an AmI environment. *Composite services* assemble a set of functionalities in relation to user tasks, and thus are closer to user actual goals. Service composition has widely been addressed in the Web Service field. Existing composition frameworks [8] enable expressing and enacting complex service compositions. However, they rely on explicitly named services, which are not discovered dynamically. On the contrary, the Semantic Web Services (SWS) approach [32] is a step toward dynamic service discovery and composition [9, 40], where intelligent systems try to build composite services from abstract user requirements with or without manual selection of services. SWS leverage knowledge representation techniques, with ontologies describing a domain in a formal manner, and AI planning methods to make composition systems more autonomous.

Although, Web Services are a key implementation technology of the SOA paradigm, the main standards defined to implement the SOA paradigm (i.e., WSDL, UDDI, SOAP) emphasize interoperability rather than the capability to accommodate seamless changes at runtime. Frameworks based on ontologies, such as METEOR-S [42], also lack flexible mechanisms for the distribution of information about services as they require the adoption of shared ontologies that impose the distribution policy. Regarding composition, Business Process Execution Language (BPEL) is the de-facto standard [1]. It takes a workflow-oriented approach to the coordination of cooperating services and provides a good solution for the design-time composition of heterogeneous components wrapped as WSDL services. However, runtime identification of partner services is not addressed and thus the degree of dynamism and flexibility is limited.

1.2.4 Overview

Based on the above discussion we give in the following table an overview of the relative advantages/shortcomings of the three middleware design paradigms regarding their support to the development of adaptive systems (Table 1.1).

Table 1.1 Pros and cons of the three middleware design paradigms

Paradigm	Pros	Cons
Reflective middleware	• A system can modify its structure and behavior at runtime • Achieves variable system size to suit different devices and environments • Changes made to the self-representation are immediately mirrored in the underlying system's actual state and behavior (causally connected)	• Conceptually contrary to the encapsulation principle • Usually heavy computational load and low performance • Dynamically tuning the scope of changes based on adaptation semantic information is still an open issue
Aspect-oriented programming	• Supports the concept of separation of concerns • Separates cross-cutting aspects during development time (e.g., security, logging) • Combines principles of object-oriented programming and computational reflection	• May give large memory footprint • Programming in terms of aspects is not easy • The order of insertion may be relevant if the corresponding aspects are not independent (composition of aspects) • Does not support adequate point cut descriptions to capture join points based on context data and business-level semantics
Service-oriented architecture	• Modular design appropriate for adaptation and reconfiguration • Late binding of services • Composite services can be defined from simple ones • Main standards defined to implement SOA provide support for interoperability	• Automatic service composition is not trivial • Main standards defined to implement SOA provide limited capability to accommodate seamless changes at runtime • Limited degree of dynamism and flexibility

In this work, we describe an approach based on the SOA paradigm. Besides SOA a novel mechanism is proposed to achieve different kinds of adaptation centered upon the management of knowledge, which is encoded in multi-layered ontologies, which are used by intelligent agents.

1.3 ATRACO Architecture

Ambient intelligence (AmI) is a paradigm that puts forward the criteria for the design of the next generation of UbiComp environments [37]. In this context we have introduced the *Ambient Ecology* (AE) metaphor to conceptualize a space populated by connected devices and services that are interrelated with each other, the environment and the people, supporting the users' everyday activities in a meaningful way [14].

In the context of the EU funded R&D project ATRACO [16] we aim to extend the AE concept by developing a conceptual framework and a system architecture that will support the realization of adaptive and trusted AEs which are assembled to support user goals in the form of *Activity Spheres* (ASs). Our approach is based on a number of well-established engineering principles, such as the distribution of control and the separation of service interfaces from the service implementation, adopting a SOA model combined with intelligent agents and ontologies. Agents support adaptive planning, task realization, and enhanced human–machine interaction while ontologies provide knowledge representation, management of heterogeneity, semantically rich resource discovery, and adaptation. ATRACO ASs are dynamic compositions of distributed, loosely coupled, and highly cohesive components that operate in dynamic environments.

Therefore the architecture and the system we propose operate in an AmI environment, which is populated with people and an AE of devices and services. Our basic assumption is that the AE components are all autonomous, in the sense that (a) they have internal models of their properties, capabilities, goals, and functions, and (b) these models are proprietary and "closed", that is, (a) they are not expressed in some standard format and (b) they can only be changed by the owner components. However, each component can be queried and will respond using a standardized protocol.

1.3.1 ATRACO World Model

The concepts discussed below constitute a critical subset of the ATRACO conceptual framework defined for building AmI applications (Table 1.2).

The basic terms and concepts of the ATRACO world model are encoded in the ATRACO Upper Level Ontology (ULO). In general, ontology is used as the means to share information among heterogeneous parties in a way that is commonly

Table 1.2 ATRACO main concepts and corresponding descriptions

Concept	Description
Ambient Ecology (AE)	The set of heterogeneous artifacts with different capabilities and provided services that reside within an intelligent environment (IE).
Activity Sphere (AS)	It is formed to support an actor' specific goal. An AS represents both the model and the realization of the set of information, knowledge, services, and other resources required to achieve an individual goal within an IE. The concept of AS is a "digitization" of the concept of "bubble" used by the psychologist Robert Sommer [39] to describe a temporary defined space that can limit the information coming into and leaving it.
Intelligent environment (IE)	A territory that has both physical properties and offers digital services. It is the container of AE. ASs are instantiated in an IE using the resources provided by its AE.
Artifact	A tangible object which bears digitally expressed properties; usually it is an object or device augmented with sensors, actuators, processing, networking unit, etc., or a computational device that already has embedded some of the required hardware components.
Actor	Any member of AE capable of setting and attaining goals by realizing activities. Within the AE actors are users or agents.
Goal	Each actor may have its own set of goals and plans to achieve them. A goal is described as a set of abstract tasks, which is described with a task model.
Task model	It may be *abstract* or *concrete*. An abstract task model describes what should be done, without details of how it should be done or by the use of what kind of modality; these are described in the corresponding concrete model. The abstract task model may also contain several decomposition rules modelled as a set of subtasks.
Local Ontology (LO)	Each member of the AE stores locally descriptions of its properties, services, and capabilities. It is a sub-class of the class Ontology.
Sphere Ontology (SO)	The SO results from the LO of those AE members that are required to achieve the AS's goal based on the resolution of its task model. Apart from device and service ontologies, it may contain user profiles, agent rule bases, and policies. It is another sub-class of the class Ontology.
Agent	A software module (is a kind of actor) capable of pursuing and realizing plans in order to achieve specific goals based on tasks. It includes three types of agents: Task agent (e.g., Fuzzy systems based Task Agent or FTA), who manipulates sensors and actuators in order to realize specific tasks; *Planning Agent* (PA), who resolves an abstract task hierarchy into concrete tasks using the resources of the AE; and *Interaction Agent* (IA), who manages user–system interaction using a mixed-initiative dialogue model.
User	The actor that uses the available services and devices in order to perform a task. When a user performs a task, this can be subdivided into different activities. Users use devices, which provide them with services. Devices run these services in a physical environment (context). Users use these services according to personal conditions (user profile) and within a physical context.
Aim	It is attributed to a user; it is decomposed into a set of interrelated goals, which are distributed to the components of the AS.

(continued)

Table 1.2 (continued)

Concept	Description
Policy	Actors specify high-level rules for granting and revoking the access rights to and from different services. Examples of policy ontologies are privacy policy ontology, interaction ontology, and conflict resolution policy ontology.
Service	The entity which describes the service offered by a device.
Device	The entity that has physical/digital properties and offers a specific service.
Resource	A resource can be the *space*, an *entity*, or a *component*, such as managers (e.g., Ontology Manager, Sphere Manager) or other basic components.

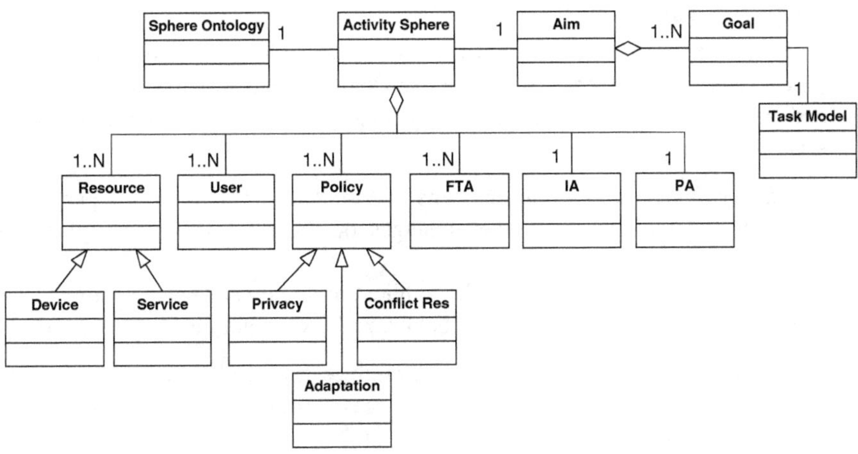

Fig. 1.3 Activity Sphere domain model (part of ATRACO ULO)

understood [20]. An ontology is a network of concepts and entities, which can be associated with different types of relations (the most common being the hierarchical association, or is a relation). More concrete (or domain) ontologies contain also instances of these entities with specific properties and values. More powerful ontologies contain constraints and rules that cause inferences for the entities. Figure 1.3 illustrates in UML representation the AS domain model which is also encoded as ontology in the ATRACO ULO.

1.3.2 System Requirements

For the requirement analysis and design of the ATRACO architecture we followed a process where initially application scenarios were defined and application requirements from a user perspective were identified. In addition as a separate process we defined high-level requirements on system perspective. Then initial requirements were used as input for a process of abstraction that allowed identifying a set of challenges that the architecture has to address, in order to frame further design [16]. These challenges are organized in the following categories:

Challenge 1: Assemble/Dissolve The first challenge has naturally to do with the formation and the dissolution of ATRACO applications (ASs). ASs encapsulate the ambient ecology resources that are necessary to serve the goal for which the AS has been created. ATRACO supports adaptation and trust requirements of ASs by integrating into the AS services of the system components that develops.

Challenge 2: Adaptability Adaptability implies that an AS should attempt to continuously provide expected behavior by adapting to unexpected conditions such as changes to the resources constituting the system or changes in the behavior of the user.

To this effect, the ATRACO system components are defined that adapt task-based usage of the sphere to the changing user behavior, environment conditions, and context. ATRACO implements mechanisms that support adaptability in several forms:

- *Activity Sphere adaptation*, in terms of *Structural adaptation*: the persistent achievement of the goal when changes on the type or cardinality of the available resources occur.
- *Behavioral adaptation*, where the application logic is changed as a result of changes in the user and/or device behavior. This category is specialized as:
  - *Artifact adaptation* (the system examines how an artifact can adapt its model of operation in reaction to changes in the device characteristics, e.g., handling a partial failure of a heater) and
  - *User behavior model adaptation* (an agent will learn and adapt its rule base to face the changes in the user desires and preferences by monitoring the user actions, e.g., the user decides to read in bed, therefore requires her bedside lamp to be on instead of her reading light).
- *User interaction adaptation* specifies adaptation interacting with the user using different devices/modalities depending on available resources, environment characteristics, tasks, and user profile.
- *Network adaptation* to allow the uniform and transparent access to devices and services present in the networked environment supporting the realization of Activity Spheres across a mixture of heterogeneous networks.

Challenge 3: Semantic heterogeneity A basic assumption is that an AmI space is available to host an ambient ecology, and devices and services are inherently heterogeneous and contain heterogeneous descriptions of their capabilities and services in the form of local ontologies. Thus, in order to achieve collaboration among them, firstly one has to deal with these forms of heterogeneity. However, the issue raised by the heterogeneity of ontologies and how to achieve semantic interoperability between systems using different ontologies is a challenge.

In ATRACO, the approach that is followed in order to address this challenge is to research, develop, and test theories of ontology alignment to achieve task-based semantic integration of heterogeneous devices and services.

To this effect, a Sphere Ontology is defined and an Ontology Manager administrates its use by performing ontology updating, ontology querying, and ontology matching services.

Challenge 4: Trustworthiness The interactions in the Activity Sphere should be trustworthy. The ambient ecology will behave in a dependable manner and will not adversely affect information, other components of the system or people.

To this effect, policies and rules are defined in the ontology and mechanisms are defined for the management of the identity of service requestors and service providers as well as access control on services and context information.

Summarizing the above discussion, ATRACO architecture should provide:

- Support for realizing user goals (Activity Spheres), by resolving abstract tasks to a workflow of concrete tasks;
- Support for executing workflows by applying service composition and control policies in the form of rules (obligation policies);
- Support for establishment and management of associations between service clients and service providers (as described in task workflows);
- Support for maintaining the Sphere Ontology which contains the contextual knowledge necessary to realize the concrete tasks;
- support for ontology alignment and lookup;
- Support for adaptation of the given tasks according to the user desires and behavior (personalization and learning over-time);
- support for use of heterogeneous network capabilities for communication (network adaptation);
- Support for discovery of services, devices, networks, and resources;
- Support for usage of services offered within ATRACO infrastructure or by third parties (e.g., external Web Services);
- Support for privacy enforcement and access control through policies;
- Support for the possibility of adapting the user interaction depending on available interactive devices and objects;
- Support for management of user profiles and preferences;
- Support for gathering, processing, and distribution of context information.

In the next section we outline the ATRACO system design that accommodate the system requirements and then we discuss in more detail the service composition framework for deploying adaptive workflows in IEs to achieve structural adaptation of ATRACO applications, which is the focus of this presentation.

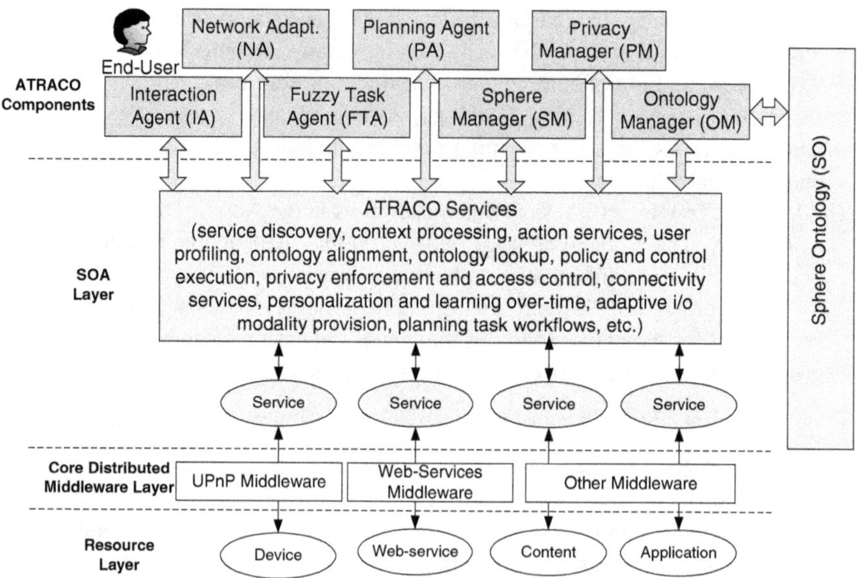

Fig. 1.4 ATRACO architecture

1.3.3 System Design

In ATRACO, we propose a combination of the SOA model with agents and ontologies (Fig. 1.4). We adopt SOA both at the resource level to integrate resources, such as devices, sensors, and context in applications and at the system level to combine ATRACO services that provide adaptation and trust features into applications. ATRACO aims to empower users with the ability to interact in environments with many resources such as devices (UPnP devices), Web Services, content (music/video file, contacts) and applications (e.g., media player) using adaptive user interfaces. The functionality in these environments is exposed as semantically rich services which an actor (either a user or an agent) can discover and then compose to form ATRACO Activity Spheres.

Each service is associated with at least one semantic description which shields the actor from the complexity of the Resource Layer realization and makes it easy for the actor to employ these services in accomplishing interesting and useful tasks. Figure 1.4 shows a conceptual layered view of the ATRACO architecture. The ATRACO infrastructure consists of SOA services. On the one hand, these context-aware services are built on “Core distributed middleware” and rely on “Network and resources” layer. On the other hand, the ATRACO infrastructure supports basic services such as context management and reasoning, communication management, user profiling and service (discovery), as well as adaptation and privacy services that form the basis for ATRACO systems (i.e., ASs).

The ATRACO architecture consists of ontologies, active entities, passive entities, and the user who as the occupant of the IE is at the centre of each AS. Active entities are agents and managers. The role of the ATRACO agents is to provide task planning (Planning Agent or PA), adaptive task realization (Fuzzy systems-based Task Agent or FTA) and adaptive human–machine interaction (Interaction Agent or IA). The PA encapsulates a search engine that exploits hierarchical planning and partial-order causal-link planning to select atomic services that form a composite service (workflow) [3]. One or more FTAs oversee the realization of given tasks within a given IE. These agents are able to learn the user behavior and model it by monitoring the user actions. The agents then create fuzzy-based linguistic models which could be evolved and adapted online in a life learning mode [43]. The IA provides a multimodal front end to the user. Depending on a local ontology it optimizes task-related dialogue for the specific situation and user [35]. The IA may be triggered both by the FTA and by the PA to retrieve further context information needed to realize and plan tasks by interacting with the user. On the other hand, ontologies complement agents regarding adaptation by tackling the semantic heterogeneity that arises in IEs by using ontology alignment mechanisms to generate the so-called, Sphere Ontology (SO). There are two main kinds of ontologies: local ontologies, which are provided by both active and passive entities and encode their state, properties, capabilities, and services and the SO, which serves as the core of an AS by representing the combined knowledge of all entities [38].

The Sphere Manager (SM) and Ontology Manager (OM) components are responsible for the formation, adaptation, and evolution of the user applications (modelled in ATRACO as ASs) and will be further examined in this paper. In the current version of the system there is also a Privacy Manager (PM) that provides a set of privacy enhancing techniques in order to support privacy in an adaptive and individualized way. Finally, devices in the IE that may come from heterogeneous networks (e.g., LonWorks, ZigBee, Z-Wave, etc.) and services (e.g., Network Time, VoIP, Real Time Streaming, etc.) are accessed transparently through a service representation layer exporting them to the ATRACO clients as UPnP services. This layer is implemented in the network adaptation (NA) component [33].

1.4 Adaptive Workflows and Structural Adaptation

In many respects, a composite service can be modelled as a workflow [36]. The definition of a composite service includes a set of atomic services together with the control and data flow among the services. Similarly, a workflow is the automation of a business process, in whole or part, during which documents, information, or tasks are passed from one participant to another for action, according to a set of procedural rules [22]. Workflows have been used to model repeatable tasks or operations in a number of different industries including manufacturing and software. In recent years, workflows have increasingly used distributed resources and Web Services through resource models such as grid and cloud computing. In this section, we argue that workflows can be used to model how various services should interact with one another as well as with the user in IEs depending on available resources, environment characteristics, user tasks, and profile.

In this section we describe how SOA can support AS adaptation. The structural adaptation (a form of polymorphism) is possible because the workflow model represents abstract services and binding to real devices can be accomplished at runtime. ATRACO-BPEL, a streamlined version of BPEL, has been defined as the specification language to describe workflows of abstract services.

1.4.1 Scenarios

In order to test our framework and to illustrate how workflows can be used to fit user interaction with an IE, as well as the structural adaptation mechanism of ASs, we use two simple scenarios. The first example corresponds to an AS that supports the realization of goal named *"Feel comfortable upon arrival at home."*

Martha arrives at the door of her smart apartment. The system recognizes her, through an RFID card, and opens the door. On entering the space the system greets Martha by saying "Welcome home" and then when she has entered the living space the lights and A/C are switched on and brightness and temperature are automatically adjusted according to her profile, season, and time of day, to make her feel comfortable. Martha then sits at the sofa to relax and after a while, the system asks "Would you also like some music?" Martha responds positively and the music plays (according to predetermined preferences). Following this, the system asks "Would you like to view yesterday's party photos?" Martha responds positively and a rolling slide show appears in a picture frame in front of her. After a while, Martha gets up, walks towards the window and opens it. Fresh air pours into the room. Temperature level drops. Brightness level increases. Some of the lights are automatically switched off, in an attempt to maintain the previous level of brightness in the room. After a while, the A/C is switched off because of the open window. Suddenly, the picture frame goes off! The system finds a proper replacement and as a result, photos are displayed in the TV set, while Martha is informed on the event.

The second example corresponds to an AS that supports the realization of goal named *"Studying AS."*

Suppose that John is using a number of objects to support the studying activity at his writing desk, according to his profile (his preferable level of light, temperature, etc.). In this case, John has set as a goal to study. This goal can be decomposed in a hierarchy of abstract tasks that constitute a task model for the goal: sit on a chair, move the chair in proximity to the desk; take the book; place the book on top of the desk; turn on the light. In the AmI environment an AS is formed to support the specific goal, by using four artifacts, a lamp, a chair, a desk and a book. The application logic can be stated as follows: when the chair is occupied and it is near the desk and the book is open on the desk, the lamp is turned on (reading activity has been inferred). The implementation of such a task specification can be represented as a graph of connected services provided by the artifacts.

Furthermore, John can move in the room and change his reading spot at the sofa. This causes an adaptation in the configuration of the Studying AS since a new artifact (sofa) is added and one is removed (desk). Another implication of this mobility is that the light service will adapt to the new reading spot. While reading at the desk the desk light is used, and when he moves to the sofa the lamp near the sofa is used. This implies that device selection for instantiating/adapting an AS depends on user location.

Since workflows are essentially graphs of activities, it is useful to express those using UML activity diagrams. Figure 1.13 describes the sequence of activities for the example scenario. Note that the tasks "AdjustLights," "AdjustAC," "ShowPhotos," and "PlayMusic" can run in parallel and therefore they have been enclosed in a fork-join block. Note also that the exception events are not part of the workflow description but they are handled by the corresponding ATRACO active entities.

1.4.2 Late Binding

We have developed a service composition mechanism which includes three phases: *task workflow planning*, *dynamic service binding*, and *execution management and control* as illustrated in Fig. 1.5.

The planning problem can be stated as "discover an execution path of services (tasks) given some state of the world to achieve a goal." In ATRACO, we use a library of abstract plans which model specific user goals. An abstract plan contains a sequence of abstract services which are actually ontological descriptions of service operations that cannot be directly invoked, but will be resolved by the SM during runtime. Having an abstract service workflow description, which is given in a

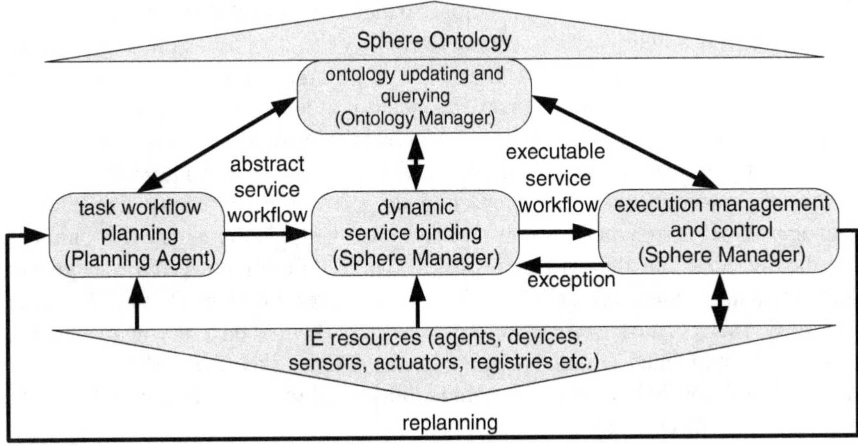

Fig. 1.5 Late service composition process in ATRACO

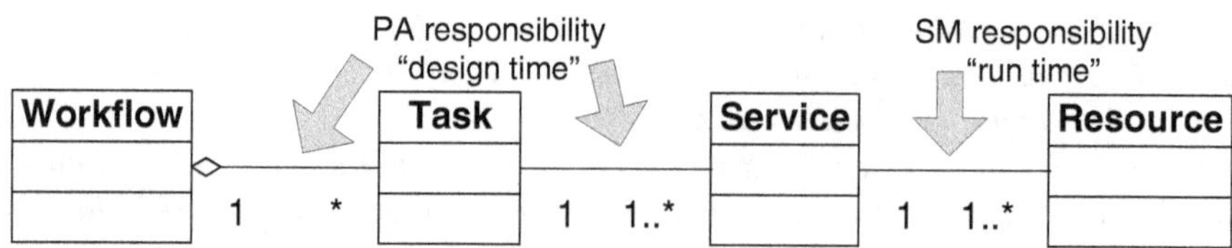

Fig. 1.6 Conceptual model for dynamic service binding

BPEL-like language, the dynamic service binding module of the SM applies a semantic-based discovery mechanism and uses information about available services and context to discover suitable services or devices in registries able to perform each abstract service. The output of this process is an executable service workflow. In the execution management and control phase the SM executes and continuously monitors the deployed services and the termination condition of the workflow.

This adaptation has been inspired by the subtype polymorphism found in the object-oriented programming paradigm [2]. The concept is that we can adapt the instantiation of the AS to different environments provided that a late binding mechanism is in place that determines the exact resources that will be used in the AS (i.e., the specific artifacts; "the lamp in the corner"). The different resources that may be involved only need to present a compatible interface to the clients (i.e., in our case, a UPnP interface). Figure 1.6 gives a conceptual view of the dynamic service binding process. A workflow is mapped into a number of tasks and a workflow task is mapped into one or more abstract services. In addition, each service would also require certain physical resources for its implementation. Mapping of the task to the services can be specified at design time by the PA as per users' functional requirements. However, mapping of the service to the actual human and physical resources is done at runtime, in keeping with service orientation. This dynamic binding is therefore dependent on the context in which the binding occurs.

In the absence of the Sphere Ontology, which has not been yet instantiated, the SM implements a lightweight Resource Discovery Protocol for artifacts or eEntities (eRDP) where the term resource is used as a generalization of the term service. eRDP is a protocol for advertisement and location of network/device resources with a semantic description. The assumption here is that there is a local ontology to describe the services/resources that each artifact can provide and as such assist the service discovery mechanism. In order to support this functionality, an Ontology Manager (OM) is assumed present that provides methods that query this ontology for the services that the artifact provides. The details of the eRDP design and implementation can be found in [15]. The matching resources are returned by eRDP and the SM selects the best set of device(s)/service(s) based on a scoring mechanism that will be explained later. Subsequently, the SM invokes the OM to create the Sphere Ontology (SO) which will include links to all the relative devices to the AS that have been discovered.

After service binding the SM starts any interaction task in conjunction with the IA and also any FTA task and executes the workflow preserving the precedence constraints or the conditions that are specified in the workflow. At runtime a

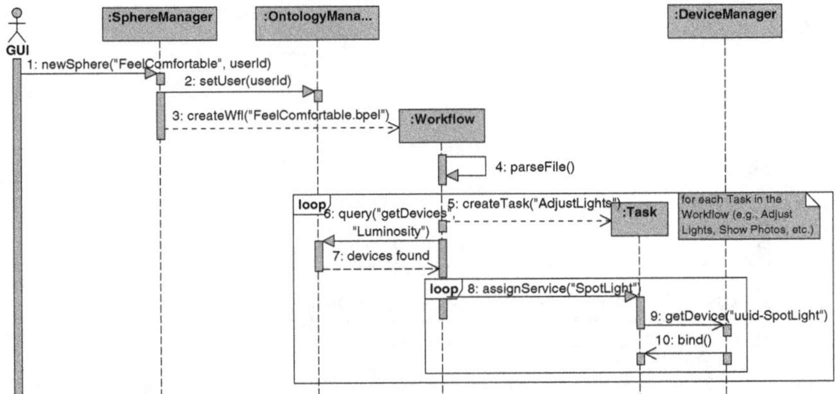

Fig. 1.7 Example AS instantiation and binding of devices to services

workflow object aggregates a number of Task objects where each object represents a task in the workflow. The services that this task requires for running are divided into input and output services and are connected with the appropriate resources. The resources that are bound to the Task object can be either devices that the Task directly controls (i.e., input sensors and actuation devices) or agents, such as the IA or the FTA. In either case the Task object is informed on the status of the resource and operates according to the pattern specified by its type. The sequence diagram in Fig. 1.7 shows the basic interaction of the software components during the instantiation of the "Feel Comfortable" AS, which employs the dynamic service binding process mentioned earlier. In the diagram, this process is implemented by the methods used inside the two loops.

The Task object "AdjustLights" is assigned to the FTA component to generate adaptive models for the individual devices/artifacts and for the user behaviors. Figure 1.8 illustrates the initialization of the FTA component to control the room lights in the example AS. The FTA is initialized by passing the input/output, light level related devices as well as light controls which are in turn retrieved from the Sphere Ontology which has been populated with the required ontologies during the AS instantiation. In addition, if the user profile stores initial light preferences (for example from previous executions of the FTA) these can be passed to the FTA in the form of a rule base.

In addition, the SM handles exception events that affect the configuration of the AS. For example, exceptions during the execution of the workflow, such as disconnection or failure of devices trigger an adaptation of the workflow by rebinding services to alternative devices. Context changes during the execution of the workflow may invalidate preconditions that were valid during the workflow instantiation. For example, if the user changes location and a follow-me property has been defined for a display service, then the execution state needs to be updated and a new display service instance to be scheduled. In order to achieve workflow adaptation, replanning capabilities may be required by the PA. Replanning comes

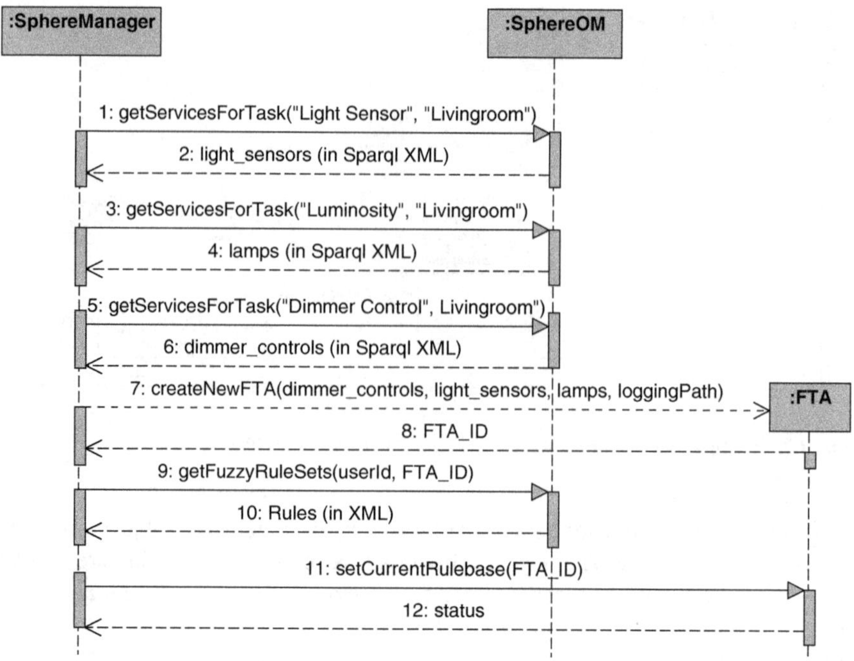

Fig. 1.8 FTA initialization for the AdjustLights task of the "Feeling Comfortable" AS

into play when the dynamic binding fails during workflow execution or update. When replanning is requested a new planning problem is defined with the services that are actually available, and the PA solves the problem and delivers a new workflow.

During AS instantiation in IEs there could be multiple devices or services providing similar functionality from which the system will have to choose. Thus, the ATRACO system must provide mechanisms for selection between similar devices or services and decide which of them is the most suitable to participate in the AS. Device selection is based on criteria such as: task suitability, efficiency (as device's proximity to the user, quality of the service or device and stability), user distraction (the inconvenience a user experiences when the system selects different groups of devices than those that the user prefers or is used to use for a specific task), and conflicts with other tasks (more details are given in the Appendix). For calculating the rank for each device we use a scoring mechanism that is similar to that proposed in [28] and is based on multi-attribute utility theory (MAUT). The overall rank of a device given a specific task is defined as a weighted sum of its evaluation with respect to its relevant orthogonal value dimensions (attributes). For ATRACO the relevant value dimensions are scores for task suitability, efficiency, negative of user distraction, and negative of conflicts with other tasks.

A ranking policy defines weights between zero and one for each of the above metrics. The scoring policies are defined per task (or task category) by the user and give priority to some of the metrics. For example, if a task is urgent, the suitability and efficiency ranks must have priority over user distraction, and inter-task conflicts. The weights are normalized to add up to one. The rank of a given device D according to policy P is computed as the dot product of the vector weights specified by the policy with the vector of scores for each one of the metrics. Applying MAUT, the device rank is computed as shown in (1.1).

$$DR(D,TP) = \sum_{i=1}^{4} w_i(TP) * D(m_i) \quad (1.1)$$

where DR is the overall rank of device D according to ranking policy TP for the task T, $w_i(TP)$ is the weight of metric i according to policy TP and $D(m_i)$ is the rank of device D for the metric m_i.

For the task suitability and efficiency we have $D(m) = DS(m)$, while for user distraction and inter-task conflict that have a negative meaning we have $D(m) = 1 - DS(m)$, where $DS(m)$ is the device's score for the metric normalized from 0 to 1.

1.4.3 Ontology Manager

The Ontology Manager (OM) component provides an interface to the SM to access AS related data, including personal and contextual information, represented in ontologies. The OM provides methods for querying and modifying User Profile Ontologies, Device Ontologies, the Privacy and Policy Ontology, and the eventual Sphere Ontology (SO) that emerges from the alignment of all the previous ontologies. The ontology alignment process can be described as: given two ontologies, each describing a set of discrete entities (which can be classes, properties, rules, predicates, or even formulas), find the correspondences, e.g., equivalences or subsumptions, holding between these entities. Thus, under the request of the SM, the OM produces ontology alignments, responds to queries regarding the state or properties of sphere resources, and creates inferences in order to enrich the SO as specified in [38].

The OM has been developed as a wrapper around the Jena Framework (http://jena.sourceforge.net/). The OM interface provides comprehensive and simple methods for creating an RDF/OWL-based ontology, importing and removing other RDF/OWL- based ontologies, updating the ontology at runtime, querying of the ontology using SPARQL, and saving the modifications in OWL files. Ontology alignment has been applied by using the Java Alignment API (http://alignapi.gforge.inria.fr/align.html). After the alignment, inference and querying is performed on a grid of imported ontologies, given the alignment points that have been produced using OWL class and individual equivalence assertions.

public String queryForSparqlXML(String query, boolean autoPrefix, String queryType)
Performs a query to the ontology. autoPrefix determines if OM will try to resolve known prefixes and the queryType can be ASK or Select. Returns the results in SparqlXML format.
public String[] getFuzzyRuleSets(String userId, String FTA_ID) throws Exception
Returns in XML format the stored fuzzy rulesets that match the given userId and FTA_ID. Used by SM to retrieve stored fuzzy rule sets during initialization of the corresponding FTA.
public String getServicesForTask(String serviceDescription, String location) throws Exception
Returns a Sparql XML string containing technical parameters for each device and service that matches the serviceDescription description tag and is within the location specified by the second parameter. Used by SM for resolving tasks to specific devices, services, actions, variables and values.

Fig. 1.9 A sample of the OM interface

Figure 1.9 illustrates a small sample of the OM interface that is used, for example, to query an ontology using SPARQL syntax, and methods related to the User Profile Ontology, e.g., for importing and exporting rules from the FTA.

A number of ontologies have been developed for and used in the prototype for the representation of AS high-level concepts (Fig. 1.3), devices and their services, and users and their profile information.

The User Profile Ontology holds personal information about the user. It consists of a local, private OWL ontology file that contains the actual user information in the form of individuals and assertions and a publicly accessible (via HTTP) ontology that contains the generic classes, properties, and restrictions that describe a user profile. Currently User Profile Ontology contains assertions about the Social Profile (name, nickname, email, address, etc.), the location, the activities (Goals and Plans), and the preferences of the user (in the form of stored fuzzy rule sets).

Figure 1.10 illustrates part of an instance of a device ontology for one of the spot lamps used in the prototype. The service concept represents an abstract service that the device can provide enriched with descriptive tags, e.g., a lamp can provide lighting service. In general, a device may offer more than one service and thus more service instances may be defined. The `StateVariable` concept represents the abstract states of the corresponding service. It encapsulates the linguistic variable and labels that are required by the FTA for the creation of adaptive device models. The device ontology includes technical characteristics and information about communication with the device in the context of a UPnP environment. Finally, the name, the owner, the location and physical properties of the device are included.

1.4.4 ATRACO-BPEL Workflow Specification

BPEL defines a model and a grammar for describing the behavior of a business process based on interactions between the process and its partners. It allows for creating complex processes by creating and wiring together different activities that can, for

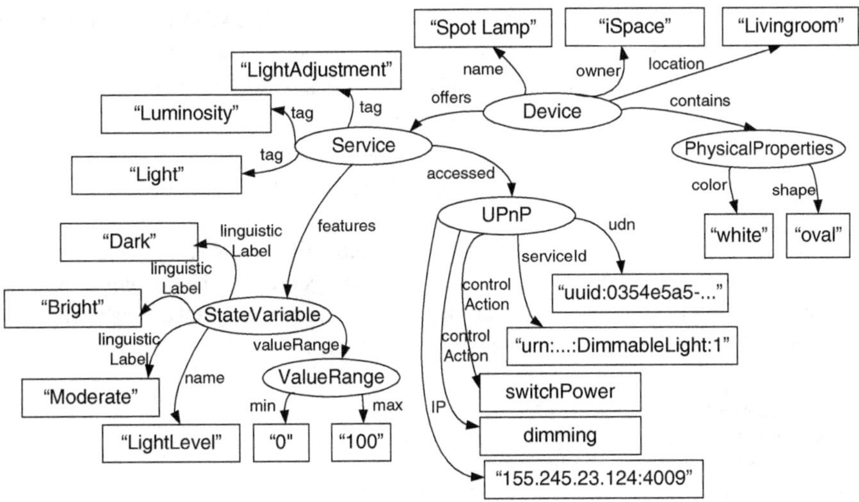

Fig. 1.10 Part of an instance of a device ontology for a spot lamp

example, perform Web Services invocations (<invoke>), waiting to be invoked by someone externally (<receive>), generate a response (<reply>), manipulate data (<assign>, throw faults (<throw>), or terminate a process (<exit>). In our case, the business process represents the process model of an AS and the partners can take the form, either of a service of a simple device, or the service of an ATRACO agent. While BPEL is a suitable language for describing workflows, an ATRACO workflow description presents requirements that cannot be completely covered by BPEL. This is due to the following:

1. BPEL partners (partnerLinks) are bound statically to specific Web Services. In the context of ATRACO, however, services are not bound at design time but dynamically during the execution of the workflow. Thus, there is a need to describe services in the workflow by their semantics which mainly define ontological related searching terms (for example, "Luminosity" for a light service).
2. The limitation of the one-to-one mapping of services between communicating partners, supported by BPEL. On the other side, ATRACO tasks may need to handle two or more services that provide input or output to the task.
3. BPEL supports a single coordinator that executes the orchestration logic. ATRACO workflows normally are centrally handled by the Sphere Manager which implements the workflow execution engine; however, a more distributed scheme can also be followed by sharing parts of the workflow with collaborating agents (e.g., IA and FTA). This collaboration sets some special requirements in the description of the workflow.

Given the above requirements a variant of BPEL, called ATRACO-BPEL, was defined in order to provide those ATRACO specific features needed in order specify workflows. In the following we explain how using the ATRACO-BPEL formalism an example task is bound with the appropriate service(s). The task *AdjustLights* is associated with the partnerLink *AdjustLightsPL* as part of the orchestration logic section:

```
<bpel:invoke
  name="AdjustLights" partnerLink="AdjustLightsPL">
</bpel:invoke>
```

The partnerLink *AdjustLightsPL* has an input role called *ATRACO:lightStatus* and an output role (partnerRole) called *ATRACO:triggerLight*. The *Continues* type denotes that the execution of the activity is to be treated as a task that is running continuously, i.e., the workflow does not wait its termination.

```
<bpel:partnerLink
  name="AdjustLightsPL"
  partnerLinkType="ATRACO:Continuous"
  myRole="ATRACO:lightStatus"
  partnerRole="ATRACO:triggerLight">
</bpel:partnerLink>
```

The input role *ATRACO:lightStatus* denotes the appropriate abstract service that must be bound to fulfill the role (Luminosity) along with any other application specific details that are needed for its operation, e.g., the task will be monitored by an ATRACO agent for learning user behavior with respect to light adjustments and all found light devices are to be used.

```
<ATRACO:role
  name="lightStatus" type="input" Agent="yes" IAmode ="none">
  <ATRACO:service semantics="Luminosity" trigger="Low" reset
      ="none" quantity="all" rules="">
  </ATRACO:service>
</ATRACO:role>
```

The corresponding definition for the output role will be:

```
<ATRACO:role
  name="triggerLight" type="output" Agent="yes" IAmode="
      withAgent">
  <ATRACO:service semantics="Actuate Light" trigger="On"
      reset="Off" quantity="all" rules="">
  </ATRACO:service>
</ATRACO:role>
```

In ATRACO-BPEL each partnerLink role is specialized as an ATRACO:role which is a new definition in ATRACO-BPEL. In each ATRACO:role the attributes listed in Table 1.3 are defined.

Each ATRACO:role envelopes a set of services that are bound to it. Each role can have more than one abstract service. If the role type is input then the activity waits for all the services to deliver their result before proceeding. If the role type is output then, upon activity completion, all the services enveloped in this role are triggered. For each abstract service-specific attributes are defined, providing the necessary support for device discovery and service operation. Table 1.4 summarizes the service-specific attributes in ATRACO-BPEL.

Table 1.3 ATRACO:role semantics in ATRACO-BPEL

Attribute	Semantics
Name	The name of the role.
Type	Denotes the type of the role. Accepted values are **input/output**.
Agent	This attribute defines whether the task is monitored by an ATRACO agent or not. Accepted values are **yes/no**.
IAmode	Specifies the interaction mode with the ATRACO Interaction Agent. Accepted values are: **none** no interaction is needed; **pure** this value is used to indicate that a single interaction with the user through a dialog interface (spoken, tangible or software) needs to be provided either to provide a message or to receive an input for the system from the user in a form of question; **direct** this value is used when the IA needs to create an interface for an output device; **withAgent** this value is used to indicate that there is a need to find proper user inputs for the Agent monitored tasks.

Table 1.4 Service-specific attributes in ATRACO-BPEL

Attribute	Semantics
Semantics	The semantics of the service as a set of keywords—these are used to find the specific device that can be bound to this abstract service.
Trigger	**input role**: denotes a linguistic value that triggers the service. **output role**: denotes a linguistic value passed to the service.
Reset	The reset state (linguistic value) that the service should apply in the case that the activity cannot be performed.
Quantity	A number that defines how many devices providing this service are needed for the specific activity. If the value is "**all**" then all found devices are used.
Rules	Any special constraints need to be met for binding the corresponding device(s).
IAdlg	This attribute is associated with the direct or pure interaction modes with IA in order to give it the proper interaction dialog type. Examples of accepted values are: GreetingMessage, LightInstructions, GrantGuestAccess, MusicQuestion, MusicControl, PhotoFrameQuestion, SlideshowControl.

1.5 Deployment

In order to test the AS adaptation mechanisms we have implemented an experimental prototype in the AmInOffice testbed. The AmInOffice is a testbed developed in the premises of Dynamic Ambient Intelligent Systems Research Unit at RACTI (daisy.cti.gr) and consists of a variety of sensors deployed in the office environment, a set of smart objects that support office tasks and the appropriate network infrastructure. In order to implement the above scenario we have set up AmInOffice with the following devices:

- An RFID reader near the door of the office to read RFID tags
- Two light sensors each one reading light level in a different spot in the office
- Two ceiling lamps controlling the ambient light
- Two lamps one placed at the desk and one near the sofa
- Speakers connected to the main PC for playing music and producing vocal messages
- A smart chair (eChair) able to sense if someone is sitting on it.
- A smart sofa (eSofa) that can sense if someone is sitting on it and at which spot (left or right).
- Smart books (eBooks). Apart from smart readers (eReaders) this includes a typical book instrumented with bending sensors that can sense if the book is opened or closed.
- A smart desk (eDesk) that can sense objects on it and near it.

Figure 1.11 illustrates how the devices have been placed in the AmInOffice.

The ATRACO components that implement the necessary functionality in order to support AS formation and adaptation based on mechanisms discussed in this work are the Sphere Manager (SM) and Ontology Manager (OM). Interaction with other ATRACO components such as Planning Agent and Privacy Manager is assumed and requires the interfaces specified in [17]. Third-party tools have been also used for

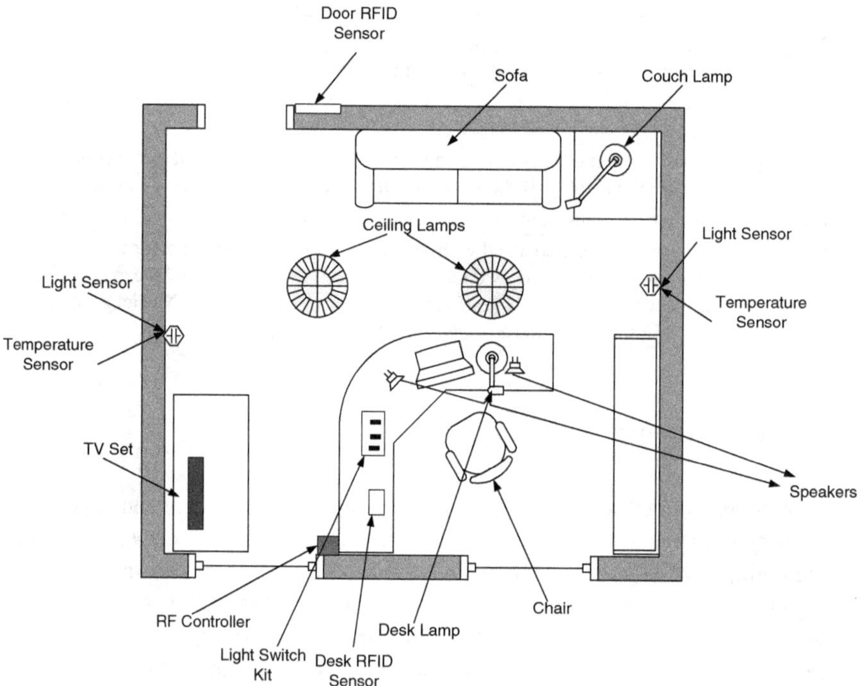

Fig. 1.11 AmInOffice setup for the experimental prototype

performing alignment. The ontologies for all the artifacts used have been developed and a semantically rich UPnP device ontology was developed to support workflow-driven inclusion of UPnP compatible devices in a sphere. In the experimental prototype we have tested the following functionality:

- *Sphere initialization*: Initiate an AS through an ATRACO-BPEL file. Test workflow creation and execution.
- *Late binding of the devices*: Bind abstract services needed for each task to specific devices that exist in AmInOffice in collaboration with the OM at runtime.
- *Runtime application behavior*: Validate that the running tasks correspond to the scenario of the experiment.
- *Handling of adaptation events*: Test system response to adaptation events that affect the configuration of the AS categorized in the following types:
  - *User location change*: test system reaction when the location of the user associated with the AS changes.
  - *Resource not available*: test system reaction when a device bound to a task fails. Check if the system can find an appropriate replacement.
  - *New resource (service/device)*: test system reaction when a new device relevant to the task that is running is available.
  - *New person*: test system reaction (in terms of security and privacy) when a new user is recognized by the system

Figure 1.12 illustrates in the form of an activity diagram the main tasks to be executed by the Sphere Manager component in order to handle each one of the above adaptation events. The starting point for running an AS is the generation of the corresponding workflow. Workflows are described in ATRACO-BPEL, but they can be represented in a more user friendly way with activity diagrams. The diagram in Fig. 1.13 illustrates the workflow for the "Feel Comfortable" AS of the example scenario. The diagram is annotated with labels from the source file in an attempt to close the gap between the high-level view of the diagram and the low-level view of the file. For example, the annotation in each box shows the activity type in the main sequence and the task name, the ontological searching term, as well as which ATRACO component, besides SM, has responsibility for running parts of this task.

The technical requirements for the deployment and testing of the ATRACO system include: the runtime versions of the ATRACO components with the specified service interfaces; the devices serving the scenarios, wrapped as UPnP devices; the domain and resource ontologies; the workflows specifying the tasks in each AS; and various third-party runtime libraries. The deployment of the system has been done in two IE testbeds using scenarios similar to the one discussed in this paper.

The implementation technologies and tools used are based on open frameworks and are compatible with the SOA paradigm. Java is the main programming language and UPnP enhanced with semantic descriptions [41] is used as the communication middleware for the integration of devices and services, instead of Web Services. OWL has been used for the development of the ontologies as it provides a strong logical reasoning framework for the expression and enforcement of ATRACO policies and rules.

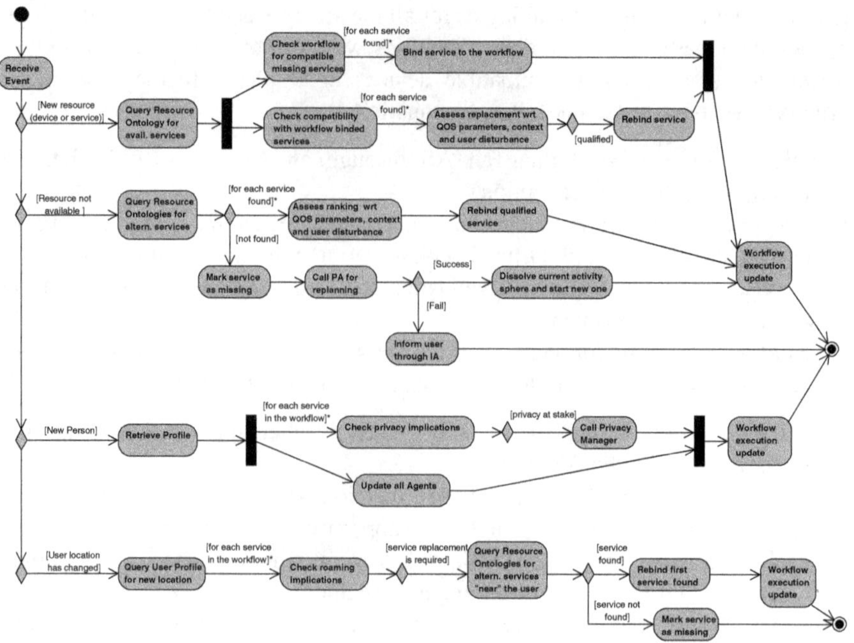

Fig. 1.12 Adaptation events handling

Although there are available (open source) execution engines for BPEL "programs" in ATRACO we need to build a layer upon such engines as a proxy in order to process the parts of the workflow description that are ATRACO-specific. In addition, most engines do not allow for dynamic binding and discovery of services. To address this limitation, the framework uses the SM as a proxy to communicate with service registries to obtain operational descriptions (e.g., UPnP or WSDL files) and instantiate services. This is achieved by encapsulating service search parameters in ATRACO-BPEL (see Table 1.4) as an input to the dynamic service binding process.

When the user changes position in the room the ATRACO system is notified for that change. While this location context can be provided either by using motion detection devices, or specific services such as Ubisense (used in iSpace), for our experiment we emulated such a device by using a WoZ interface and selecting the appropriate location. When the SM receives a location change event, it queries OM for the new location. Then for each service that is bound to the active task it checks if there are any requirements for device replacement. This is done by querying the OM with the new user location context. If the device that OM returns is not equal to the currently bound then it proceeds with service replacement for the appropriate task. The sequence diagram in Fig. 1.14 shows the exact messages that are exchanged for the task "Reading" when the user changes location to the sofa.

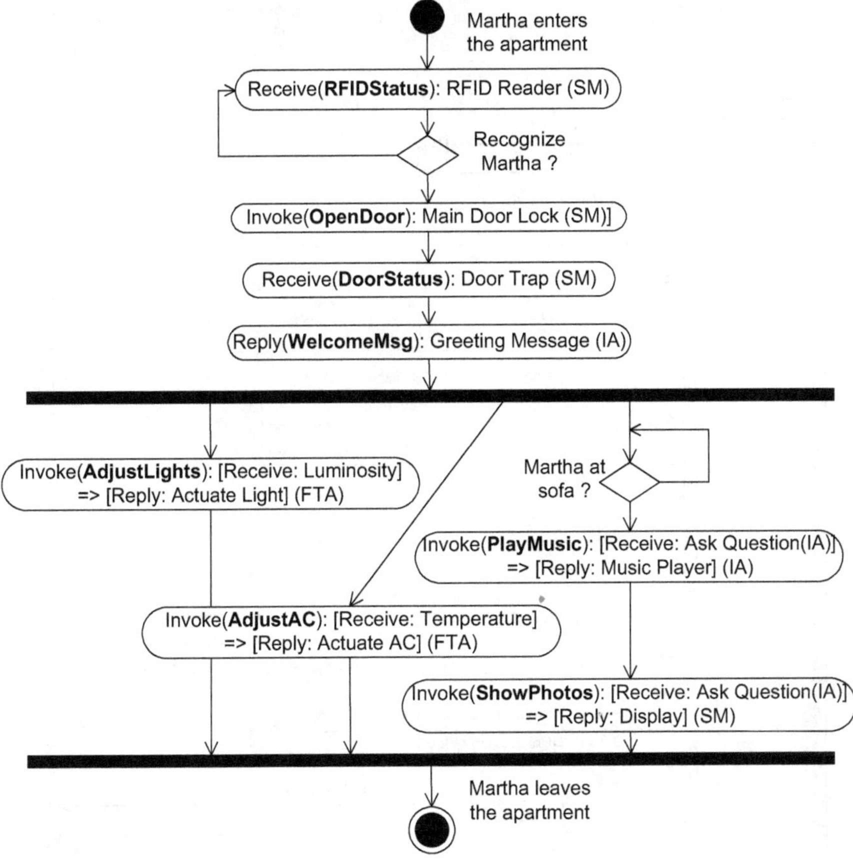

Fig. 1.13 Annotated activity diagram for the "Feel Comfortable" AS

1.6 Discussion

The SOA approach appears to be a convenient architectural style towards meeting one of the key objectives of the ATRACO project that is the need for adaptable and reconfigurable systems. Analyzing contemporary software technologies complying with the SOA architectural paradigm, such as OSGi, UPnP, and the Web Services architecture appears that current software technologies do not meet the adaptability and interoperability requirements for the ATRACO project.

In the first case SOA provides little support on how adaptive services can be used to allow people to interact with an AmI environment in a seamless and unobtrusive manner. In other words, research into service composition has mainly focused on the composition mechanism rather than on guiding composition to enable the user to perform activities in the way they wish to do. A challenge here is how to automate the service composition process, so that the service offered to users appears to be

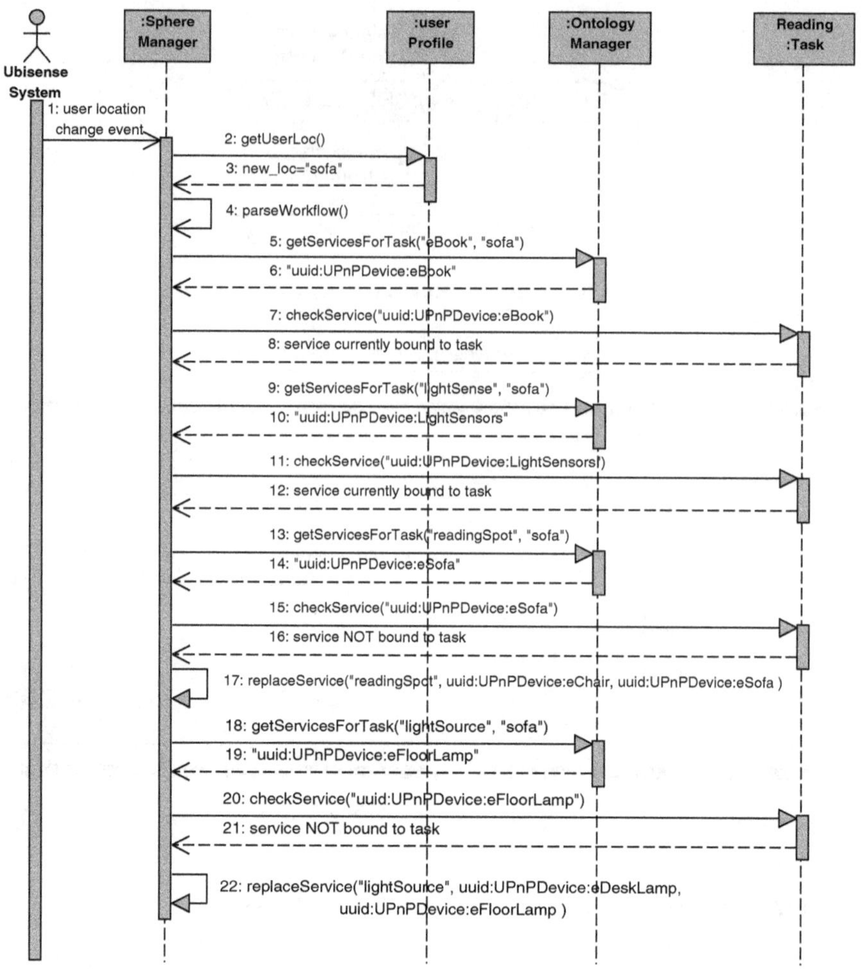

Fig. 1.14 Replacing the light providing device as a result of user mobility

adaptive, in the sense that the service provided changes dynamically according to the task the user wishes to perform and the context in which they wish to perform it.

In the second case, current solutions provide little support for semantic-based interoperability, hence dealing with interaction between services based on syntactic description for which common understanding is hardly achievable in an open environment. The latter issue may be addressed using semantic modeling through ontologies. Ontologies can provide an extensible and flexible way of expressing the basic terms and their relations in a domain, task or service. However, the issue that can be raised by the heterogeneity of ontologies and how to achieve semantic interoperability between systems using different ontologies remains a challenge. In ATRACO the approach that is followed in order to address this challenge is to research, develop and test theories of ontology alignment to achieve task-based semantic integration of heterogeneous devices and services. This issue is examined thoroughly in a separate chapter.

ATRACO concrete plans are described as mentioned previously by workflow specifications using an extension of BPEL (ATRACO-BPEL). Normally, workflow management systems have not been used for dynamic environments requiring adaptive behavior. Typically an intranet-based workflow system executes, by using a collection of services that are owned and managed by the same organization. In this setting, service interruptions are rare and typically they are scheduled during system maintenance. On the contrary, in ATRACO we require adaptive workflows which need to react to varying environmental conditions. This transition from the static to dynamic and adaptive nature of workflows increases the runtime complexity of the management system, since the coordination mechanism must become more fault tolerant. On the other side our approach is viewed as collaborative problem solving approach where a set of autonomous agents work together to achieve a common goal. Our general idea then is that since a workflow describes the relationship between services and if an agent is represented by such a service, then the relationship between the agents would be possible to specify. Following such a combined agent-based and SOA approach means that a workflow could be used to establish the initial relationships of the ATRACO components. Applications can be specified then first with a workflow description using ATRACO-BPEL that defines the most common scenario and fault conditions. Once the basic system has been deployed, the agents could be working proactively so they can adapt to unforeseen circumstances and automatically handle the extension of the workflow description.

1.7 Conclusion

ATRACO supports the deployment and execution of applications that need to be adapted and reconfigured in dynamic environments. The need for adaptation and reconfiguration calls for a modular design approach, which the SOA paradigm tends to provide. Following this architectural style, each device provides services through which other components can obtain information or control its behavior. When an application has to be adapted, either during application transition to a new environment or when a device running a service fails, a description of the structure of the application, which is modelled as a workflow of abstract services, is used by an adaptation module which makes use of ontologies, context information, and defined policies to generate a new structure for the adapted application. The agent approach complements the SOA modular and flexible infrastructure by providing high-level adaptation to user's tasks, as an intelligent control layer above SOA (See Table 1.5 in the Appendix).

Appendix

Table 1.5 Metrics for device selection

Metric	Description
Task suitability	This quantifies, as a percentage, how well a specific device or service is suitable for a specific task. Its value is calculated based on the semantic relevance between the task's description as it is presented in an abstract plan, and the device's or service's description provided by the device's ontology. For example, for a service of providing light, both a lamp and a computer monitor can be candidates. The fact that the lamp's ontology states as primary purpose of the device the supply of light while the monitor's ontology state the light emission as a secondary attribute gives the lamp a higher score for this metric.
Efficiency	This metric measures the efficiency of a device or service for a certain goal. It expresses how well or to what degree the device is able to contribute in achieving a goal. Its value is calculated over a combination of other measures such as the device's proximity to the user (based on location info from User Profile Ontology (UPO) and device ontology), the quality of the service or device (as inferred from the specifications of the device that are encoded in its ontology) and the stability that quantifies how well a device will be able to perform a task to completion. The exact measures that participate in the calculation of efficiency are depended on the nature of the task and are derived from the policies encoded in the ATRACO ontologies. For example, if the goal is to provide enough light for the user to read a book for an hour, a lamp located closer to the user has a strong potential to be selected. But if its light is weaker than the minimum needed for reading and another lamp exists a little more far, but in the range of the user, and can provide the desirable light level, the later should get a higher efficiency score. In a similar way if a device runs out of battery and will not last to achieve fully the goal its efficiency score should be discounted.
User distraction	User distraction expresses the inconvenience a user experiences when the system selects different groups of devices than those that the user prefers or is used to use for a specific task. User's preferences and habits are expected to be stated at the UPO or inferred from it. For example, if a lamp with strong light is available near the user as he reads his book, but the user has expressed (directly or indirectly) its preference to use a specific one with weaker light when reading at this part of the room, the system should penalize the first lamp by increasing its user distraction score.
Conflicts with other tasks	This quantifies the number of other running tasks that will be blocked by selecting a device. For example, if a monitor is currently used for watching a film, its conflicting score for selecting it for the task of displaying incoming emails should be greater than zero because it will obstruct the film watching task.

References

1. Alves, A., Arkin, A., Askary, S., Barreto, C., Ben, Curbera, F., Ford, M., Goland, Y., Guízar, A., Kartha, N., Liu, C.K., Khalaf, R., König, D., Marin, M., Mehta, V., Thatte, S., van der Rijn, D., Yendluri, P., Yiu, A.: Web services business process execution language version 2.0. Technical Report, OASIS Web Services Business Process Execution Language (WSBPEL) TC (2007). http://docs.oasis-open.org/wsbpel/2.0/OS/wsbpel-v2.0-OS.html
2. Armstrong, D.J.: The quarks of object-oriented development. Commun. ACM **49**, 123–128 (2006). doi:http://doi.acm.org/10.1145/1113034.1113040. http://doi.acm.org/10.1145/1113034.1113040
3. Bidot, J., Schattenberg, B., Biundo, S.: Intelligent planner. Technical Report, University of Ulm (2010)
4. Birrell, A.D., Nelson, B.J.: Implementing remote procedure calls. ACM Trans. Comput. Syst. **2**, 39–59 (1984). doi:http://doi.acm.org/10.1145/2080.357392. http://doi.acm.org/10.1145/2080.357392
5. Blair, G.S., Coulson, G., Andersen, A., Blair, L., Clarke, M., Costa, F., Duran-Limon, H., Fitzpatrick, T., Johnston, L., Moreira, R., Parlavantzas, N., Saikoski, K.: The design and implementation of open orb 2. IEEE Distrib. Syst. Online **2**, 2001 (2001). http://portal.acm.org/citation.cfm?id=1435643.1436507
6. Brønsted, J., Hansen, K.M., Ingstrup, M.: Service composition issues in pervasive computing. IEEE Pervasive Comput. **9**, 62–70 (2010). doi:http://dx.doi.org/10.1109/MPRV.2010.11. http://dx.doi.org/10.1109/MPRV.2010.11
7. Capra, L., Blair, G.S., Mascolo, C., Emmerich, W., Grace, P.: Exploiting reflection in mobile computing middleware. SIGMOBILE Mob. Comput. Commun. Rev. **6**, 34–44 (2002). doi:http://doi.acm.org/10.1145/643550.643553. http://doi.acm.org/10.1145/643550.643553
8. Chakraborty, D., Joshi, A.: Dynamic service composition: state-of-the-art and research directions. Technical Report, University of Maryland, Department of Computer Science and Electrical Engineering (2001)
9. Dustdar, S., Schreiner, W.: A survey on web services composition. Int. J. Web Grid Serv. **1**, 1–30 (2005). doi:10.1504/IJWGS.2005.007545. http://portal.acm.org/citation.cfm?id=1358537.1358538
10. Emmerich, W.: OMG/CORBA: An object-oriented middleware. In: Marciniak, J.J. (ed.) Encyclopedia of Software Engineering, pp. 902–907. Wiley, New York (2002). http://www.cs.ucl.ac.uk/staff/w.emmerich/publications/Encyclopedia
11. Erl, T.: Service-Oriented Architecture: Concepts, Technology, and Design. Prentice Hall PTR, Upper Saddle River, NJ (2005)
12. Frei, A., Popovici, A., Alonso, G.: Eventizing applications in an adaptive middleware platform. IEEE Distrib. Syst. Online **6**, 1– (2005). doi:10.1109/MDSO.2005.20. http://portal.acm.org/citation.cfm?id=1069591.1069686
13. Gjørven, E., Eliassen, F., Lund, K., Eide, V.S.W., Staehli, R.: Self-adaptive systems: a middleware managed approach. In: SelfMan, pp. 15–27 (2006)
14. Goumopoulos, C., Kameas, A.: Ambient ecologies in smart homes. Comput. J. **52**, 922–937 (2009). doi:http://dx.doi.org/10.1093/comjnl/bxn042. http://dx.doi.org/10.1093/comjnl/bxn042
15. Goumopoulos, C., Kameas, A.: Smart objects as components of ubicomp applications. Int. J. Multimedia Ubiquit. Eng. Special Issue on Smart Object Systems **4**(3), 1–20 (2009). http://www.sersc.org/journals/IJMUE/vol4_no3_2009/1.pdf. SERSC Press
16. Goumopoulos, C., Kameas, A., Hagras, H., Callaghan, V., Gardner, M., Minker, W., Weber, M., Bellik, Y., Meliones, A.: Atraco: adaptive and trusted ambient ecologies. In: Proceedings of the 2008 Second IEEE International Conference on Self-Adaptive and Self-Organizing Systems Workshops, pp. 96–101. IEEE Computer Society, Washington, DC, USA (2008). doi:10.1109/SASOW.2008.13. http://portal.acm.org/citation.cfm?id=1524875.1525041

17. Goumopoulos, C., Calemis, I., Togias, K., Kameas, A., Pruvost, G., Wagner, C., Meliones, A., Wiedersheim, B., Bidot, J.: Integrated component platform for prototype testing and updated specification and design report. Technical Report, Computer Technology Institute, ATRACO ICT 1.8.2 216837 D7 (2010)
18. Grace, P., Truyen, E., Lagaisse, B., Joosen, W.: The case for aspect-oriented reflective middleware. In: Proceedings of the 6th International Workshop on Adaptive and Reflective middleware: Held at the ACM/IFIP/USENIX International Middleware Conference, ARM '07, pp. 2:1–2:6. ACM, New York, NY (2007). doi:http://doi.acm.org/10.1145/1376780.1376782. http://doi.acm.org/10.1145/1376780.1376782
19. Grimes, R.: Professional Dcom Programming. Wrox Press Ltd, Birmingham, UK (1997)
20. Gruber, T.R.: Toward principles for the design of ontologies used for knowledge sharing. Int. J. Hum.-Comput. Stud. **43**, 907–928 (1995). doi:10.1006/ijhc.1995.1081. http://portal.acm.org/citation.cfm?id=219666.219701
21. Hayton, R.: Flexinet open orb framework. Technical Report, APM Ltd, Poseidon House, Castle Park, Cambridge, UK (1997)
22. Hollingsworth, D.: Workflow management coalition - the workflow reference model. Technical Report, Workflow Management Coalition (1995)
23. Kiczales, G.: Aspect-oriented programming. ACM Comput. Surv. **28** (1996). doi:http://doi.acm.org/10.1145/242224.242420. http://doi.acm.org/10.1145/242224.242420
24. Kiczales, G., Hilsdale, E., Hugunin, J., Kersten, M., Palm, J., Griswold, W.G.: An overview of aspectj. In: Proceedings of the 15th European Conference on Object-Oriented Programming, pp. 327–353. Springer, London (2001). http://portal.acm.org/citation.cfm?id=646158.680006
25. Kiczales, G., Rivieres, J.D.: The Art of the Metaobject Protocol. MIT Press, Cambridge, MA (1991)
26. Kon, F., Román, M., Liu, P., Mao, J., Yamane, T., Magalhã, C., Campbell, R.H.: Monitoring, security, and dynamic configuration with the dynamictao reflective orb. In: IFIP/ACM International Conference on Distributed systems platforms, Middleware '00, pp. 121–143. Springer, New York, Secaucus, NJ (2000). http://portal.acm.org/citation.cfm?id=338283.338355
27. Kon, F., Costa, F., Blair, G., Campbell, R.H.: The case for reflective middleware. Commun. ACM **45**, 33–38 (2002). doi:http://doi.acm.org/10.1145/508448.508470. http://doi.acm.org/10.1145/508448.508470
28. Kumar, R., Poladian, V., Greenberg, I., Messer, A., Milojicic, D.: Selecting devices for aggregation. In: WMCSA, pp. 150–159. IEEE Computer Society, Los Alamitos, CA, USA (2003). doi:http://doi.ieeecomputersociety.org/10.1109/MCSA.2003.1240776
29. Ledoux, T.: Opencorba: a reflective open broker. In: Proceedings of the Second International Conference on Meta-Level Architectures and Reflection, Reflection '99, pp. 197–214. Springer, London, UK (1999). http://portal.acm.org/citation.cfm?id=646930.710404
30. Maciel da Costa, C., da Silva Strzykalski, M., Bernard, G.: An aspect oriented middleware architecture for adaptive mobile computing applications. In: Proceedings of the 31st Annual International Computer Software and Applications Conference - Volume 02, COMPSAC '07, pp. 81–86. IEEE Computer Society, Washington, DC, USA (2007). doi:http://dx.doi.org/10.1109/COMPSAC.2007.59. http://dx.doi.org/10.1109/COMPSAC.2007.59
31. Maes, P.: Concepts and experiments in computational reflection. SIGPLAN Not. **22**, 147–155 (1987). doi:http://doi.acm.org/10.1145/38807.38821. http://doi.acm.org/10.1145/38807.38821
32. McIlraith, S.A., Son, T.C., Zeng, H.: Semantic web services. IEEE Intell. Syst. **16**, 46–53 (2001). doi:http://dx.doi.org/10.1109/5254.920599. http://dx.doi.org/10.1109/5254.920599
33. Papadopoulos, N., Meliones, A., Economou, D., Karras, I., Liverezas, I.: A connected home platform and development framework for smart home control applications. In: Proceedings of the 7th IEEE International Conference on Industrial Informatics (INDIN09) (2009)
34. Pawlak, R., Seinturier, L., Duchien, L., Florin, G.: Jac: A flexible solution for aspect-oriented programming in java. In: Proceedings of the Third International Conference on Metalevel Architectures and Separation of Crosscutting Concerns, pp. 1–24. Springer, London, UK (2001). http://portal.acm.org/citation.cfm?id=646931.710426

35. Pruvost, G., Kameas, A., Heinroth, T., Seremeti, L., Minker, W.: Combining agents and ontologies to support Task-Centred interoperability in ambient intelligent environments. In: Proceedings of the 2009 Ninth International Conference on Intelligent Systems Design and Applications, ISDA '09, pp. 55–60. IEEE Computer Society, Washington, DC, USA (2009). doi:http://dx.doi.org/10.1109/ISDA.2009.195. http://dx.doi.org/10.1109/ISDA.2009.195
36. Rao, J., Su, X.: A survey of automated web service composition methods. In: Cardoso, J., Sheth, A. (eds.) Semantic Web Services and Web Process Composition. Lecture Notes in Computer Science, vol. 3387, pp. 43–54. Springer, Berlin (2005). http://dx.doi.org/10.1007/978-3-540-30581-1_5
37. Remagnino, P., Foresti, G.L.: Ambient intelligence: a new multidisciplinary paradigm. IEEE Trans. Syst. Man Cybern. Part A **35**(1), 1–6 (2005)
38. Seremeti, L., Goumopoulos, C., Kameas, A.: Ontology-based modeling of dynamic ubiquitous computing applications as evolving activity spheres. Pervasive Mob. Comput. **5**, 574–591 (2009). doi:10.1016/j.pmcj.2009.05.002. http://portal.acm.org/citation.cfm?id=1630161.1630223
39. Sommer, R.: Personal Space: The Behavioral Basis of Design. Prentice Hall Trade, Englewood Cliffs, NJ (1969)
40. Sycara, K., Paolucci, M., Ankolekar, A., Srinivasan, N.: Automated discovery, interaction and composition of semantic web services. J. Web Semant. **1**(1), 27–46 (2003)
41. Togias, K., Goumopoulos, C., Kameas, A.: Ontology-Based representation of upnp devices and services for dynamic Context-Aware ubiquitous computing applications. In: International Conference on Communication Theory, Reliability, and Quality of Service, pp. 220–225. IEEE Computer Society, Los Alamitos, CA, USA (2010). doi:http://doi.ieeecomputersociety.org/10.1109/CTRQ.2010.44
42. Verma, K., Sivashanmugam, K., Sheth, A., Patil, A., Oundhakar, S., Miller, J.: Meteor-s wsdi: A scalable p2p infrastructure of registries for semantic publication and discovery of web services. Inf. Technol. Manag. **6**, 17–39 (2005). doi:10.1007/s10799-004-7773-4. http://portal.acm.org/citation.cfm?id=1047575.1047628
43. Wagner, C., Hagras, H.: Toward general type-2 fuzzy logic systems based on zslices. Trans. Fuz Sys. **18**, 637–660 (2010). doi:http://dx.doi.org/10.1109/TFUZZ.2010.2045386. http://dx.doi.org/10.1109/TFUZZ.2010.2045386
44. Wollrath, A., Riggs, R., Waldo, J.: A distributed object model for the java system. Comput. Syst. **9**(4), 265–290 (1996)
45. Yang, Z., Cheng, B.H.C., Stirewalt, R.E.K., Sowell, J., Sadjadi, S.M., McKinley, P.K.: An aspect-oriented approach to dynamic adaptation. In: Proceedings of the First Workshop on Self-Healing Systems, WOSS '02, pp. 85–92. ACM, New York, NY (2002). doi:http://doi.acm.org/10.1145/582128.582144. http://doi.acm.org/10.1145/582128.582144

Chapter 2
Adaptive Networking

Apostolos Meliones, Ioannis Liverezas, and Dimitrios Economou

Abstract Intelligent environments have been commercially available already over 10 years without becoming such a mass product as expected. The expectations of potential users of mentioned solutions have not been fulfilled yet due to missing globally accepted standards causing interoperability problems of different hardware and software components, complexity of configuration and use, lack of universal service consideration, and insufficient ROI for private residence owners. Clearly, a stronger emphasis is needed on device adaptation, usability, and scalability, which can seamlessly accommodate new IE services. Although several research efforts have addressed the development of IEs through networking existing devices and resolving interoperability issues with the help of middleware, there has been little work on specifying at a high level of abstraction how such abstraction services would work together at the application level taking into account in a combined dynamic way the heterogeneous networks and services. This chapter presents a framework for Network Adaptation in IEs using OSGi and UPnP technology allowing the uniform and transparent access to devices and services present in the networked environment and supporting the realization of activity spheres across a mixture of heterogeneous networks. Our work specifically focuses on the issue of heterogeneity at the network level and defines adaptation as the systemic mechanism in order to deal with this issue. The proposed network adaptation framework has been implemented within the ATRACO project and supports an ambient ecology trial hosted in the iSpace in Essex demonstrating a number of futuristic user activity spheres.

A. Meliones (✉)
Department of Digital Systems, University of Piraeus, Piraeus, Greece

inAccess, 12 Sorou Str., 15125 Maroussi, Athens, Greece
e-mail: meliones@unipi.gr

I. Liverezas • D. Economou
inAccess, 12 Sorou Str., 15125 Maroussi, Athens, Greece
e-mail: iliverez@inaccess.com; decon@inaccess.com

© Springer International Publishing Switzerland 2016

S. Ultes et al. (eds.), *Next Generation Intelligent Environments*,
DOI 10.1007/978-3-319-23452-6_2

2.1 Introduction

Intelligent environments have been commercially available already over 10 years without becoming such a mass product as expected. The expectations of potential users of mentioned solutions have not fulfilled yet due to missing globally accepted standards causing interoperability problems of different hardware and software components, complexity of configuration and use, lack of universal service consideration, and insufficient ROI for private residence owners. The main reason is that smart environment networks consist of a large variety of content sources (e.g., sensors), multiple information carriers (wired and wireless media), and communication standards which lead to problems of interoperability, administration, and reducing the ease of use. At the same time, there is significant interest in home networking today, which stems from the availability of low-cost communication technology and readily available access to content and services from various sources and suppliers, extending to accommodate a number of smart applications, including management, control, and security, in the home environment. Therefore a stronger emphasis needs to be set on device adaptation, usability, and scalability, which can seamlessly accommodate new smart environment services.

Several research efforts have addressed the development of an Ambient Intelligent Environment (AIE) through networking existing devices and resolving interoperability issues with the help of middleware. The NGN@Home ETSI initiative [12] facilitates interoperability between the various home network end devices and various home hub technologies and will provide a standardized approach to Next Generation Networks at home and in home intelligent device technologies. TEAHA [19] addressed networked home control applications and their complementarity to audio-visual networked applications via an interoperable middleware having a hardware centric view for creation of universal solutions. EPERSPACE [4] concentrated on creating a home platform to link different devices at home to an interoperable network, to provide these devices information about what you or your friends need and make the system respond accordingly. AMIGO [1] aimed to develop ontology-based middleware and interoperability of devices, artifact and services. e-SENSE [18] concentrated on developing methods, tools and a test platform for the design, implementation, and operation of smart adaptive wireless networks of sensing components. HOME2015 [6] is a multidisciplinary research programme aiming to create future systems and technologies for future homes. MUSIC [11] develops a software development framework that facilitates the development of self-adapting, reconfigurable software that seamlessly adapts to dynamic user and execution context in ubiquitous computing environments. In [17] a home network is described as a set of interconnections between consumer electronic products and systems, enabling remote access to and control of those products and systems, and any available content such as music, video, or data. The elements of the above definition are connections, access, and control. One point, implicit in the above definition, that must be emphasized is ease of use. For those involved in the consumer electronic industry, it is understood that the *home networking* implies fundamental simplicity.

Other efforts present solutions based on web services technology to solve the interoperability problem in smart home environments [13], permit users or agents to aggregate and compose networked devices and services for particular tasks [8], discuss how recent technological progress in the areas of visual programming languages, component software and connection-based programming can be applied to programming the smart home [7], introduce supporting infrastructures moving the burden of connectivity away from the end devices to the system core [3], enhance the OSGi standard to integrate many existing home protocols and networks with emphasis on realizing an effective converged service gateway architecture for smart homes [9, 21], and propose hardware architectures for ubiquitous computing in smart home environments where the devices may autonomously act and collaborate with each other using agent-based service personalization software architectures [22]. Other approaches to modelling and programming devices for intelligent environments model devices as collections of objects [13], as Web services [10], and as agents [14]. The lack of a de-facto standard middleware for distributed sensor-actuator environments has been identified by many researchers as one of the key issues limiting research on intelligent environment and the proliferation of intelligent environments from research environments to their deployment in our everyday lives, as for instance in [16] presenting a robotic middleware as glue between sensors, actuators, and services for complex ubiquitous computing environments. Early attempts to adapt heterogeneous environments consisting of dissimilar networks and computing devices primarily address primitive system architectural issues and reconfiguration principles [23], but broader usability of the developed approaches is limited for AIE applications. Moreover, there has been little work on specifying at a high level of abstraction how such abstraction services would work together at the application level taking into account in a combined and dynamic way the heterogeneous networks and services.

One of the main objectives of the ATRACO project [2] is to research and design an architecture and provide a system specification that will lay the foundations for the development of adaptive and trusted ambient ecologies. The proposed system operates in an AIE, which is populated with people and an ambient ecology of devices and services. People have goals, which they attempt to attain by realizing a hierarchy of interrelated tasks. For each user goal and the corresponding task model, an Activity Sphere (AS) is initialized, which consists of all software modules, services, and other resources necessary to support the user in achieving the goal. The ATRACO system architecture enables meaningful integration of relevant services and information during runtime and accomplishes that in a privacy-sensitive manner. ATRACO adopts a service-oriented architecture, combined with (a) intelligent agents that support adaptive planning, task realization, and enhanced human–machine interaction and (b) ontologies that provide knowledge representation, management of heterogeneity, semantically rich resource discovery, and adaptation using ontology alignment mechanisms. An ATRACO system supports adaptation at different levels, such as the changing configuration of the Ambient Ecology (AE), the realization of the same AS in different AIEs, the realization of tasks in different contexts, and the dynamic interaction between the system and the user.

In ATRACO, Network Adaptation (NA) is one of the key project dimensions allowing the uniform and transparent access to devices and services present in the networked environment and supporting the realization of activity spheres across a mixture of heterogeneous networks. Unlike other past and recent research efforts briefly presented above, our proposed framework specifies how network abstraction services should work together at the application level taking into account in a combined and dynamic way the heterogeneous networks and services present in the AIE. Our work additionally focuses on the issue of heterogeneity at the network level, concerning both devices and services, and defines adaptation as the systemic mechanism in order to deal with this issue.

NA in ATRACO offers a wealth of new exciting AIE experience on top of existing broadband service bundles. By exploiting the capability of continuous connectivity of an AIE to the Internet, it allows the provisioning of a set of services focused on the AIE offering at the same time a new experience for AIE management and control. Furthermore, it gives the operator the opportunity to expand its business by offering new customer services, such as security, safety, surveillance, energy management and monitoring, comfort and way of living management, and many other value-added services. NA can also interface a wide variety of sensors and actuators by multiple suppliers according to the user needs. NA is fully scalable and new services can be seamlessly accommodated using a sophisticated service development framework. Other benefits include the generation of alerts and user notification to always be aware of the AIE condition, the time-scheduling of certain functions for carrying out time-dependent tasks (e.g., switch on/off electrical devices), the creation and management of scenes and scenarios to easily adapt the AIE according to the situation (e.g., scene creation for watching movies by applying the desired lighting), as well as the recording and presentation of historical data and statistics to provide the user with useful information that can be further exploited for his own benefit (e.g., energy saving, user behavior adaptation, etc.). The following sections present in detail the network adaptation framework.

2.2 Intelligent Environment Network

An intelligent environment platform usually consists of different software components running in different network elements such as GUI clients for mobile phones and PC, portal server for remote login and user authentication and gateway software for accessing various home devices like automation equipment and network cameras. The user is able to have control over his environment either from inside using a PC or remotely using a PC or a mobile phone. Remote access is usually provided through a dedicated portal server after the user has authenticated himself. The connections on the portal server are encrypted while the connection between the home gateway and the portal is initiated from the gateway and is encrypted as well.

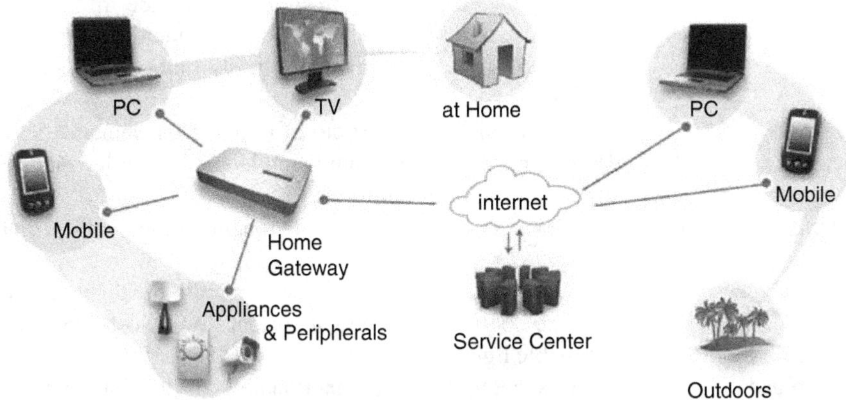

Fig. 2.1 Intelligent environment network architecture

This mechanism, on one hand, provides the user with advanced protection from harmful Internet attacks and on the other hand offers simple and easy installation since it does not need any network configuration on the modem/router. Figure 2.1 depicts the intelligent environment network architecture, addressing a smart home, contributed to the ATRACO project.

The smart home platform contributed to the ATRACO project provides a set of services focused on the home and offers a new experience for home automation and management, including security, safety, surveillance, energy management and monitoring, comfort and way of living management, and other exciting smart living services. The networking technology supported by the smart home platform can be easily installed without the need for extra wiring. A wide variety of COTS sensors and actuators can be seamlessly integrated in the home network according to the user needs. Main platform benefits include the generation of alerts and user notification through email, phone, or SMS to always be aware of home condition, the time-scheduling of certain functions for carrying out time-dependent tasks (e.g., switch on/off electrical devices), the creation and management of scenarios to adapt the home environment according to the situation (e.g., scene creation for watching movies by applying the desired lighting), and the processing of persistent data in order to provide the user with useful information (e.g., towards energy saving).

The intelligent environment network is a network integrating wired and wireless peripherals from multiple technologies. The user may select from a long list of wired and wireless automation peripherals to be integrated with the IE, giving unlimited capabilities in connectivity with other devices, user access and service provision. Devices have to be based on one of the automation technologies supported by the smart home platform, e.g., Sienna powerline, LonWorks PL/TP, or RF Z-Wave. Each type of device is supported using a driver per different technology adapting the functionality to an upper network independent layer. Automation peripherals are grouped depending on the functionality type:

Sensors	Devices that can sense a medium and provide information about its state (binary). This group encapsulates devices like wall mounted switches, remote controls, motion sensors, and door traps.
On/off	Devices used to open or close a circuit on demand. Such devices may be used to control appliances like coffee makers, water heaters, floor lights, etc.
Combined on/off	Devices that combine the functionality of the on/off and sensor types.
Dimmers	Devices used to open or close a circuit on demand and can also control the electrical current flow. Mostly used for dimmable lights.
Combined dimmers	Devices combining the functionality of the dimmers and sensor types.
Motors	Devices used to control two-directional motor devices. They can be used with devices like electrical shades, garage doors, valves, etc.
Combined motors	Devices combining the functionality of the motors and sensors types.
Thermostats	Devices giving feedback of a room temperature and control the central heating.

The intelligent environment platform contributed to the ATRACO project (see Fig. 2.2) integrates a set of Java tools and components that enable an OSGi-literate engineer to quickly design, develop and deploy new intelligent environment services utilizing the provided OSGi service platform and widely adopted automation technologies. The Home Controller is used to integrate connectivity with home devices of various home control technologies. The service platform embeds the use of OSGi technology in the home controller, ensuring interoperability with home devices and an easy way to integrate new home services. The base OSGi platform is extended by a set of OSGi network subsystems integrating various automation technologies supported by the home controller. The different network subsystems are interfaced in a common way through a Network Adaptation Layer. This layer adapts any network subsystem supported by the service platform, providing unified interfaces for devices from different networks, adding any specific support required for each type of device. The main intelligent environment functionality is actually offered and built by using the Network Adaptation API. Following the Network Adaptation API specification, a developer may build various applications, such as presentation layer applications (e.g., a web-based UI), monitoring applications that collect data and send them to a backbone server, and other control applications in the IE.

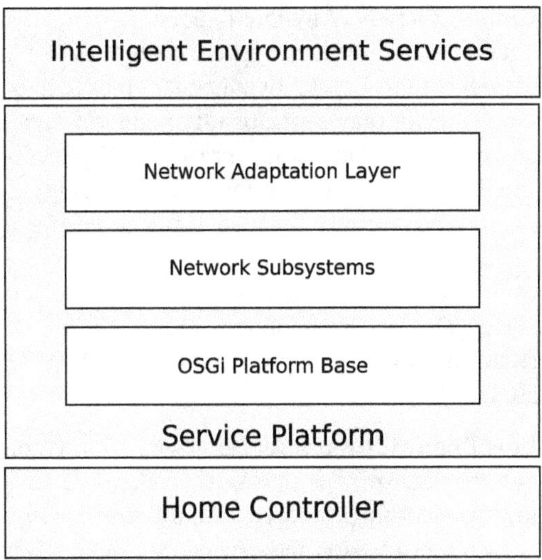

Fig. 2.2 Intelligent environment development framework

2.3 Intelligent Environment Service Design Considerations

The design and development of an intelligent environment service should take into account a number of considerations, such as embedded systems performance, other embedded system constraints, service responsiveness, low bandwidth automation networks, battery operated peripherals, etc.

Performance Smart home controllers are embedded systems having enough processing power to load featured operating systems and OSGi applications, however, they are more than far the capabilities of a desktop system. That is why it is really important to consider performance when developing intelligent environment applications. There are generally three rules to consider when developing applications for systems with resource constraints: (1) do not do what you don't really need to do, (2) do not allocate memory if you can avoid it, and (3) search for a faster implementation with the same result.

Storage Constraints Smart home controllers use flash memory components with limited write operations comparing to traditional storage components. Taking into account that the controllers' flash memory components are not modular and replaceable, they should never be used for frequent storage operations. In smart home controllers, flash memories are used to store the operating system filesystem, except the /tmp filesystem which is stored in RAM. So, any file created not under /tmp, is stored in controller's flash memory. On the other hand, any file stored in /tmp filesystem does not affect the lifetime of flash memory, however, it is lost in case of an unexpected shutdown of the controller. In case of critical information or useful log information, other resources should be considered: USB memory stick, backbone logging servers, etc.

Responsiveness Any intelligent environment service providing user feedback and interaction should be responsive, that is not hanging or freezing for significant period of time, especially when it comes to every day operations. Such non responsive behavior may be the result of either design choices not taking into account user perception or bad coding practices. Especially regarding OSGi, synchronization should be used with extreme care, especially between cooperating bundles: OSGi platform is a multi-threaded application, managing the execution of other multi-threaded applications in the same process context. Synchronization should thus be using custom techniques, ensuring that two bundles will not interoperate in a way that produces application deadlocks or long hanging periods. Especially regarding OSGi, see a presentation regarding OSGi best coding practices from OSGi alliance [5].

Low Bandwidth Automation Networks Home automation networks dealing with lighting, sensors, and HVAC systems usually utilize physical mediums using low bandwidth protocols, usually between the range of 5–80 kbps with variable error/retransmission rate per packet. While 5–40 kbps seems enough for protocols with limited data communication, we should always consider how scalable the behavior would be when the number of network nodes is increased. Basic rule is to avoid sending data to a device when this is not really needed: polling a thermostat for the temperature value, while knowing that the thermostat will send an update when the temperature changes, is not needed. In case the thermostat does not provide such an automatic update message, we should carefully consider the polling period required for a satisfying user perception or for other applications purposes: since the temperature usually will not fall significantly in 5 min, unless a door opens (and the outer temperature is below zero), the polling period could be set to 5 min. In case we would like a more frequent update, try for a 1 or 2 min periods, however, more frequent periods would likely be unreasonable: in such a case, it is likely that the device application is not intended for the use you desire and they should be replaced with a different device.

Battery Operated Peripherals Battery operated devices, e.g., motion sensors, are used in some automation networks. The battery lifetime is limited, usually specified by the manufacturer between 1 and 5 years. However, such specifications refer to wise use of the battery cycles. The three parameters that affect more the battery life are:

Sleeping period	Usually a device sleeps for some time and then wakes up to check for data available; keep the sleeping period of a device as long as possible.
Off delay	The period used until a device checks again for a change of its state, for example, a motion sensor to declare that it no longer detects motion.
Polling operations	In case each time the device wakes up it receives a number of polling commands, then the battery is exhausted sooner.

2.4 Network Adaptation Definition

In the ATRACO project we address the design of AIE systems using a service-oriented approach, in which resources in the environment provide independent, heterogeneous, loosely coupled, primitive services. The ATRACO infrastructure consists of SOA services. It supports basic services such as context management and reasoning, communication management, user profiling and service discovery, as well as adaptation and privacy services that form the basis for ATRACO ambient ecologies. NA is a set of functions and protocols allowing an AIE system to interact with network resources. The main goals of the NA layer are to allow devices and services to be used seamlessly by an AIE system and to simplify the access to networks in the AIE (usually for control, data sharing, communications, and entertainment). NA will achieve this by defining the appropriate functions and designing and implementing the needed abstraction layers.

Network resources in an AIE can be either devices or services. In general, devices include wired or wireless automation devices and entertainment devices. All these devices, intelligent or not, can use different networking technologies, thus could be part of different networks in a specific environment. For example, in a household the lights could be controlled using LonWorks technology while the motion sensors could be wireless Z-Wave for flexible and easy installation. The access from a single control point on both lights and motion sensors demands the use of two different networking protocol stacks: LonTalk and Z-Wave. This complicates the implementation of a control application since its design should cope with different protocol architectures and software structuring. The problem escalates when the control networks increase in numbers and the complexity of the control application gets maximized. The term intelligent device refers to a networked device with sufficient processing power to run network protocols. An intelligent LonWorks lamp or motor has built in the LonWorks technology so it can be controlled over the LonWorks network from another LonWorks device. Non intelligent devices can be converted to intelligent ones by connecting them on other intelligent nodes which are called network adapters.

Figure 2.3 illustrates the internal layering of the ATRACO NA component. To cope with the complexity of accessing diverse control networks, a common device representation layer is introduced. This representation layer should be architected in such a way in order to handle different networks (LonWorks, KNX, ZigBee, Z-Wave, X10, etc.) while it hides the complexity and details of each one of them. The application which should be placed on top of this representation layer, could not tell the difference between a Z-Wave lamp and a LonWorks lamp. The device representation layer creates a unique representation of each different device type across networks, which simplifies the evolution of applications and services. Beside devices there are various services which in most of the cases are services provided over IP networks. Examples of such services are the NTP (Network Time Protocol) to synchronize time to a reference time source, VoIP (Voice over IP) to implement Internet Telephony, RTSP (Real Time Streaming Protocol) to deliver voice and/or

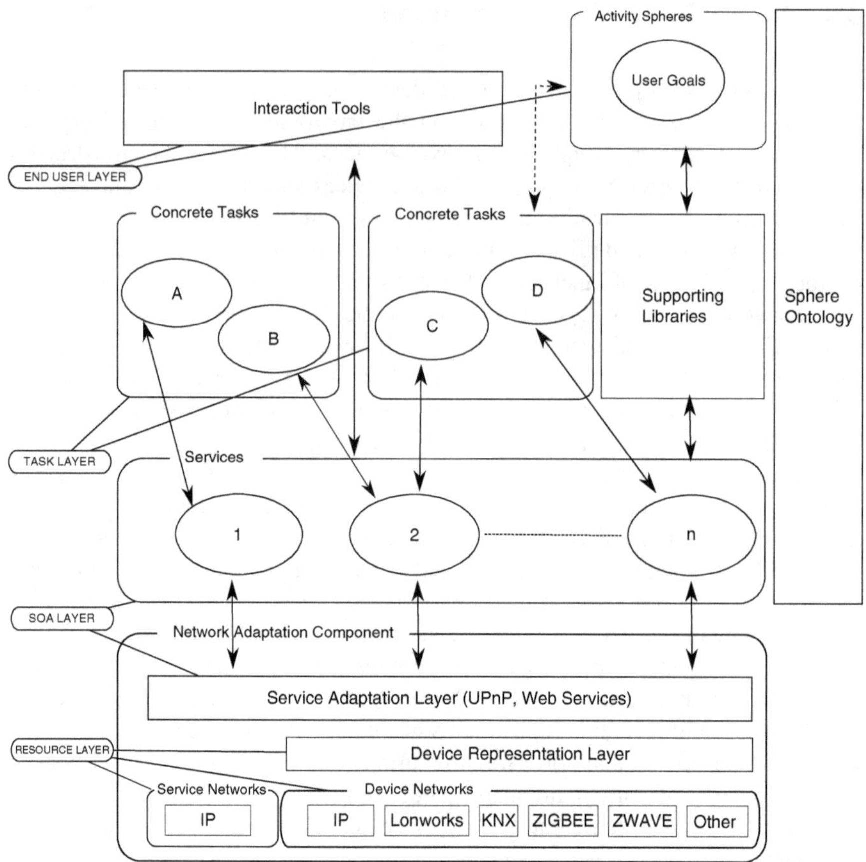

Fig. 2.3 Network Adaptation concept

video stream with real time properties, etc. Such types of services are of interest to an AIE so they must be shaped as AIE services and provided through the NA layer.

All network resources, devices, and services, should be provided in an AIE as services. The functionality in these environments is exposed as semantic services which an actor (either a user or an agent) can discover and then compose to form AIE applications. Each service is associated with at least one semantic description which shields the actor from the complexity of the Resource Layer realization and makes it easy for the actor to employ these services in accomplishing interesting and useful tasks. SOA offers one such prospective architecture where every component is either a service provider or a service consumer or both service provider and consumer. SOA may unify AIE processes by structuring large applications as an ad hoc collection of services. Different groups of applications both inside and outside an AIE system can use these services. SOA may use web services standards and web technologies and is rapidly becoming a standard approach for many information systems. However, web services face significant challenges because of particular requirements. Applying the SOA paradigm to an AIE system presents many problems, including response time, support of event-driven, asynchronous parallel applications, complicated human interface support, reliability, etc. The following sections describe the technology that ATRACO has chosen to implement its NA SOA architecture.

In ATRACO, NA is one of the key project dimensions allowing the uniform and transparent access to devices and services present in the networked environment and supporting the realization of activity spheres across a mixture of heterogeneous networks. The NA layer basically consists of two sub-layers; the common Device Representation Layer and the Service Adaptation Layer. The NA Layer is implemented using a mixed architecture, including a centralized home controller able to seamlessly integrate various different devices from different technologies around different home automation and control space and a variety of modules, some of which may be able to work in a distributed architecture. The Device Representation Layer provides an abstraction of the AIE devices to OSGi services that can be used by other AIE components. The service adaptation layer focus is first to provide the Device Representation Layer devices as OSGi services to the other AIE entities. On top of that, OSGi device representations are eventually transformed to UPnP services. Another focus of the Service Adaptation Layer is the UPnP adaptation of IP services participating in an AIE system. With the assumption that the available services are web services it would be straightforward to wrap them as UPnP services or services of any other technology of choice (e.g., R-OSGi [15]). It might even be possible to fully automate this procedure in a service agnostic way. UPnP seems to cover most of the aspects that are specified for the Service Adaptation Layer, such as ease of network setup, device discovery, self announcement and advertisement of supported capabilities.

2.5 Network Adaptation Guidelines

An ATRACO system supports adaptation at different levels, such as changing configuration of the ambient ecology, realization of the same activity sphere in different AIEs, realization of tasks in different contexts, and dynamic interaction between the system and the user. The design and specification of the NA component is done having the following adaptation guidelines in mind:

- Unified access on devices that belong to different networks (e.g., LON, Z-Wave, etc): The proposed NA framework defines common device types that share common variables across networks.
- Interoperability amongst multiple networks: The proposed NA framework allows events from any network to trigger actions to any network.

- Services can use alternative resources for task completion: The proposed NA framework employs a device registry where the available devices can be registered and queried by identifier, property value, constraints, and interface.
- Dynamic discovery for new networks and devices in an Ambient Intelligence (AmI) space ecology: The proposed NA framework features a Device/Driver manager to assure dynamic support for new networks and devices.
- Representation of device resources as services in a SOA middleware in order to be consumed by other AIE components.
- Representation of legacy Internet services in order to be consumed by other AIE components.
- Relaying of existing Web Services: An AIE is able to use a web service provided by the NA component without worrying about service availability, accessibility, and quality (e.g., NA provides single access to the AIE for a weather report service being able to evaluate availability, accessibility, and quality of many existing weather report web services provided by Internet sites).

2.6 Network Adaptation Framework

The NA middleware is a technological solution we have used for prototyping a typical AIE which seamlessly blends IP networking with a wealth of multimodal home automation functionality. NA provides uniform access to the controlled devices (including the full range of sensors and actuators) through an adaptation layer mapping all different network domains in the AIE space to the IP level, and some basic services for task execution and event management. Within ATRACO, NA contributes to the realization of activity spheres under the orchestration of the Sphere Manager (SM). NA is able to represent the integrated AIE to the ATRACO ontology level maintaining local device and policy ontologies and collaborating with the Ontology Manager (OM) to respond to queries regarding the state or properties of devices and during ontology alignments to propagate context changes to the sphere ontology.

NA integrates a set of Java tools and components that enable to quickly design, develop, and deploy services in an AIE space utilizing the provided OSGi service platform and widely adopted automation technologies. A home controller is used to integrate connectivity with devices of various home control technologies, such as LON PL, LON TP, Z-Wave (RF), X10, etc. The service platform embeds the use of OSGi technology in the home controller, ensuring interoperability with devices in the AIE and an easy way to integrate new services. The different network subsystems are interfaced in a common way through a device representation layer, known as NA-OSGi or ROCob layer, providing unified device representations. The main NA functionality is actually offered through the device representation middleware. User applications may include presentation layer applications, monitoring applications that collect and process data and send them to a backbone server or trusted recipients, home control and pervasive applications.

2.6.1 *Why OSGi?*

The key reasons for choosing OSGi as the framework to implement the device representation layer of the NA middleware are summarized in the following list:

- Reduced development complexity—Developing with OSGi technology means developing bundles: the OSGi components. Bundles are modules. They hide their internals from other bundles and communicate through well-defined services. Hiding internals means more freedom to change later. This not only reduces the number of bugs, but also makes bundles simpler to develop because correctly sized bundles implement a piece of functionality through well-defined interfaces.
- Ability to reuse ready made bundles in application development.
- Ease of component installation and management via a standardized API (e.g., using a command shell, a TR-69 or OMA DM management agent, a cloud computing interface, etc.). The standardized management API makes it very easy to integrate OSGi technology in existing and future systems.
- Ability of dynamic system updates. Bundles can be installed, started, stopped, updated, and uninstalled without bringing down the whole system.
- Availability of a dynamic service model allowing bundles to find out what capabilities are available on the system and adapt the functionality they can provide making code more flexible and resilient to changes. This requires that the dependencies of components need to be specified and it requires components to live in an environment where their optional dependencies are not always available.
- Simplicity of the OSGi API, powerful security model, portability across different execution environments, and wide use.

2.7 Device Representation Layer

NA acts in device representation multiple device types functioning using different technologies. In such environments devices of different types and communication technologies should be interoperable. NA needs a way of controlling and representing seamlessly devices of the same type (e.g., a Smart Light), even if they use different communication technology. Moreover, it needs a flexible way of supporting new device types and technologies, through the use of quick, efficient, and modular interfaces.

2.7.1 *Architecture Description*

Figure 2.4 depicts all blocks currently available in the device representation architecture, which constitute a working software release. This architecture can easily be extended to support other technologies, such as X-10 for example, with

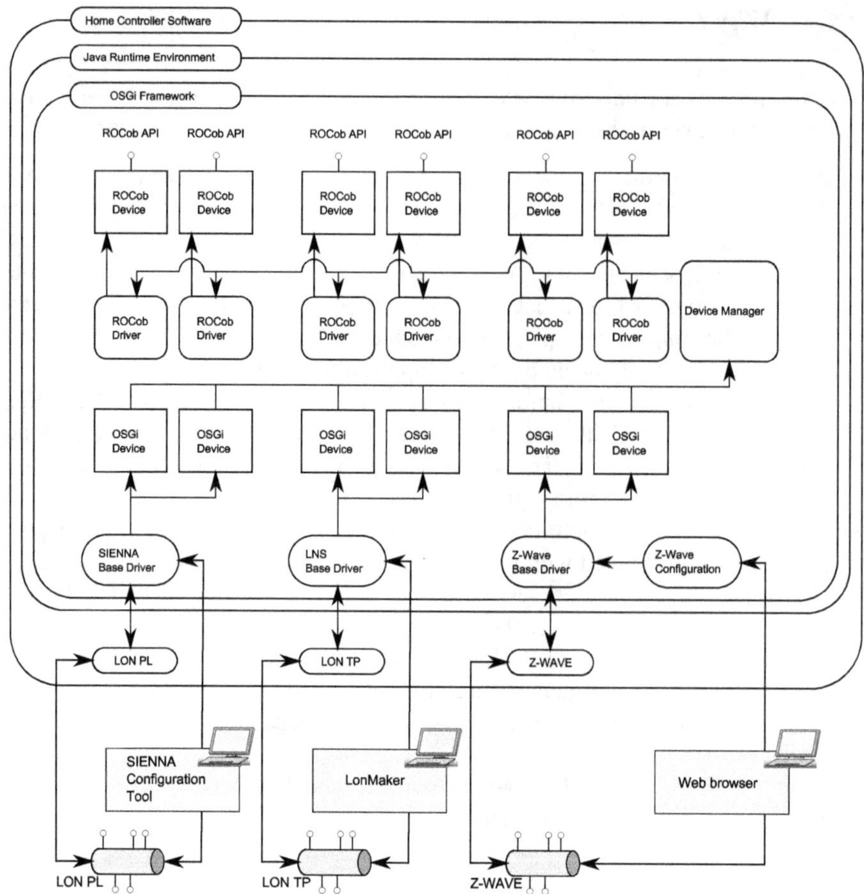

Fig. 2.4 Device representation layer architecture

the development of the corresponding device representation layer drivers. Moreover, the device representation layer API can be easily extended maintaining backwards compatibility, to support new functions in order to cover the functionality of new devices. In the following we include an overview that describes the functionality of each block (generic input/output, internal functionality) and a reference to relevant standards that the block complies to.

LON PL LonWorks is a home network protocol supported by the home controller. The physical layer as the PL implies is the powerline network of the house. LonWorks is seven layer protocol (OSI). The LON PL block is a Linux driver for accessing the neuron chip which runs the lower four LonWorks layers. All these layers are stacked in a commercial firmware provided by Echelon (the company who invented LonWorks) named MIP (Microprocessor Interface Program). The Linux driver is the software that communicates with MIP and transfers data back and forth from the neuron chip to the microprocessor memory.

LON TP It is also about LonWorks but the physical layer here is the Twisted Pair especially called FTT-10. LON TP block is the same Linux driver like LON PL regarding the source code. At runtime LON TP is a different instance and communicates with a MIP running inside a FTT-10 neuron chip. LON TP and LON PL can coexist in the same home controller.

Z-Wave It is a standard serial Linux driver which allows user programs accessing Z-Wave firmware running on a microcontroller located on a daughter-board which is connected internally on a serial port of the Home Controller. Z-Wave is an RF technology from Zensys that has a five layer protocol stack. The layers are the physical (RF), MAC (CSMA/CA), Transfer, Routing, and Application. The first 4 layers are implemented inside the Z-Wave microcontroller. Apart from the protocol layers, the Z-Wave microcontroller provides a serial API so that microprocessor hosted applications have access on the Z-Wave network.

Network Base Drivers The network base drivers are OSGi software elements that basically implement all or part of a control network protocol like LonWorks or Z-Wave. These drivers based on the provided configuration information coming from the network configuration tools, create software representations of the real network devices. These representations are in fact software objects which are called OSGi devices and provide a general interface for other OSGi applications to make actions and take feedback from the real network devices. A device object represents some form of a device. It can represent a hardware device, but that is not a requirement. Each network driver registers the associated device services in the framework in order device manager to find them and attach them to appropriate driver service.

Sienna Especially for the PL network the Home Controller has adopted the SIENNA protocol variation which provides a lightweight and reliable communication protocol for powerline networks. The Sienna Base driver is an OSGi base driver which implements the SIENNA protocol on top of the standard LNS driver. For simplicity the LNS driver doesn't appear as a different block.

LNS The LNS base driver is an OSGi base driver which implements the three upper layers of the LonWorks protocol stack. The LNS base driver is a third party proprietary block provided by Echelon. The LNS base driver has been modified in order to support the hardware LON TP and PL interface on the Home Controller.

Z-Wave The Z-Wave base driver is an OSGi base driver which implements the Z-Wave serial API and allows other OSGi applications to have access over the Z-Wave network but also to information which is locally stored in the Z-Wave microcontroller.

Sienna Configuration Tool The Sienna Configuration tool is a third party proprietary software which comes from Secyourit, through which the installer is able to setup a network of SIENNA LON PL devices. After the installation is completed this tool generates an XML file that contains valuable network information like the names, addresses, and the types of the devices. This XML file is further used by the Sienna Base Driver in order to instantiate the sienna OSGi devices.

LonMaker LonMaker is the network configuration tool provided by Echelon for setting up a LNS network. This is a professional tool indented to be used by the installer. After the installation is completed, LonMaker generates an XML file that contains network information like the device names, addresses, network variables, etc. This XML file is used by the LNS Base Driver so that the latter to instantiate the LNS OSGi devices.

Z-Wave Configuration The Z-Wave configuration service is responsible to setup a network of Z-Wave devices through a friendly user interface. This interface guides the user via a step-by-step procedure and after the configuration has been completed it transfers all valuable network information to the Z-Wave network base driver.

OSGi Device An OSGi device is a representation of a physical device or other entity that can be attached by a Driver service. OSGi devices that represent physical devices are instantiated by Network Base Drivers.

Device Manager The device manager is responsible for initiating all actions in response to the registration, modification, and unregistration of Device services and Driver services. The device manager detects the registration of Device services and coordinates their attachment with a suitable Driver service. All available Driver services participate in a bidding process. The Driver service can inspect the Device properties to find out how well this Driver service matches the Device service. The highest bidder is selected. The selected Driver service is then asked to attach the Device service. If no Driver service is suitable, the Device service remains idle. When new Driver bundles are installed, these idle Device services must be reattached.

ROCob Driver The ROCob drivers are in principle refinement drivers that translate the network-specific device interface to a common interface. This common interface named as ROCob API is used by other OSGi or non OSGi applications in order to communicate with devices in the control networks. Each ROCob driver registers itself in the framework registry expressing its interest for OSGi devices by setting-specific criteria. The ROCob driver driven by the Device Manager bids for the device of interest and finally if it is the highest bidder amongst other ROCob drivers it uses the provided device services.

ROCob Device The ROCob device is a representation of a physical device but at a higher abstraction layer compared with the respective OSGi device. ROCob devices are instantiated by ROCob Drivers.

ROCob API The ROCob API is a common interface that hides the specific network interface and gives a general interface for each different device category. For example, the API calls for controlling an actuator or receiving events from a sensor are the same either the devices belong to the LON PL network, the LON TP network or the Z-Wave network.

2.7.1.1 The LonWorks Subsystem

The LonWorks Base Driver communicates with the available devices and with the aid of an external Network Configuration XML file it creates the corresponding LonWorks OSGi Base devices. This external file can be produced by a network built with LonMaker with a utility that exports a LonWorks network image to an XML file, suitable for use by the smart home platform.

The LonWorks Base Driver communicates with the available devices and with the aid of an external Network Configuration XML file it creates the corresponding LonWorks OSGi Base devices. This external file can be produced by a network built with LonMaker with a utility that exports a LonWorks network image to an XML file, suitable for use by the smart home platform.

These OSGi Base devices can be used by various clients, such as the ROCob drivers or the Devbrowser bundle. The clients can communicate with the LonWorks Base Devices either with standard LonMark SNVTs and SCPTs provided by the OSGi-stdtypes package or with custom ones.

The LonWorks Subsystem (Fig. 2.5) is supported in the smart home platform with the LonWorks OSGi stack, which comprises of the following bundles:

LonMark OSGi Device Access API This bundle provides an API with which the various LonWorks ROCob drivers communicate with the LonWorks OSGi Base Devices that are offered by the Base Driver for LonWorks devices bundle. This bundle offers the "language" to read/write network variable values and communicate with the LonWorks devices and LonWorks network interfaces.

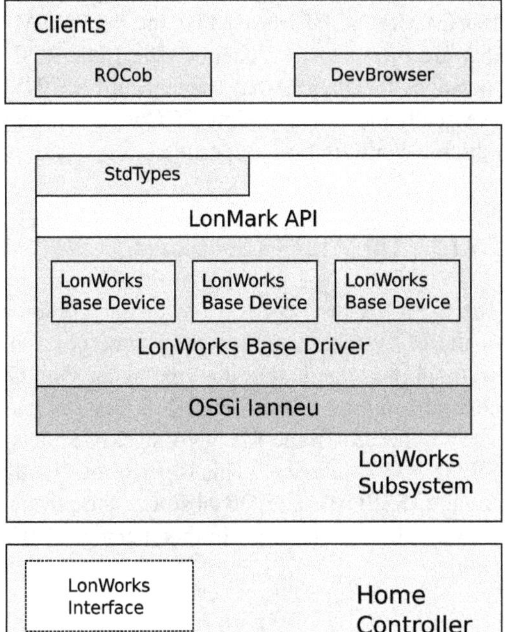

Fig. 2.5 The LonWorks network subsystem in the home controller

LonWorks Base Driver Bundle This bundle implements the LonMark API and is the heart of the LonWorks support in the home controller. The base driver keeps and maintains the LonWorks network image. Each LonWorks operation is performed through this bundle. Moreover, it supports the LonWorks over IP protocol and the LNS Pass-Through mode which it can be used for network creation. The base driver for LonWorks devices (also called LonWorks bundle) is loaded with a special XML file that holds the information about the LonWorks network. Based on the information that this file provides, the LonWorks bundle creates and registers LonWorks Base OSGi Devices that correspond to the real devices in the LonWorks network. These devices are then used by the ROCob LonWorks Composite Driver in order to dispatch them to the appropriate ROCob drivers and create the relevant ROCob devices.

LonWorks Device Browser This bundle is a debugging tool for base device services and presents a hierarchical explorer tree view of the LonWorks network. If the device has input network variables, the current values of the input network variables are displayed in text boxes (without formatting applied). The input and output network variables can be read, and the input network variables on the device can be written using the Device Browser.

Neuron Network Interface Driver This bundle takes up the responsibility to communicate with the native LonWorks interfaces of the home controller. It is the interface between the operating system's LonWorks interface driver and the LonWorks bundle.

LonMark SNVT/SCPT Formatters This bundle provides the values for the standard LonMark SNVTs and SCPTs, as these are described in detail in the LONMARK SCPT Master List and the LONMARK SNVT Master List documents that are provided by Echelon. The LNS ROCob drivers use this bundle to form messages for devices that use standard SNVTs and SCPTs. However, devices and eventually the corresponding ROCob drivers may use custom NV types and network variable configuration properties.

2.7.1.2 The SIENNA Subsystem

The SIENNA protocol is a power line communication protocol which sits on top of the LonWorks communication protocol. On the home controller, the SIENNA protocol implementation uses the underlying OSGi LonWorks implementation for communicating with the SIENNA Devices and on top of it builds its own logic. The Connected Home SIENNA stack abstracts the physical SIENNA devices into SIENNA OSGi devices. This is performed by the SIENNA Stack Base-driver OSGi Bundle (SSBOB). SSBOB abstracts the physical SIENNA Network in a way that is usable by the end user. In this manner, SSBOB ends up to understand two basic entities:

GE The GE is a group of devices that have the same G and E address values as defined by the SIENNA Protocol. These devices may offer either actuator or sensor type functionality.

GEType The GEType is a group of devices inside a GE that have the same functionality type, actuator or sensor. This means that a GE can contain at most two GETypes. Furthermore, there must be reasonable address assignment between the devices, so that devices with incompatible functionality are not grouped together. If this happens, then a GEType with the safest possible common functionality is created.

The SIENNA stack base driver OSGi Bundle is the interface between the SIENNA devices in the power line network and the upper OSGi layers in the smart home platform. Thus, it creates OSGi services that represent the actual SIENNA or groups of SIENNA devices. On its one end, SSBOB transforms LonWorks messages to SIENNA Messages, both in the incoming and the outgoing direction. For example, LonWorks NV updates that refer to the available SIENNA Devices are translated to the corresponding SIENNA updates and outgoing commands to the SIENNA devices are transformed to the appropriate LonWork messages. On the other end, SSBOB provides OSGi services to an upper layer OSGi clients, such as the SIENNA Control OSGi bundle. To describe the whole process in the OSGi framework, in the beginning, SSBOB registers the GEs as services. Then, the GETypes can be extracted from the GEs. The GETypes offer the capability to control and receive status updates from the physical SIENNA devices. The task of extracting the GETypes is performed by the clients of SSBOB. Such clients are, for example, the SIENNA Control OSGi bundle, which provides a testing tool for the SIENNA Network, and the SIENNA ROCob drivers (Fig. 2.6).

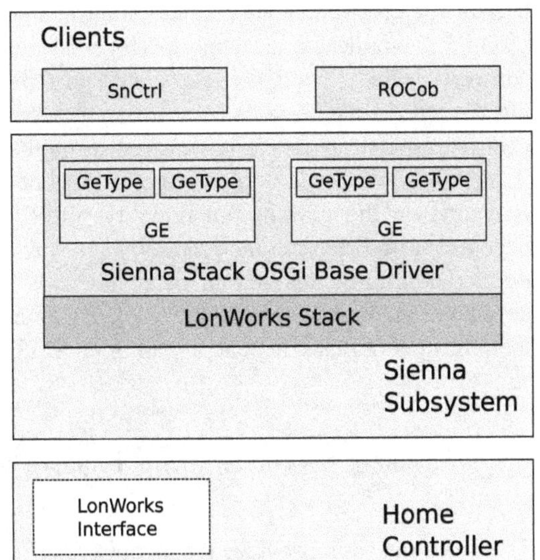

Fig. 2.6 The SIENNA network subsystem in the home controller

The integration of the SIENNA technology is achieved with the use of the SIENNA Stack Base-driver and SIENNA Control OSGi Bundles.

2.7.1.3 The Z-Wave Subsystem

The smart home platform integrates the Z-Wave technology and provides installation, maintenance, control, and monitoring capabilities over Z-Wave devices. The Z-Wave integration is achieved by an embedded Z-Wave interface on the home controller and an extension to the OSGi service platform to provide Z-Wave device services. The architecture of the Z-Wave subsystem is described in the following picture.

The home controller features a Zensys ZM3102 Z-Wave RF module. The Zensys ZM3102 module uses ZW0301 Zensys chips to integrate Z-Wave technology, providing a Serial Application Interface over a custom Zensys serial protocol. The home controller integrates the Z-Wave technology with the Z-Wave Base Driver and the Z-Wave Shell OSGi bundles.

The Z-Wave Base Driver is an OSGi bundle offering application interfaces for creating, managing and controlling Z-Wave networks. It enables the user to create, maintain, control, and monitor networks of Z-Wave devices. Using this bundle a developer may provide services for performing new devices registration, configuration, association and removal, as well as most installation and maintenance Z-Wave actions. A developer may also use the Z-Wave bundle in order to provide control, monitor, and simple test scenarios execution on Z-Wave devices.

Z-Wave Shell is an OSGi bundle offering both a command-line shell tool over the Z-Wave Base Driver OSGi bundle enabling management and control actions on Z-Wave networks. It enables the user to use the interface of the Z-Wave OSGi bundle for creation, maintenance, control and monitoring of Z-Wave networks. Using this bundle the user may perform a number of tasks by typing commands: add new devices, configure, associate and remove existing ones, as well as most installation/maintenance Z-Wave tasks; the user may also use the Z-Wave Shell to control, monitor and perform simple tests on the devices.

The Z-Wave Shell is also a good practical—yet not complete or detailed—reference on the current functionality offered by the Z-Wave OSGi bundle. It exposes most offered functionality in a way that is understood by a potential developer prior to reading in detail the Z-Wave API documentation: either the developer targets at a new installer tool, user interface for devices control or a simple monitoring tool that will send a mail or an SMS notification to a user (Fig. 2.7).

2.7.2 Device Representation Layer API Specification

The OSGi platform provides the level of required flexibility to add support for new technologies on the runtime, along with other dynamic features. The NA layer we

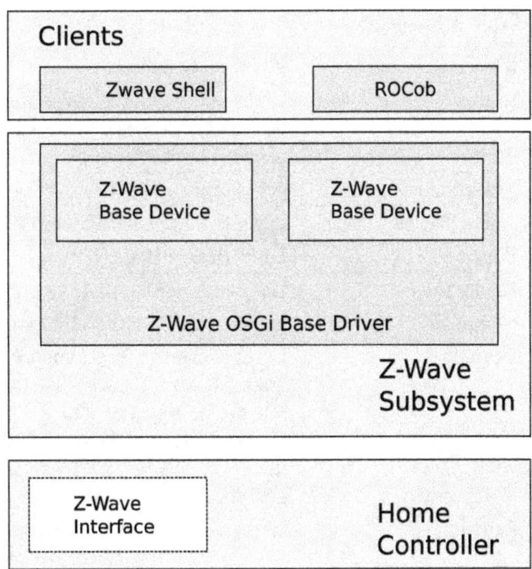

Fig. 2.7 The Z-Wave network subsystem in the home controller

present in the following sections is a set of JAVA classes and interfaces, bind with OSGi services, which are targeting on adding a level of transparency between the higher level services acting on the OSGi layer and the real physical devices. The physical device (OSGi) drivers must implement a number of the NA interfaces, in order that bundles built on top of the NA layer can acquire and use them seamlessly to any underlying physical device technology (Table 2.1). The main ingredients of the NA API are the NA Device and Function interfaces. Each OSGi implemented device driver that takes care of the communication specifics between the physical devices and the OSGi framework finally generates and registers to the framework a NA-OSGi Device. A similar term for NA-OSGi is ROCob and we use these similar terms interchangeably in the chapter. The registration must be done in a well-defined way (see below). Each registered NA-OSGi Device implementation must implement one or more Function interfaces. Using the methods defined in these interfaces other NA-OSGi aware services will be able to communicate with the NA-OSGi Devices and thus with the physical devices in the underlying network.

2.7.2.1 NA-OSGi Device Interface

A service must implement this interface to indicate that it is a NA-OSGi Device. Services implementing this interface give the system components the opportunity to discover them and retrieve all required information for managing and performing certain actions. Every service that has the intention of being registered as a NA-OSGi Device must conform to the semantics specified by the Device interface of the NA-OSGi API. In detail:

Table 2.1 Specification summary of the NA-OSGi API

Interface	Fields and methods
Alarm	Irrelevant, Level (E), Message (E), NoAlarm, Type (E), Communication Error, Device Inaccessible, Device Specific, Intrusion, Maintenance, Tamper Attempt
AlarmState	getAlarmLevel, getRelevantAlarmTypes
AnalogMeter	Level (E), getAnalogMeterLevel
Battery	Level (E), getBatteryLevel
BinarySensor	Current State (E), State Off (E), State On (E), getState
Composite	getEndpointCount, getEndpointFunctions(index), getEndpoint(index)
Device	Level, Function (E), Functions (S), Type (S), UniqueId (ES), Event Topic, Binary Sensor (S), Door Trap (S), Flood Detector (S), Gas Valve (S), Key Fob (S), Lamp (S), Motion Sensor (S), Shade (S), Smoke Detector (S), Switch (S), Tank Level Meter (S), Thermostat (S), Water Valve (S), Unknown (S), getDefaultName, getFunctions, getType, getUniqueId
Function	Alarm (DE), Analog Meter (DE), Battery (DE), Binary Sensor (DE), Composite (DE), Level Absolute (DE), Level Absolute Timed (DE), Level Relative (DE), Level Relative Control (DE), Level Relative Timed (DE), Switch (DE), Thermostat (DE), Function Classes, Function Names, Count, Device Error, Driver Error, Invalid Args, Invalid State, Network Error, Queued, Stack Error, Success, Transmitted
LevelAbsolute	Current Level (E), Max (E), Min (E), getLevel, setLevel
LevelAbsoluteTimed	setLevel
LevelRelative	startLevelChange, stopLevelChange
LevelRelativeControl	Current Direction (E), Down (E), Stopped (E), Up (E), getLevelChangeDirection
LevelRelativeTimed	startLevelChange
ROCState	getDeviceId, getFunctionId, getProperties
Scene	Actions Count Error, Function ID Error, getActionCount, getDeviceId(index), getFunctionId(index), getProperties(index), getSceneId
SceneManager	addScene, getScene, getSceneIds, removeScene, runScene
SceneTrigger	add, getTrigger, getTriggerIds, remove
Switch	setState
Thermostat	Auto (E), Cool (E), Heat (E), Off (E), Mode (E), Setpoint (E), Temperature (E), getActualTemperature, getOperationMode, getTemperatureSetPoint, setOperationMode, setTemperatureSetPoint
Trigger	getDeviceId(index), getFunctionId(index), getProperties(index), getSceneIds, getTriggerDeviceCount, getTriggerId

E Event Property, *S* Service Property, *DE* Device.Function Event Property

OSGi Registration Classes This property should be a list containing the NA-OSGi Device interface class name and the class names of all NA-OSGi Function Interfaces implemented by this NA-OSGi Device (see below).

Device Unique ID The value of this registration property denotes the Unique ID among the NA-OSGi Devices registered in the framework. This property is aimed to be used by high level services to distinguish the NA-OSGi Devices. This value should remain the same for every distinct NA-OSGi Device, if it needs to be re-registered in the future. This is because higher lever services may have generated data associated with this ID. This property can be acquired using the getUniqueId() method of the NA-OSGi Device interface.

Device Functions The value of this registration property is a list of the NA-OSGi Functions that are supported (implemented) by this registered NA-OSGi Device. This property is commonly used by tracker services to filter the search on a specific NA-OSGi Function type. This list can also be acquired by calling method getFunctions() of the NA-OSGi Device interface.

Device Type The value of this registration property denotes the underlying physical device type. This property's value can also be acquired using the NA-OSGi Device method getType().

The registered NA-OSGi Devices need to establish a two way communication between other NA-OSGi services. In the following we describe how other services can handle the NA-OSGi Devices leading to the control of the underlying physical devices and how the NA-OSGi Devices can notify other NA-OSGi services for events such as device property updates.

The registered NA-OSGi Devices can be tracked by other services using the defined by the OSGi framework methods. The properties that escort the NA-OSGi Devices during their service registration can be used to help services tracking the desired devices. Once a NA-OSGi Device service is tracked, it can be controlled using the methods of the NA-OSGi Function interfaces that it implements. Then, the service could be casted to the appropriate interface and use its implemented methods.

In the general case, almost every NA-OSGi Device will eventually need a way to notifying other NA-OSGi aware services for device property updates, derived by the underlying physical device (like a lamp turning on), or for any other useful reason. The NA-OSGi Devices use the OSGi r4 EventAdmin specification to deliver events to the other Framework services. These types of events must conform to the following:

Event Topic All the NA-OSGi Device Events must start with the "NA-OSGi" prefix to inform the OSGi Event Handlers that this is a NA-OSGi Event. This can be used for event filtering. The event topic must end with the NA-OSGi Interface name that triggered this event generation. For instance, if an event was generated by a NA-OSGi Device that implements the NA-OSGi Switch Interface, the full event topic would be NA-OSGi/Switch.

A NA-OSGi Device may implement more than one NA-OSGi Interfaces. In that case, the suffix of the OSGi Event Topic will be the Interface name associated with the property, the value of which is published for notification to the Framework. For instance, consider a NA-OSGi Device that stands for a Dimmer physical device and implements the Switch and the LevelAbsolute NA-OSGi Interfaces, to support the on/off and dimming functionality of this dimming physical device. Suppose that this device turns off (i.e., changes its state) and the NA-OSGi Device needs to notify the framework for this change. The property CURRENT STATE that was modified lies within the NA-OSGi Switch Interface. So, the Event topic would be "NA-OSGi/Switch." If the NA-OSGi Device needed to notify for a change in the brightness level of the dimmer device as well, it should create one more event for the property CURRENT LEVEL of the LevelAbsolute Interface with a topic like: "NA-OSGi/LevelAbsolute."

Device Function This property must accompany the NA-OSGi Events. The allowed values for this property derive from the NA-OSGi Function interface. It provides information about the NA-OSGi Interface type that triggered the Event. It has the exact same meaning as the suffix of the OSGi Event Topic of the NA-OSGi Events. Although this information is redundant, it can be used for Property-based filtering or alternatively to be used in "switch" statements.

Device Unique ID This is the same property that accompanies the NA-OSGi Device to the framework registration.

In addition to these properties, every NA-OSGi Event must be accompanied with one or more properties that carry the actual information for this event. In the general case, the properties that are used depend on the NA-OSGi Function that triggered the event. These properties are required to exist in the NA-OSGi Event. For instance, the CURRENT STATE property mentioned in the example above is a part of the NA-OSGi Switch Interface. Any other property that could be of interest to the event handlers may be added as well.

2.7.2.2 Function Interface

This interface is not intended to be implemented by any class. Its reason of existence is to provide a number of fields of the available NA-OSGi Function interfaces (that should be implemented by the NA-OSGi Devices), represented as primitive integers along with helper collections that map these integers to strings like friendly names or class names. These collections and fields are meant to help the developer at the NA-OSGi Device registration, event generation, and handling. With these functional profiles a NA-OSGi device can represent the functionality of the underlying physical device. The enumeration of the available Function interface representation is the following:

Composite	Interface commonly used for physical devices with multiple endpoints.

Binary Sensor	Interface commonly used for sensor-like devices, like motion sensors, etc.
Level Absolute	Interface used for physical devices with level properties like dimmers, etc.
Level Absolute Timed	Interface used for physical devices with level properties like dimmers with an extra timing functionality.
Level Relative Control	Interface used for devices with direction semantics, with read only values.
Level Relative	Interface used for physical devices with direction semantics like shades etc.
Level Relative Timed	Interface used for physical devices with direction semantics like shades, etc., with an extra timing functionality.
Switch	Interface used for on/off physical devices like lights, simple home appliances, etc.
Battery	Interface which is commonly used for representing a physical battery.
Alarm	Interface which is commonly used for alarming devices.
Analog Meter	Interface used for devices that measure analog inputs like pressure and temperature.
Thermostat	Interface which is commonly used for representing a physical Thermostat device.

2.7.2.3 Scene Interface

The scene-related interfaces intend to provide a service layer for scene (task) execution. In general, a scene can be considered as a collection of NA-OSGi Devices that will execute a number of actions on these devices when a certain event occurs. An alarm service is included, with which the device representation layer can send alarm events to other NA-OSGi components or higher layers. It makes use of a set of defined alarm levels, types and properties, to cover a wide range of alarm events.

The Scene Interface is used to define the collection of the NA-OSGi Devices that take part in a scene, along with a number of properties per Device that will lead to the appropriate actions on the Device when the Scene is executed. Each Scene implementation has a unique ID among the scenes and holds a collection of NA-OSGi Device IDs. For each of these IDs it holds a Dictionary of properties, the data of which depend on the NA-OSGi Device Function interfaces the specific NA-OSGi Device implements. Moreover, it holds a Function ID per NA-OSGi Device. This Function ID is used from the Scene implementation when the Scene is executed to lead to the appropriate NA-OSGi Device method call on the specific device. At the same time, the properties provided by the Dictionary per NA-OSGi Device are used to determine the method's arguments.

For instance consider a Scene with one NA-OSGi Device that implements the Switch interface and we want this device to turn on upon the Scene execution. The function ID for this device should be "NA-OSGi.Function.SWITCH" and the property within the Dictionary for this device should be: "Switch.CURRENT STATE=Switch.STATE OFF." When this Scene is to be executed, "setState(Switch. STATE OFF)" should be called on this NA-OSGi Device, which will lead to the physical device to turn off.

2.7.2.4 Scene Manager Interface

The implementation of this interface works as a Scene manager service. The role of this service is to hold a collection of scenes provided by other OSGi bundles and execute them upon request. The implementations of this interface should register to the OSGi framework under the class name "NA-OSGi.SceneManager," for other bundles to track and use it.

2.7.2.5 Trigger Interface

For a scene to execute, a certain event must occur. These events are property modifications of the NA-OSGi Devices, like a Binary Sensor at STATE ON, which can be tracked using the OSGi r4 EventAdmin specification as described before. This interface is introduced to be used for that reason. It holds a collection of NA-OSGi Device IDs that take part to the triggering of a scene (or scenes) execution. It also holds a collection of the scene IDs to be executed when a certain event is occurred. It further holds the event that should occur to trigger the scenes' execution. For each of the NA-OSGi Device IDs that take part on this a Dictionary of properties (the data of which depend on the NA-OSGi Device Function interfaces implemented from that specific NA-OSGi Device) is hold. The event is supposed to activate the trigger when these properties of the NA-OSGi Device are met. If just one of the NA-OSGi Devices in the collection of this Trigger meets these property values, the scenes must be executed. Moreover, a NA-OSGi Function ID is held for each NA-OSGi Device in the collection. This is supposed to be used on the NA-OSGi Device event filtering.

For instance, suppose we want a number of scenes to be executed when a Binary Sensor's state changes to STATE ON. This Binary Sensor is described by a NA-OSGi Device implementing the Binary Sensor interface. The Dictionary for this Device in the Trigger should hold the property "BinarySensor.CURRENT STATE=BinarySensor.STATE ON" and the Function ID for this Device should be "Function.BINARY SENSOR."

2.7.2.6 Scene Trigger Interface

The implementations of this interface are used to manage implementations of the Trigger interface. In other words, the implementations of this interface should be OSGi services that hold a collection of Triggers added by other OSGi bundles.

The implementations must be registered to the OSGi framework under the class name NA-OSGi.SceneTrigger, for other bundles to track and use to add or remove triggers. They should also take care of the evaluation of the triggers that hold and the scene execution of the scene IDs associated with these triggers.

2.7.2.7 Alert Service

This utility service is designed to provide a simple alerting service to the other NA-OSGi aware services in the framework, based on custom Rules. The alerts make use of the alarm characteristics. This service listens for NA-OSGi Device events and evaluates a series of user defined Rules to finally create the specific NA-OSGi alerts. This service is most commonly used when analog meters are used, like depth meters. For instance in fuel tanks, you may need to have an alert when the fuel level drops below a specific threshold. This alert could be used by a NA-OSGi service to create a notification (e.g., SMS) to the user.

The rules used by the NA-OSGi Alert service are based on the NA-OSGi Device-specific properties. The user creates a number of rules based on these specific properties and the evaluation is done by the service by listening for NA-OSGi Device property changes from the NA-OSGi Device events. There are two ways to evaluate a set of properties for a NA-OSGi Device (AND, OR). The NA-OSGi alert service uses the OSGi r3 Configuration Admin Specification to create new Rules. The properties that need to be set in order to create a new Rule include the NA-OSGi Device IDs that take part in this rule, the message that will accompany the alert when generated, the level of the alert, the properties to be evaluated, the evaluation type, as well as a bounce filter and a delay used for avoiding bouncing effects.

2.8 Service Representation Layer

Services are the basis of distributed computing across the Internet. A service consumer locates a service and invokes the operations it provides. As ATRACO has adopted the SOA model each component could be either a service provider or a service consumer or both consumer and provider.

An ATRACO system will use the Network Adaptation component to access network resources. Resources can be either devices or services, but both can be accessed within ATRACO as services. In order to do that, Network Adaptation defines an internal upper layer called service adaptation layer on top of which all resources are viewed and accessed as services.

Devices organized in device types according to their functionality will expose through the service adaptation layer a list of access methods in the form of services. From the service point of view a device can be analyzed as a list of state variables (or parameters) and a list of methods or functions that either read the value or change the value of these state variables.

Other types of services which can be found and accessed in the Internet are of significant interest to an ATRACO system and thus they should be provided in the ATRACO environment in a proper shape in order to be consumed from the ATRACO entities. These services could be already implemented as Web Services or may have a completely different architecture and access interface. In either way, the Network Adaptation Layer should intervene in order to reshape the service to follow the selected ATRACO service architecture acting as a service proxy to existing Web Services and enforcing authentication and security policies.

The NA-OSGi layer provides an abstraction layer for a variety of underlying networking technologies. Looking from the perspective of higher layers, it can be either used directly to implement intelligent environment services or a presentation layer, but it can also act as a common base for extending interoperability with services provided by other well-established technologies, like UPnP and R-OSGi.

The upper layer of NA includes the Service Representation Layer, avoiding that way a direct interface of the service adaptation to different underlying networks. In an ATRACO environment where NA is used to access devices participating in multiple different networks the integration of the service representation layer with the device representation layer minimizes the complexity of the implementation and allows future additions of networks in a seamless way. This layer converts the available devices from the device representation layer as well as other wrapped legacy services to services of a common technology of choice, e.g., UPnP or R-OSGi. As a result, devices and services can consume or be consumed by other devices and services available in an AIE environment.

The service representation layer supports zero-configuration networking. An automation device from any vendor and supported technology, once registered in the NA component, dynamically announces its name, conveys its capabilities upon request, and learns about the presence and capabilities of other devices or services.

UPnP seems to cover most of the aspects that are specified for the Service Adaptation Layer, such as ease of network setup, device discovery, self announcement and advertisement of supported capabilities. Currently the Network Adaptation framework supports UPnP and integrates it with the OSGi Service Platform. On top of that, a set of NA-OSGi to UPnP wrappers have been developed that eventually transform the NA-OSGi services to UPnP services.

2.8.1 UPnP Adaptation

The UPnP Device Architecture specification provides the protocols for a peer-to-peer network. It specifies how to join a network and how devices can be controlled using XML messages sent over HTTP. The UPnP specifications leverage Internet protocols, including IP, TCP, UDP, HTTP, and XML. The OSGi specifications address how code can be downloaded and managed in a remote system. Both standards are therefore fully complimentary. Using an OSGi Service Platform

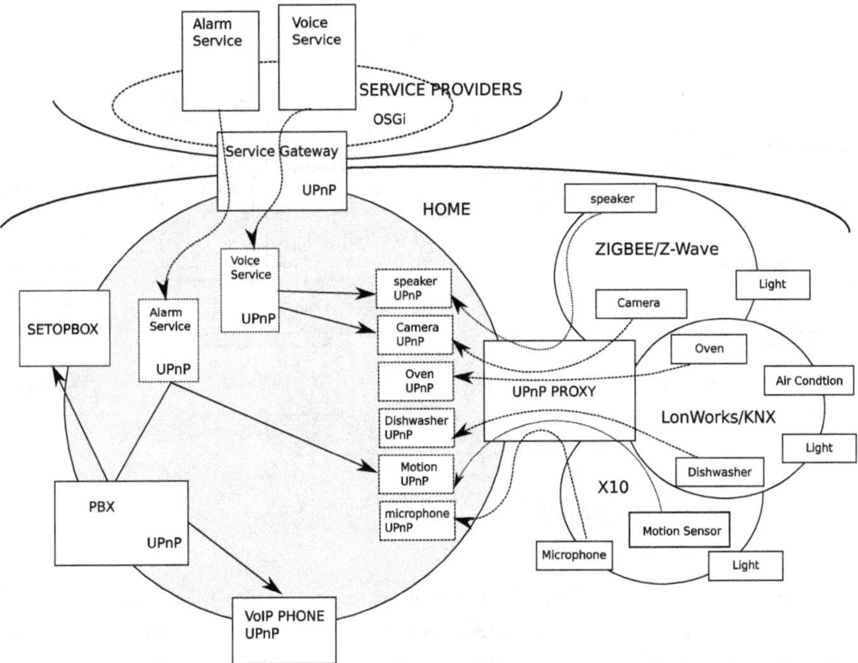

Fig. 2.8 Example of UPnP proxy virtualization function

to work with UPnP enabled devices is therefore a very successful combination allowing the development of OSGi bundles that can interoperate with UPnP devices and UPnP control points.

By using UPnP above the NA-OSGi layer, the functionality of services for standard home control devices (lighting, shades, heating, etc.) can be combined with other spheres of interest, such as audio-visual devices or other web services transformed into UPnP services (network clocks, weather report, news services, and others). This combination opens a new wide layer for offered services or complex presentation designs, accommodating and re-innovating the modern lifestyle inside a smart environment.

NA provides a UPnP virtualization of the interfaced physical devices, which eventually belong to heterogeneous non IP networks. Figure 2.8 illustrates a generic view of a networked AIE. UPnP proxies bridge IP networks with non IP networks representing at the same time devices belonging to non IP networks as UPnP entities. Figure 2.9 illustrates the virtualization function of the UPnP proxy. The proxy knows how to communicate with the Z-Wave microphone. For that reason it uses a special library that encodes, over the Z-Wave API, microphone commands to start/stop recording as well as to control the gain and the sampling rate. In order higher perceptual components, such as the voice recognizer, to make use of the microphone, the proxy represents it as a UPnP device exporting appropriate actions

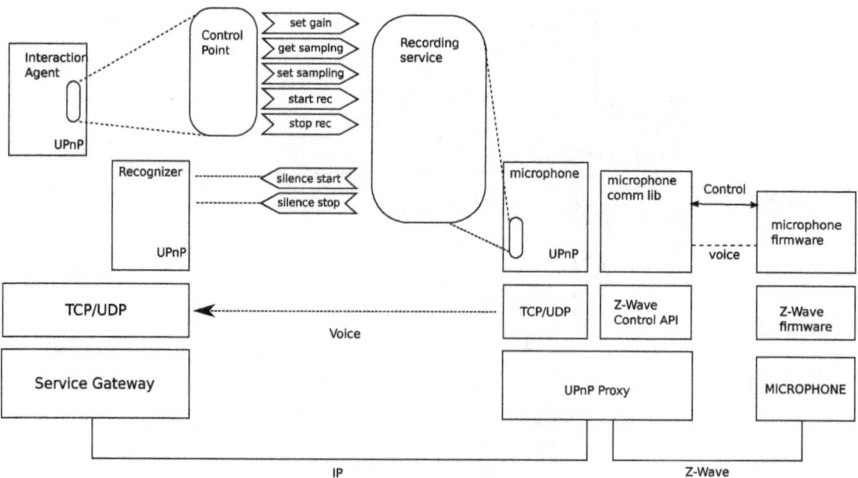

Fig. 2.9 Generic NA view of an AIE

for remote invocation (set gain, set sampling, start rec, etc.). On the other hand taking advantage of the silence detection feature of the microphone, proxy sends appropriate events triggering that way the voice processing at the recognizer side.

Considering the device types, functional profiles, and event properties defined in the device representation layer as well as the availability of a standard OSGi UPnP Driver implementation it is straightforward to develop OSGi Wrapper services that export NA-OSGi devices as UPnP devices. The NA layer provides UPnP device services to the AIE network via the UPnP wrappers. Those services expose the relative UPnP device and service description XMLs. Clients that intend to use the offered services can use third party programming APIs, based on the UPnP library on use, e.g. CyberLink (Java), CyberDomo (UPnP), Intel (C#). A side advantage of this procedure is that a single UPnP wrapper can provide UPnP services for a device class, e.g. on/off switch, irrelevant to the technology that the switch uses (LON, Z-Wave, SIENNA), due to the abstraction offered by the NA-OSGi layer. The NA UPnP wrapping architecture is briefly illustrated in Fig. 2.10, while Fig. 2.11 depicts the NA device layers (in parallel for exemplary LON and Z-Wave networks), from physical device to UPnP device.

A Wrapper service would usually track the availability of NA-OSGi devices with certain characteristics and then register a UPnP Device service with the appropriate device and service description to represent the underlying device. The registered UPnP Device service is then tracked by the UPnP Base Driver, the bundle that implements the bridge between OSGi and the UPnP networks, and exported to the network as a UPnP device. The registered UPnP Device service should be implemented in such a way that the provided UPnP Service class implementations will correspond to the functions supported by the NA-OSGi device. Similarly each UPnP Service should be implemented in such a way that the provided UPnP Action

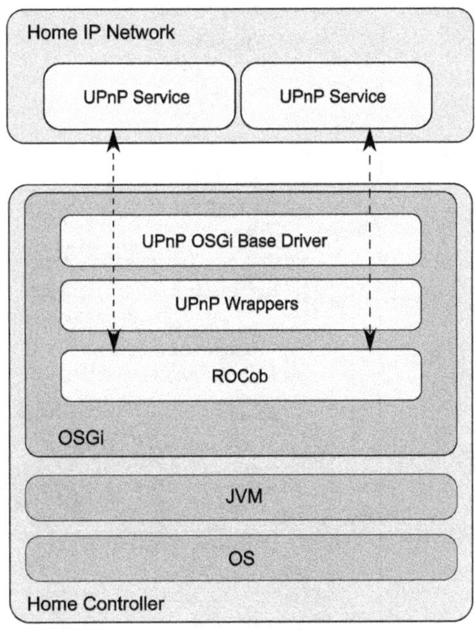

Fig. 2.10 UPnP wrappers overview

class implementations will correspond to the Java methods defined in the associated NA-OSGi function interface. In the reverse path the NA-OSGi event properties defined in each supported NA-OSGi function should be represented as UPnP State Variable in the corresponding UPnP Service class implementation and appropriate notify UPnP Event calls should be made to registered UPnP Event Listener services for each NA-OSGi event received.

Device adaptation can also be used on the reverse path to control devices and web services already present in the AIE provided by other sources. The way to do this is to import them in the NA framework, transform them to ROCob devices and operate them like local ROCob devices. This way, various UPnP services tracked by the OSGi UPnP base driver can be registered with the OSGi framework. Then, appropriate refinement drivers can refine these devices to NA-OSGi devices. Since the imported UPnP devices have their NA-OSGi equivalent in the OSGi framework, they can be combined with the rest NA-OSGi devices and exploit advanced features of the NA-OSGi layer. For example, they can be used seamlessly in scenes creation and execution.

2.8.1.1 General Exported UPnP Device Model

A NA-OSGi UPnP Wrapper refines a NA-OSGi device to the appropriate UPnP device. Then, with the aid of the UPnP base driver, this device is advertised in the network. The UPnP wrapper consists of two main components:

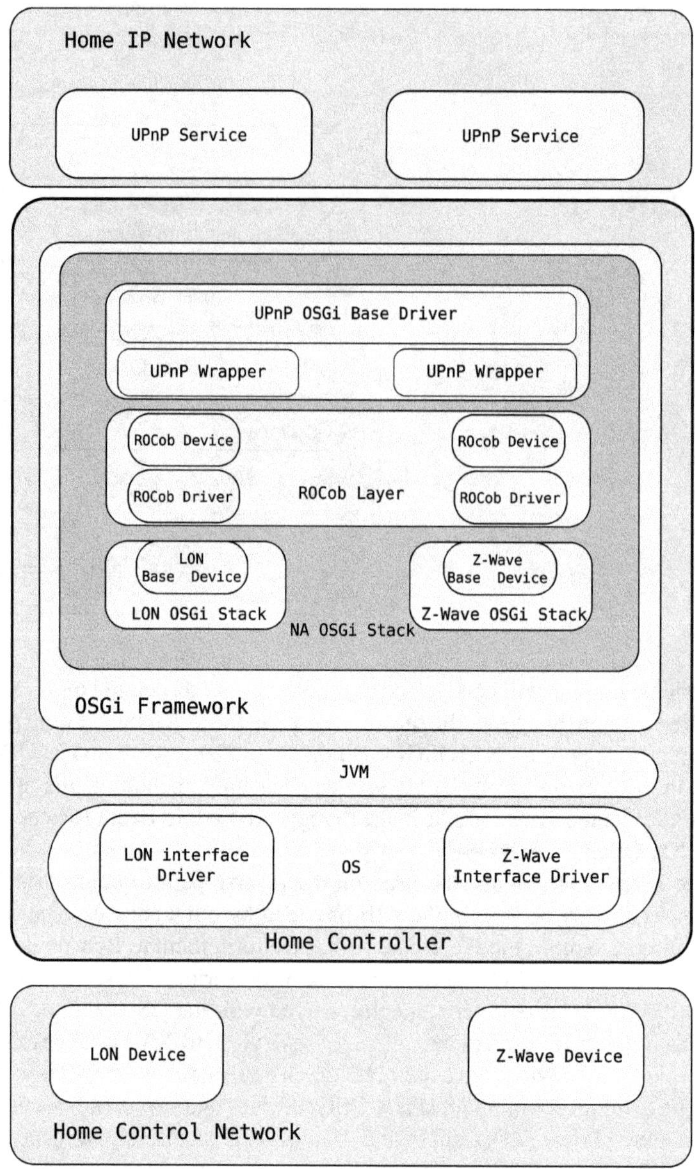

Fig. 2.11 Device adaptation layers, from physical device to UPnP device

- The NA-OSGi to UPnP driver, that tracks the available NA-OSGi devices, selects the suitable devices according to an LDAP filter and creates the corresponding UPnP devices.
- The NA-OSGi to UPnP device, which maps the NA-OSGi device functionality to the corresponding UPnP device functionality.

Table 2.2 UPnPDevice service registration properties

Property name	Property type	Property value
DEVICE CATEGORY	String	UPnP
objectClass	String	org.OSGi.service.upnp.UPnPDevice
UPnP.device.friendlyName	String	A characteristic term for the device plus the value returned by com.inaccessnetworks.rocob. Device.getDefaultName()
UPnP.device.manufacturer	String	inAccess Networks
UPnP.device.modelDescription	String	ROCob—device type—UPnP Wrapper
UPnP.device.UDN	String	A standard or a custom device UDN, according to the device type and the UPnP specification
UPnP.export		Not Required
UPnP.device.type	String	A standard or a custom device type definition, according to the UPnP specification
UPnP.service.id	String[]	Array of Strings with the provided functions, e.g. "urn:upnp-org:serviceId:SwitchPower.001"
UPnP.service.type	String[]	Array of Strings with the supported UPnP service types by the wrapper, e.g.: "urn:schemas-upnp-org:service:SwitchPower:1"

For each tracked device a new service is registered implementing the UPnPDevice interface and the registration properties listed in Table 2.2. The UPnPDevice instance will provide a UPnPService for each corresponding NA-OSGi function supported by the NA-OSGi device. The wrapper's architecture creates a pool of UPnP services that correspond to NA-OSGi functions. This way wrapped UPnP devices can select and combine services from this pool to offer the corresponding functionality from NA-OSGi to UPnP layer. This design can lead to rapid development, since replication of common functionality is avoided.

2.8.1.2 Switch UPnP Wrapper

Since the NA-OSGi Switch function matches exactly the UPnP Forum standardized SwitchPower service, the wrapper should eventually export to the network a UPnP device that utlizes this standard UPnP service. Therefore the UPnPService instance should provide two UPnPAction implementations corresponding to *GetStatus* and *SetTarget* actions defined in SwitchPower:1 as well as two UPnPStateVariable implementations corresponding to *Status* and *Target* state variables defined in SwitchPower:1 (returned by UPnPService methods getActions() and getStateVariables(), respectively).

The *GetStatus* UPnPAction.invoke() implementation should call the NA-OSGi Switch.getState() method, while the *SetTarget* UPnPAction.invoke() implementation should call the NA-OSGi Switch setState() method. In both cases the wrapper service must convert NA-OSGi values to UPnP values. Since NA-OSGi Switch function does not have the concept of two state variables, this must be implemented by the wrapper explicitly. The *Status* UPnPStateVariable should have the value returned by the last NA-OSGi Switch.getState invocation or the last NA-OSGi event received. *Target* UPnPStateVariable should just hold the requested value from the last *SetTarget* UPnPAction.invoke() call.

NA-OSGi Switch function defines a single event property Switch.CURRENT STATE and a NA-OSGi event is raised containing this property every time the underlying physical device changes state. Therefore the *Status* UPnPStateVariable is evented and the wrapper bundles should maintain a list of the registered UPnPEventListener implementations (that requested events for this state variable) and call the notifyUPnPEvent callback every time a ROCob event is received. If at least one UPnP control point in the network subscribed for events of this state variable, then the UPnP Base Driver should have an appropriate UPnPEventListener registered and propagate a UPnP Event Notification to the network every time notifyUPnPEvent is called. The *SwitchPower* service description can be found at the UPnP Forum web site [20]. Listing 2.1 presents the UPnP profile for the Switch device in XML description language.

```
<root xmlns="urn:schemas-upnp-org:device-1-0">
  <specVersion>
    <major>1</major>
    <minor>0</minor>
  </specVersion>
  <URLBase>http://192.168.2.253:4004</URLBase>
  <device>
    <deviceType>urn:com.inaccessnetworks:device:Switch:1</deviceType>
    <friendlyName>BinarySwitchZW0000001205</friendlyName>
    <manufacturer>inAccess Networks</manufacturer>
    <manufacturerURL>http://www.inaccessnetworks.com</manufacturerURL>
    <modelDescription> Binary Switch UPnP Wrapper</modelDescription>
    <modelName>ROCob Binary Switch</modelName>
    <modelNumber>1.0</modelNumber>
    <modelURL>http://www.inaccessnetworks.com</modelURL>
    <serialNumber></serialNumber>
    <UDN>uuid:UPnPROcobZW0000001205</UDN>
    <UPC></UPC>
    <serviceList>
      <service>
        <serviceType>urn:schemas-upnp-org:service:SwitchPower:1</serviceType>
        <serviceId>urn:schemas-upnp-org:serviceId:SwitchPower:1</serviceId>
        <SCPDURL>/service/0/gen-desc.xml</SCPDURL>
        <controlURL>/service/0/ctrl</controlURL>
        <eventSubURL>/service/0/event</eventSubURL>
```

```
      </service>
    </serviceList>
    <presentationURL></presentationURL>
  </device>
</root>
```

Listing 2.1 The UPnP profile for the Switch device in XML description language

2.8.1.3 More UPnP Wrappers

Binary Sensor This function is not directly mapped to any standardized UPnP service. Since the binary sensor only sends events with its current state, a UPnP service is needed that can read the binary sensor status or can get notification events when the status is modified. The ideal service to handle this task is a subset of the SwitchPower service. Therefore, a new service has been created that uses one state variable, called *Status*, and a corresponding action to get its value called *getStatus*.

Door Trap UPnP wrapper resembles the functionality of the binary sensor. Their difference lies in the device description. This is necessary to serve better the presentation layer, to allow correct visualization of the devices, as well as to allow expandability in the future.

Motion Sensor UPnP wrapper, just like the door trap wrapper, currently follows the functionality of the binary sensor. This means that it implements only the binarysensor UPnP service. The corresponding profile will be possibly enhanced in the future, when devices with advanced features become available. Such features may be sensor sensitivity and sensor reset timeout.

Illumination Detector UPnP wrapper takes care to represent a NA-OSGi illumination detector in the UPnP layer. This is achieved by utilizing the AnalogMeter UPnP service, which corresponds to the ANALOG METER NA-OSGi subfunction. The AnalogMeter service includes one StateVariable of type float, and an action that returns the value of this variable, called GetAnalogMeter. The measured value corresponds to the percentage of the maximum value. The UPnP device reads the analog meter level value either by polling the NA-OSGi device, using the GetAnalogMeterLevel action, or by serving incoming events that are generated by the NA-OSGi device.

2.8.1.4 Device/Service Control Example

This paragraph presents a device control example in ATRACO (see Fig. 2.12). When an ATRACO client needs to use a Z-Wave Switch connected to the NA component, it will send a UPnP device discovery message to the ATRACO IP network. The OSGi UPnP base driver of the NA component will send the description of the available UPnP devices, including the switch in question, which is handled by the relative

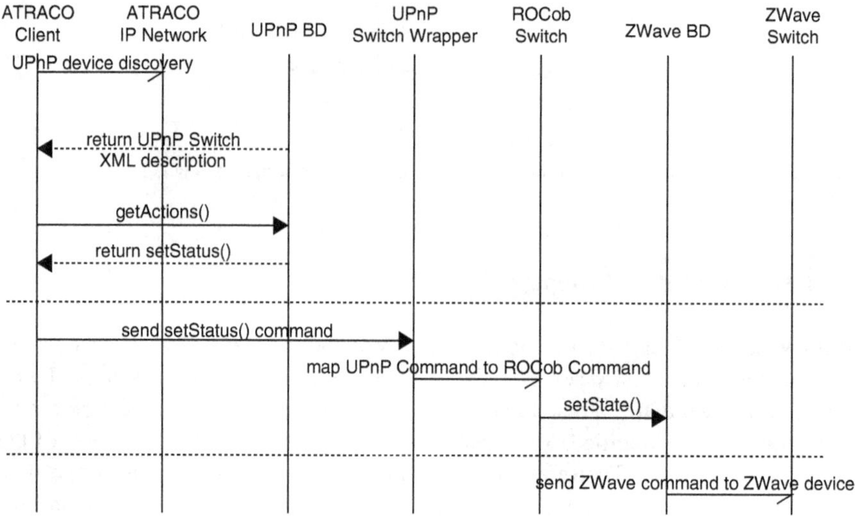

Fig. 2.12 Device control through NA

OSGi UPnP wrapper. Then the client can query from the UPnP wrapper the available actions of the device, and send, for example, the ON command to the wrapper. The wrapper, using the ROCob device ends up to send the command to the Z-Wave Base driver of the NA component. The latter will finally send the Z-Wave command to the Z-Wave switch.

2.8.1.5 Device Event Propagation Example

Figure 2.13 depicts device propagation in ATRACO. When an event from a Z-Wave door trap connected to the NA component occurs, this event is propagated through NA to the various ATRACO clients. The event is first handled by the Z-Wave Base Driver of the NA component, which propagates it to the corresponding ROCob device. The ROCob device then sends an event to the OSGi framework, which is handled by the relative UPnP wrapper. The wrapper translates the event to a corresponding UPnP event. This event is broadcasted to the ATRACO IP Network and listeners for this event will receive it.

2.8.2 Simple Task Execution Manager

The Simple Task Execution Module (STEM) is an OSGi-based UPnP service that based on a simple or nested condition evaluation generates a UPnP event about the result of the evaluation and even executes a set of actions on one or more

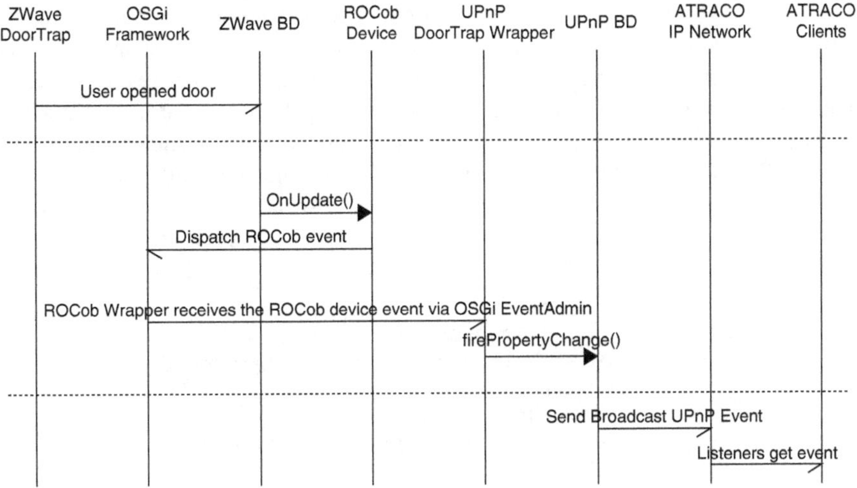

Fig. 2.13 Device event propagation through NA

UPnP devices when the evaluation is true. STEM basically operates on UPnP devices/services in a home network. Architectural wise, although STEM could be a simple OSGi service, it follows a network approach and is a UPnP service itself, so that other UPnP services in the same network can use it. It is planned however to provide an OSGi service API in the near future.

The two main concepts of STEM are the scene and the trigger. A scene is a set of actions that can be performed in one or more devices. The actions are described using unmodified UPnP identifiers of the participating devices, so that users of STEM can easily construct them. For functional reasons, each scene has its own unique ID. Thus, it is possible to create, modify, and delete a scene.

A trigger is a set of rules, that when the total evaluation turns true an event is generated that notifies that this specific trigger is true. If the trigger is associated with one or more scenes, then these scenes are executed. A rule is set of simple or nested AND and OR conditions. The conditions compare state variable values of a UPnP device with state variable values of the same or other UPnP device or with arbitrary values of the same UPnP data type. This is achieved with a recursive logical evaluation function. Just like the scenes, triggers also have a unique ID that is used for their creation, modification, or removal or association/deassociation with a scene.

In order to be able to evaluate conditions, STEM subscribes to the services of the devices that are included in the conditions of the various triggers. When an updated value for a state variable that is used in a condition arrives, STEM re-evaluates the condition and as a result the whole trigger. If the total evaluation turns true, an event containing the trigger id and the status true value is generated and sent to the listeners that have subscribed to STEM's service. If the result of the evaluation is false, an event with the trigger id and the false status value is sent.

STEM by itself is not a UPnP control point, so it is not aware of the available devices. The OSGi UPnP base driver is responsible for discovering available UPnP devices and updating them or removing lost ones. Using the OSGi API of the OSGi UPnP base driver, STEM finally becomes aware of the various devices. When a STEM user adds a scene or a trigger, these are validated against the existing UPnP devices. If a device included in a scene or trigger does not exist, or a wrong service or action or state variable is used, then the trigger/scene is discarded. Listing 2.2 shows an example of a scene description XML which handles two binary lights.

```
<xml>
  <scene id="scene1">
    <device id="uuid:UPnPROCobZW 0000000702">
      <service id="urn:schemas-upnp-org:serviceId:SwitchPower
          :1">
        <action id="SetTarget">
          <statevar id="NewTargetValue" type="boolean" value
              ="true"/>
        </action>
      </service>
    </device>
    <device id="uuid:TestLight+004fecde">
      <service id="urn:schemas-upnp-org:serviceId:SwitchPower
          :1">
        <action id="SetTarget">
          <statevar id="NewTargetValue" type="boolean" value
              ="false"/>
        </action>
      </service>
    </device>
  </scene>
</xml>
```

Listing 2.2 An example of a scene description XML which handles two binary lights

Listing 2.3 shows an example of a condition description XML; *ge*, *eq*, and *nt* are condition operands meaning greater equal, equal and not, respectively.

```
<xml>
  <trigger id = "testTrigger">
    <logic operator="OR">
      <condition operator="ge">
        <lcomparator type="upnp" device="uuid:org-upnp:
            testdevice1" service="urn:org-upnp:testservice"
            statevar="level"/>
        <rcomparator type="value" vartype="Integer" value
            ="12"/>
      </condition>
      <logic operator="AND">
        <condition operator="eq">
          <lcomparator type="upnp" device="uuid:org-upnp:
              testdevice2" service="urn:org-upnp:testservice"
               statevar="Status"/>
```

```
        <rcomparator type="upnp" device="uuid:org-upnp:
            testdevice3" service="urn:org-upnp:testservice"
             statevar="Status"/>
      </condition>
      <condition operator="nt">
        <lcomparator type="upnp" device="uuid:org-upnp:
            testdevice4" service="urn:org-upnp:testservice"
             statevar="Status"/>
        <rcomparator type="value" vartype="Boolean" value="
            true"/>
      </condition>
    </logic>
  </logic>
 </trigger>
</xml>
```

Listing 2.3 An example of a condition description XML

Figure 2.14 demonstrates a simple task execution manager example, where a self managed service add triggers that depend on events from UPnP devices A,B and scenes that operate on devices C and D. When events from devices A and B arrive, if the trigger is evaluated true, the associated scene is executed and actions are performed on devices C and D. The task execution manager runs in the OSGi framework, but since it is a UPnP service too, it has also a part in the home IP network.

2.9 Network Adaptation in the ATRACO System Prototype

The ATRACO project has implemented a prototype of an ATRACO system in order to test its proposed ambitious architecture. It consists of the main components supporting adaptation specified and prototyped by the project, namely Network Adaptation, Structural Adaptation, User Behavior Adaptation, Semantic Adaptation, and User Interaction Adaptation, as well as several basic components for controlling the environment (e.g., control of lights, HVAC, music player). Regarding NA, both in terms of device and service adaptation, a set of tools has been created in order to demonstrate the use of NA in real ATRACO environments. This set includes supporting a variety of devices as well as creating all the necessary services that combine the elements found inside an ATRACO ecosystem in order to realize a user goal.

In principle, a user goal, represented as an Activity Sphere (AS), is matched with available device local ontologies, the AS is decomposed into a hierarchy of abstract tasks represented as a workflow, and the Sphere Manager (SM) forms the AS for the specific user goal. The SM connects to device registry to get connectivity data of the devices through the help of the NA component which provides access to home peripherals and a mapping of all different network domains in the AIE to the IP level. Then, the SM initiates the various components (Ontology Manager,

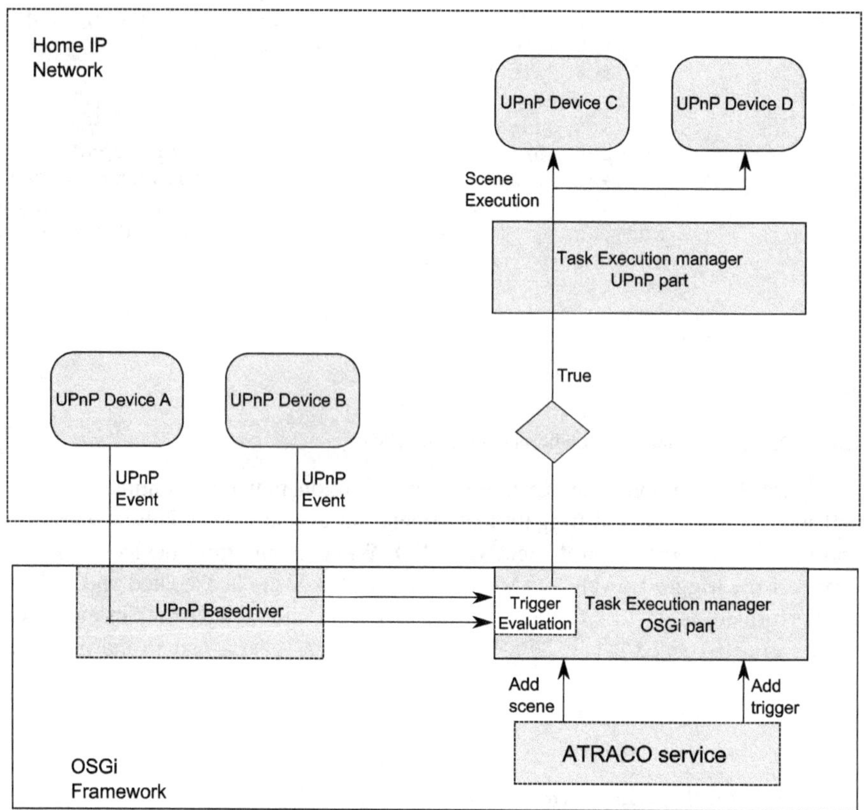

Fig. 2.14 Task execution manager example

Interaction Agent, and Fuzzy Task Agent) and performs dynamic service binding in order to execute the workflow. When an action is required to take place in the AIE, like starting the music player or switching on/off HVAC, the SM invokes the NA layer to change the state of the devices and the services. Through the NA, the SM also generates events that can be used by other components of the system, e.g. continuously sending light levels to the FTA adapting the light. The devices involved in an AS communicate over wired/wireless network domains, overlaid with TCP/IP and OSGi/UPnP middleware programmed in Java. The use of Java as the development platform facilitates our system deployment on a wide range of devices including mobile phones and PDAs.

Using the proposed Network Adaptation framework, several device UPnP wrappers for binary sensors, door traps, motion sensors, lights, switches, on/off smart plugs and illumination detectors, as well as services supporting task execution, have been implemented within ATRACO to support the complex trial hosted in the iSpace in Essex demonstrating three activity spheres (AS) for concept validation and prototype evaluation. These activity spheres include an entertainment AS involving tasks for watching TV, reviewing photos, listening to music, reading and playing on a game console, a work AS involving tasks for surfing the Internet, writing documents and reading, as well as a sleep AS referring to end of day/sleep activities and involving tasks for watching TV, listening to music/radio and reading. NA represents the physical devices participating in the ASs as ROCob devices (device representation layer) and exports them to the ATRACO ecosystem as UPnP devices (service adaptation layer).

The role of NA is demonstrated in the ATRACO prototype with the transparent communication of devices and services residing in various different physical networks. For instance, X-10 lights connected on a server in the iSpace can be controlled by the Fuzzy Task Agent (FTA) used for artifact and user behavior adaptation, based on the feedback given by a Z-Wave illumination detector physically connected to the NA component. FTA is agnostic to the protocol used for the control of each individual device, because the devices are presented in a common way thanks to NA.

To provide a clear demonstration of the functionality of NA in an ambient environment such as an ATRACO ecosystem, we examine use cases in different activity spheres. Specifically, we will examine two use cases. The first one demonstrates how a high level preference of a user for energy efficiency is realized transparently through NA inside the context of an activity sphere related to user *entertainment*. The second use case provides the implementation of a security service during a *sleep* activity sphere by employing actual devices as well as abstract UPnP services that do not correspond to actual devices.

2.9.1 Key Components in Use Cases

Every use case is implemented with the participation of various persons, devices, and services. The most obvious of them are the persons that participate and initiate actions or become receivers of the results of adaptation actions performed by the ATRACO system.

Other less obvious components that take place in use case realization are the devices that are used in the context of an AS. This set of components is less obvious because apart from the devices that the users interact directly with, it includes other devices that have an auxiliary role. Moreover, an AS may define multiple use cases, each one of them may use a differentiated set of devices. It is important to note that the characteristics of a device, such as network protocol and physical medium, are transparent to the ambient ecosystem thanks to NA and do not affect the definition and the functionality of the use case in the activity sphere.

A key device in the ambient ecosystem is the home controller, which provides connectivity with devices of different protocols and physical mediums. This device diversity is finally abstracted in the ambient space by using suitable UPnP wrappers that bring all these devices in a common network layer (UPnP).

Apart from the obvious ones, such as the devices and persons involved, some services and functionalities get involved in almost all use cases and are transparent to the user. The most important of them is STEM, which is a network service (UPnP) that enables the execution of a set of actions on one or more devices, when certain conditions are met.

2.9.2 Entertainment Activity Sphere Use Case

The entertainment AS, as implied by its name, demonstrates functionalities and adaptations that take place during an entertainment activity involving particular tasks for watching TV, reviewing photos, listening to music, reading and playing on a game console. For the purpose of this book, we analyze Network Adaptation in the use case of enabling an energy saving capability, in case this is enabled in the user preferences.

The energy saving use case demonstrates how energy can be saved, if this is desired by the user, by combining information and usability of various ATRACO components. The user will be watching TV, and at some time will open the window to let some fresh air in. If energy saving is enabled in the user profile, the ATRACO system should stop the heating from working. Then, when the window is closed, the Interaction Agent will ask the user if the heating should be turned on again.

This energy saving service is realized by using a set of window traps, motor driven curtains and heating equipment inside the ATRACO ecosystem, as well as the *interaction agent* and mostly *STEM*. Additionally there is a software component that provides the *energy saving service* that keeps track of the user's choice of enabling or disabling the energy saving feature and the heating status and also triggers the interaction agent to ask the user if he likes to restart the heating during the *entertainment AS*. The energy saving use case is executed in parallel with the watch TV use case, which additionally involves lights, illumination sensors, various location sensors and a software component providing time and daytime information.

The purpose of the energy saving use case is to demonstrate:

- User behavioral adaptation. The lights are adjusted according to the settings that FTA has progressively learnt for each specific user.
- Sphere adaptation. The system switches from the entertainment *watch TV* sphere to the *energy saving* sphere.
- Application of preferences set via the privacy manager.
- Cooperation of various ATRACO components (FTA, SM, IA, NA)
- *Network Adaptation demonstrating that devices spanning UPnP, x10, LonWorks TP, and Z-Wave networks operate seamlessly in the ATRACO environment and helping to implement user behavioral adaptation and sphere execution.*

The SM becomes aware of the user and his preferences via the privacy manager and sets a condition for the *energy saving* sphere, which is energy saving is enabled, heating is on and window is open. Then it associates with the above condition a scene to turn off heating and send event to set accordingly the appropriate state variable in the energy saving UPnP service. According to the extended use case, the user will turn on the TV and the SM will initiate the watch TV sphere. FTA becomes aware of the active sphere and reads the light sensors data. According to the value it has read, it adjusts the ceiling lights. Optionally, FTA reads the value of a UPnP network clock, and decides whether it is daytime or not. If it is, it closes the curtain in order to prevent the natural light from making watching TV difficult (e.g., too much luminance, possible reflections, etc.). Sometime during the watch TV sphere the user may want to let some fresh air in. If the curtain is closed, he uses the interaction agent and controls the curtain (or does this with a relative switch on the wall). Then he opens the window. This sends a *window open* event from the door trap sensor. STEM receives this event, and it checks the condition set. As a result it executes the associated scene as explained before. The SM will receive the heating off event and will place a new condition on STEM, which is heating is off and window is closed and further associates this condition with a scene to ask about enabling the heating. In parallel, the light sensors, if it is day, sense the luminance difference and FTA sets the ceiling lights accordingly. FTA will readjust the dimming value of the ceiling lights if the user closes the curtain again, according to the measured luminance value by the luminance sensors. The user returns to the sofa and continues watching TV. When he decides to close again the window the door trap sensor will send a *window closed* UPnP event. STEM will receive this event and ask IA to display the *Energy Saving UI* on the TV set, which asks the user if the heating should turn on. If the answer is yes, the IA propagates the answer to the SM, which subsequently sends an event to STEM to turn the heating on.

Network Adaptation in this AS is responsible for executing actions on the devices participating in the presented use case, such as turning on/off the heating, opening/closing the motor driven curtain, adjusting the lights etc., as well as translating state changes to UPnP events sent to ATRACO components and services, such as the wireless Z-Wave events sent by the door/window traps and received by the home controller in the ambient ecosystem and via its UPnP wrappers end up as UPnP events, or events conditionally generated by STEM to the UPnP energy saving service by evaluating a set of relative rules defined on STEM by the SM through invoking appropriate STEM UPnP actions.

2.9.3 Sleep Activity Sphere Use Case

The sleep AS, as implied by its name, demonstrates functionalities and adaptations that take place when the user goes to sleep. For the purpose of this book, we analyze the use case of enabling an alarm service.

This alarm service is realized by using a set of door/windows traps and a siren device inside the ATRACO ecosystem, as well as the *interaction agent* and mostly *STEM*. Additionally there is a software component that provides the *alarm service* that keeps track of the user's choice of enabling or disabling the security feature during the *sleep AS*.

Network wise, this is a UPnP service which allows the modification of its state by the user. This change of state can be performed either by using a remote control, such as a Z-Wave keyfob or by a relative user interface, provided that the privacy manager allows the use of these devices or UIs. The first way employs STEM, which connects the actions on the keyfob with the state of the alarm service. This is achieved by translating the Z-Wave commands issued by the keyfob to UPnP events that are received by STEM, which just afterwards sends the correct UPnP messages to the alarm service. STEM knows which events to use and monitor for this action, by evaluating a set of rules defined by the SM. SM checks the availability of devices, both in terms of physical presence in the ambient space and permission to use by the privacy manager, and sets the relative rules on STEM by invoking the appropriate STEM UPnP actions.

Apart from enabling the alarm service, a new set of rules has to be added, that checks the state of the participating door/window traps (in our example wireless Z-Wave devices) and the alarm service. If the latter is enabled, and at least one sensor generates an *'ON'*, then STEM sends a new event in the ATRACO UPnP network indicating the activation of the alarm service (intrusion), and starts the siren. From a network perspective, wireless Z-Wave events are sent by the door/window traps and received by the home controller in the ambient ecosystem and via its UPnP wrappers end up as UPnP events. Since STEM uses the state of these devices for rule evaluation, it accepts them and evaluates if it should fire the alarm event, again as a UPnP message.

2.10 Lessons Learned

The Network Adaptation component provides the lowest layer of service and device communication upon which the whole ATRACO environment is built. As such it has to tackle requirements and special needs of almost every component. This broad involvement of the NA component in the project was the reason that many lessons have been learned from its development and evaluation.

From a system design point of view, the greatest lesson learned was that widely adopted protocols are not always well implemented. Thus, choosing a technology based among others on adoption and availability of existing implementations does not always ensure that it will provide the fastest and most reliable solution. For example, although both OSGi and UPnP are widely used for a lot of years already, the existing implementation of UPnP for Java and OSGi had a lot of problems. In fact the problems were so many that new UPnP protocol implementations for Java and OSGi have been developed from scratch.

Along the same lines, although software written in Java is widely known to be portable across various operating systems and architectures, it proved that this is not entirely true. Special features of the functionality and implementation of network sockets of each OS running on machines based on different computer architectures (e.g., ARM, x86, x86_64) brought up lots of problems that were very difficult to spot and resolve. Resolution of the problem for a specific pair of architecture and OS very often was breaking the functionality of another. Thus, complex network operations require deep knowledge of the specifics of the underlying OS implementation and it must not be assumed that a simple use of an API provided by the programming language is sufficient.

It is worth mentioning that deviations from the specified functionality were also met in some devices used during the prototype development. This should be expected when working with quite new devices or even not publicly released ones and thus fallback scenarios must exist. Problems of this type were affecting communication with the device, the discovery procedure or could lead to lost events. In some cases that no software workaround was possible, some devices were not used at all. Another interesting issue was that devices with similar functionality, e.g., light sensors, provided input, under the same configuration, with large deviations, requiring the corresponding drivers to be modified in order to weight the incoming data or even to avoid using specific devices in order to achieve robust functionality.

Apart from problems based on deviations from specifications, such as the ones described above, useful conclusions have been drawn regarding integration of multiple components. The first lesson in this area is that quite often integration effort is heavily underestimated. Experience drawn from the ATRACO project has proved that a lot of unexpected problems arise during integration and the more the partners are involved the more the problems appear. The severity of these problems during the project was minimized by constant communication of partners, component unit testing, and following a mixed mode of "big picture" design and step by step design and implementation. Furthermore, excessive use of team development tools and testing of one component from another component's point of view lead to faster development, API clarification, and bug resolution. The second great lesson has been learned during the last year of the prototype development, where detailed specification of use cases before implementation helped to identify and resolve quite early component race conditions and limitations.

A very useful conclusion from the evaluation of the system is that testing in the lab cannot provide the same evaluation and testing data compared to the use of the system in real-life conditions. This happens because the time spent in the lab testing a specific part of the system is usually quite limited in comparison with the exposure of the system in real-life conditions. Moreover this difference is amplified in real-life conditions where usually the system is deployed in larger scale and includes many more devices affecting performance and stability. These observations have changed the way that tests were designed and run and as a result the testing procedure was much more effective.

2.11 Conclusions

Within the ATRACO project we have specified and implemented a Network Adaptation framework in order to provide a set of functions and protocols in a SOA middleware allowing an ATRACO system to interact with the network resources. The goal of the NA layer is to provide seamless use of the devices and services to the ATRACO system and to simplify the access to networks in the AIE. The design and specification process of the NA component involves the consideration of several adaptation guidelines, such as unified access on devices that belong to different networks, interoperability between networks, task completion with alternative resources, dynamic discovery for new networks and devices, representation of legacy Internet services and relaying of existing web services. In ATRACO, the NA component meets these requirements through defining common device types across networks, allowing events from any network to trigger actions in any other network, featuring a device registry and device/driver manager as well as providing single access for web services (i.e., a weather report service) able to evaluate the availability, accessibility, and quality of multiple existing web services provided by the Internet sites.

To cope with the complexity of accessing diverse control networks, a common Device Representation Layer has been introduced. This representation layer is designed in such a way that it handles different networks while hiding the complexity and details of each one of them. For instance, an application which is to be placed on top of the device representation layer should not tell the difference between a Z-Wave lamp and a LonWorks lamp. The device representation layer creates a unique representation of each different device type across networks, which simplifies the evolution of ambient intelligent applications and services. In addition to the devices, the proposed NA framework additionally shapes various other services provided over IP networks (e.g., NTP, VoIP, RTSP, etc.) as ATRACO services. The functionality in an ATRACO environment is exposed as semantic services which an actor can discover and then compose to form ATRACO applications. After all, each service is associated with at least one semantic description which shields the actor from the complexity of the resource layer realization and makes it easy for the actor to employ these services in accomplishing interesting and useful tasks.

The NA Layer has been implemented using a mixed architecture, including a centralized home controller that is able to seamlessly integrate various different devices from different technologies around different home automation and control space together with a variety of modules, some of which may be able to work in a distributed architecture. The Device Representation Layer provides an abstraction of the network devices to OSGi services, sharing a common device format and API in OSGi context, that can be used by other ATRACO components. These OSGi services represent the functionality of the underlying physical devices, giving the architectural and application components the opportunity to discover them and retrieve all the required information for managing and performing control actions. The available functional profiles include physical devices with multiple endpoints, binary sensors, level properties like dimmers (with or without timing functionality), direction semantics like shades (with or without timing functionality), direction semantics with read only values, switches, batteries, analog meters, alarms, thermostats, etc. Moreover, the Device Representation Layer provides several utilities that intend to provide a service layer for scene execution and alerting. A scene is a collection of OSGi devices that will execute a number of actions on these devices when a certain event occurs. An alarm service allows sending alarm events to other OSGi components or higher layers. It makes use of a set of defined alarm levels, types and properties, to cover a wide range of alarm events. Furthermore, the alert service provides simple alerts to the other OSGi aware services in the framework. It listens to the device events and evaluates a series of user defined Rules to finally create the specific alerts.

The Service Adaptation Layer provides an abstraction of the Device Representation Layer devices as OSGi services so that additional external services can take advantage of the unified abstraction. On top of that, OSGi device representations are eventually transformed to UPnP services. Additionally, it provides UPnP adaptation for IP services participating in an ATRACO system. UPnP seems to cover most of the aspects specified for the Service Adaptation Layer as it provides the ease of network setup, device discovery, self-announcement and advertisement of supported capabilities.

The NA Layer additionally provides an OSGi based UPnP service for task execution, which is called STEM. Based on a simple or nested condition evaluation, STEM generates a UPnP event about the result of the evaluation and even executes a set of actions on one or more UPnP devices when the evaluation is true. In order to be able to evaluate conditions, STEM subscribes to the services of the devices that are included in the conditions of the various triggers. STEM, by itself, is not a UPnP control point, so it is not aware of the available devices. It becomes aware of the existing devices using the OSGi API of the OSGi UPnP base driver. When a STEM user adds a scene or a trigger, these are validated with respect to the existing UPnP devices.

A prototype that realizes the ATRACO architecture has been implemented, which exhibits, although in primitive form, all types of adaptation specified by the ATRACO project, including adaptation at the network level. The system has been deployed in the iSpace at the University of Essex and tested by real users, with promising results. Using the proposed Network Adaptation framework, several device UPnP wrappers for binary sensors, door traps, motion sensors, lights, switches, on/off smart plugs and illumination detectors, as well as services supporting task execution, have been implemented to support the complex trial hosted in the iSpace demonstrating three activity spheres for concept validation and prototype evaluation. NA materializes a UPnP middleware allowing the transparent communication of devices and services residing in various different physical networks. In order to provide a clear demonstration of the functionality of NA in an ambient environment such as an ATRACO ecosystem, we have presented in detail two use cases. The first one demonstrates how a high level preference of a user for energy efficiency is realized transparently through NA inside the context of an activity sphere related to user *entertainment*. The second use case provides the implementation of a security service during a *sleep* activity sphere.

References

1. Amigo project. http://cordis.europa.eu/project/rcn/71920_en.html (2008)
2. Atraco project. http://www.uni-ulm.de/in/atraco (2012)
3. Baker, C., Markovsky, Y., Van Greunen, J., Rabaey, J., Wawrzynek, J., Wolisz, A.: Zuma: a platform for smart-home environments. In: Int. Conf. on Intelligent Environments, pp. 257–266 (2006)
4. Eperspace project. http://www.ist-eperspace.org/ (2006)
5. Hargrave, B., Kriens, P.: Osgi best practices. In: OSGi Alliance Community Event, pp. 26–27 (2007)
6. Home2015, Multi-Disciplinary Research Programme in Future Home Systems and Technologies, Agency for Science, Technology and Research, Singapore (2007)
7. Jahnke, J., d'Entremont, M., Stier, J.: Facilitating the programming of the smart home. IEEE Wirel. Commun. **9**(6), 70–76 (2003)
8. Kumar, R., Poladian, V., Greenberg, I., Messer, A., Milojicic, D.: Selecting devices for aggregation. In: IEEE Workshop on Mobile Computing Services and Applications, pp. 150–159 (2003)
9. Lin, R., Hsu, C., Chun, T., Cheng, S.: Osgi-based smart home architecture for heterogeneous network. In: 3rd Int. Conf. on Sensing Technology, pp. 527–532 (2008)
10. Matsuura, K., Hara, T., Watanabe, A., Nakajima, T.: A new architecture for home computing. In: IEEE Workshop on Software Technologies for Future Embedded Systems, pp. 71–74 (2003)
11. Music project. http:///www.ist-music.eu (2010)
12. Ngn @ home. http://portal.etsi.org/at/ATNGNSummary.asp (2007)
13. Perumal, T., Ramli, A., Leong, C., Mansor, S., Samsudin, K.: Interoperability among heterogeneous systems in smart home environment. In: IEEE Int. Conf. on Signal Image Technology and Internet Based Systems, pp. 177–186 (2008)
14. Ramparany, F., Boissier, O., Brouchoud, H.: Cooperating autonomous smart devices. In: Smart Objects Conference, pp. 182–185 (2003)
15. Rellermeyer, J., Alonso, G., Roscoe, T.: R-osgi: distributed applications through software modularization. In: ACM/IFIP/USENIX 8th Int. Middleware Conference (2007)
16. Roalter, L., Kranz, M., Moller, A.: A middleware for intelligent environments and the internet of things. In: 7th International Conference on Ubiquitous Intelligence and Computing, pp. 267–281 (2010)
17. Rose, B.: Home networks: a standards perspective. IEEE Commun. Mag. **39**(12), 78–85 (2001)
18. e-Sense project. http://cordis.europa.eu/project/rcn/80695_en.html (2008)
19. Teaha project. http://cordis.europa.eu/project/rcn/74666_en.html (2007)
20. Upnp switch power service description. http://upnp.org/specs/ha/UPnP-ha-SwitchPower-v1-Service.pdf (2011)
21. Valtchev, D., Frankov, I.: Service gateway architecture for a smart home. IEEE Commun. Mag. **40**(4), 126–132 (2002)
22. de Vicente, A., Velasco, J., Marsá-Maestre, I., Paricio, A.: A proposal for a hardware architecture for ubiquitous computing in smart home environments. In: Int. Conf. on Ubiquitous Computing: Applications, Technology and Social Issues (2006)
23. Wang, Q., Cheng, L.: Awareware: an adaptation middleware for heterogeneous environments. In: International Conference in Communications, pp. 1406–1410 (2004)

Chapter 3
Ontology-Based Knowledge Management in NGAIEs

Achilles Kameas and Lambrini Seremeti

Abstract The objective of the knowledge architecture of ATRACO is to enhance communication, as well as to ensure effective knowledge sharing among ATRACO components. It is built around ontologies, Ontology Managers, and agents. Every component of an activity sphere uses an ontology to model its local knowledge and state. These ontologies will certainly be heterogeneous, but they must be used transparently in the context of any sphere. Thus, knowledge management in ATRACO is concerned with the alignment of heterogeneous ontologies, in order to produce the Sphere Ontology, which encodes the sphere knowledge.

In this chapter, we shall describe the ontology management framework developed in the context of ATRACO, which includes:

1. A set of ontologies, which includes Upper Level Ontology (ULO), User Profile Ontology (UPO), Generic Device Ontology, Privacy Policy Ontology, and Interaction Ontology;
2. The Ontology Manager, a software module that manages ontology and provides an interface to the other system components; it also manages the Sphere Ontology;
3. The Alignment Module, which is responsible for aligning two ontologies and storing the alignment in a machine readable format; and
4. A mechanism based on Category Theory responsible for propagating relations.

In Information Technology and Artificial Intelligence, ontologies are the structural framework for organizing and representing information, as a set of concepts in a domain and the relations between those concepts. In the context of ATRACO project, three types of ontologies have been developed: domain ontologies, each one describing devices, services, user profiles, and agent knowledge bases; ontologies describing tasks and policies; and an ULO describing the basic entities and relationships of the ATRACO world model.

Ontology alignment is the output of the ontology matching process that attempts to find relationships, or correspondences between entities of different ontologies, in order to produce sets of such correspondences between two or more ontologies.

A. Kameas (✉) • L. Seremeti
The Hellenic Open University and DAISy Research Unit, Patras, Hellas
e-mail: kameas@cti.gr; seremeti@cti.gr

© Springer International Publishing Switzerland 2016

S. Ultes et al. (eds.), *Next Generation Intelligent Environments*,
DOI 10.1007/978-3-319-23452-6_3

During the alignment operation, the original ontologies are kept unaltered, while the alignment results are stored separately from the ontologies themselves. In ATRACO, we applied alignment procedures in order to deal with lexical, syntactic, and semantic heterogeneity.

In order to reconcile the intrinsically different models of local knowledge expressed via ontologies, a mathematical formalism based on Category Theory serves as a solid foundation, since it allows the coexistence of heterogeneous entities (ontologies) while focusing on relationships

3.1 Introduction

Ambient intelligent environments (AIEs) are human activity spaces populated with smart communicating objects, which are able to perceive the environment, act upon it, process and store data, manage their local state, communicate and exchange data. AIEs provide an infrastructure that supports services such as networking, communication, discovery, location, and context estimation [20]. The emerging Next Generation of AIEs (NGAIEs) are being designed to inherently exhibit intelligent behavior and adaptive functionality, in order to provide optimized resource usage and support consistent functionality and human-centric operation. Each of NGAIEs "stakeholders," i.e., humans, agents, devices, services, having its own conceptualization of the world, assumes a variety of roles, in order to realize successfully its tasks, within the NGAIE context.

Intelligence, as the primary means to achieve adaptation, will appear at various levels. For example, local resource management may require decision-making mechanisms and even embedded intelligent agents. At a system scale, multi-agent systems, using semantically rich descriptions, learning mechanisms and possibly cognitive functions (such as perception, homeostasis, etc.) will be embedded in NGAIEs. But intelligence relies on knowledge; thus, different kinds of knowledge will be encoded in NGAIEs, including knowledge about the state of resources, the tasks to be achieved, the preferences of the users, and the policies to be realized.

In the general case, NGAIE resources will be heterogeneous, as they will originate from different manufacturers; consequently, they will probably use proprietary, heterogeneous information and knowledge representation schemes. Within NGAIEs, as well as among interacting NGAIEs, multiple sources of heterogeneity can appear:

- Artifacts and other engineered NGAIE components, each of which has a proprietary, usually closed, model of itself and the world.
- NGAIEs, public or temporarily private, which have their own models of their resources and services.
- Task models, expressed in various domain-dependent notations, as well as dialogue states and interfaces.
- Multimedia objects, which usually adhere to the metadata of multimedia standards; the same holds for other types of "intelligent" content.

- Networking protocols and, in general, communication schemes, which require specific descriptions of artifacts and services and usually have a restricted closed world model.
- People, who have their own individual profiles and ways of perceiving, understanding and accepting technology.

Thus, two important goals of NGAIE design are (a) to increase the amount of knowledge that is available to the system and (b) to minimize the inaccuracy of knowledge and the ambiguity regarding the interpretation of the shared information, thus enabling NGAIE components to interact successfully through a common communication channel, despite their heterogeneous representations of the world.

Ontologies can be used to address these issues through the semantics they convey. Ontology is defined as an explicit and formal specification of a shared conceptualization [19]. A "conceptualization" refers to an abstract model of some phenomenon in the world, which identifies the relevant concepts of that phenomenon. "Explicit" means that the type of concepts used and the constraints on their use are explicitly defined. "Formal" refers to the fact that the ontology should be machine readable. "Shared" reflects the notion that ontology captures consensual knowledge, that is, it is not private of some individual, but accepted by a group. Thus, ontology is a structure of knowledge, used as a means of knowledge sharing within a community of heterogeneous entities.

A straightforward solution for achieving interoperability is to develop a commonly accepted ULO for NGAIEs and verify its wide applicability over various contexts. To this end, a few upper level ontologies have been proposed, while standardization efforts by the World Wide Web Consortium have led to the development of the ontology language OWL (Web Ontology Language) [47], which is the evolution of DAML + OIL. However, the development of a single ontology, able to completely and accurately describe the key concepts of such environments, as well as the processes carried out within them, is almost infeasible, because of the engineering difficulty of soundly depicting the countless relationships and properties of the large set of concepts. Thus, we expect that multiple heterogeneous ontologies will be developed autonomously, in order to semantically describe the features and capabilities of different NGAIE components.

Consequently, semantic mediation among these different types of knowledge becomes necessary. There are many approaches and techniques, which are based on ontology operations that can be used to develop mechanisms for interfacing the heterogeneous and possibly inaccurate ontologies of the NGAIE components [9]. These include ontology alignment, mapping, and merging.

With ontology mapping, the correspondences between the two ontologies are stored separately from the ontologies and thus are not part of the ontologies themselves. The correspondences can be used, for example, for querying heterogeneous knowledge bases using a common interface, or transforming data between different representations. When performing ontology merging, a new ontology is created, which is the union of the source ontologies. The new ontology, in general, replaces the original source ontologies and captures all their knowledge. The challenge in ontology merging is to ensure that all correspondences and differences between the ontologies are reflected in the merged ontology. The first stage of both ontology mapping and merging is ontology alignment, which identifies the semantically related concepts across multiple ontologies.

In a nutshell, knowledge management in NGAIEs comprises two important strands: ontology engineering and ontology matching. In the following section, we discuss the related work on knowledge representation and management in AIEs, starting from ontology-based ubiquitous computing systems and then focusing on ontology engineering methodologies and ontology alignment approaches. In the third section we present the ontology engineering methodology we developed for the purposes of project ATRACO, followed by a presentation of a set of specific NGAIE ontologies. Then, we present the three-step ontology alignment strategy that we used in ATRACO. This strategy includes an algorithm that guides the selection of appropriate matchers and an algorithm for deciding the combination of matchers to be used. In the fifth section we briefly present a mathematical formalism based on Category Theory that can serve as the framework for describing the combination and integration of ontological objects of the ATRACO world. Moreover, a mechanism for propagating relations in this settings is proposed, in order to reveal inconsistencies in a network of aligned ontologies. Throughout these sections, we highlight the main ATRACO contribution, an ontology-based knowledge management framework, based on the notion of activity spheres, that permits to overcome the barriers of dynamic nature and heterogeneity intrinsic in NGAIEs. Finally, Sect. 3.6 concludes the chapter, while Sect. 3.7 proposes further readings.

3.2 Related Work

The challenging issues associated with the dynamic nature of NGAIEs are complex and mostly related to heterogeneity problems encountered at different levels. In this chapter, we shall discuss only those related to the heterogeneity issues present at the level of knowledge representation and management, which appear in cases where different systems utilize different types of ontologies and ontology operations.

3.2.1 Knowledge Representation and Management in AIEs

As stated in the Introduction, ontologies can be used to tackle the issues that stem at the level of knowledge representation and management from the various sources of heterogeneity that exist within NGAIEs. Already quite a few research projects have employed ontologies for this task, including Gaia [35], CoBRA [7], CoCA [12], Semantic Spaces [47], SOFIA [4], DRAGO [38], and CAMPUS [37]. These projects have developed middleware infrastructure for NGAIEs and support reasoning between ontology-based components. Most of these systems either develop a common ULO, which serves as a reference source for heterogeneous domain ontologies, or use domain ontologies that have been built using a shared vocabulary, that is, they are a priori linked with a core ontology which describes the key concepts of an NGAIE. Only projects DRAGO and CAMPUS consider completely heterogenous ontologies. In the case of DRAGO, predefined mappings are used to align the knowledge representations provided by different ontologies, while CAMPUS focuses on the automatic ontology alignment between ontologies.

In our research, we follow an approach similar to that in CAMPUS project. But, because we expect that numerous inconsistent ontologies will appear within an NGAIE, and taking into account the kind of heterogeneity present in these, we propose a multi-faceted strategy to achieve reliable ontology alignments. This strategy firstly estimates the similarity of ontologies to be aligned, and based on the outcome, automates to a different degree the alignment process (obviously, the higher the similarity, the more automated the alignment process). In this way, an alignment is guaranteed, although in the worst case, the user may be asked to serve as "trusted third party" and evaluate the proposed alignments. The next two sections are dedicated to a survey of ontology engineering methodologies and ontology alignment approaches.

3.2.2 Ontology Engineering

Ontology engineering refers to a set of activities that concern the ontology development and management processes, the ontology life cycle, the methodologies for building ontologies, and the tool suites and languages that support them. The six basic aspects to consider when creating an ontology are [19, 31, 40]:

- The content of the ontology
- The application in which it will be used
- The language in which it is implemented
- The methodology which has been followed to develop it
- The software tool used to build and edit the ontology and
- The objective principles for guiding and evaluating ontology design

Several research groups have proposed various methodologies and ontology development environments for building ontologies. In this section, we discuss several methodologies for building ontologies, each one involving different activities, depending on the intended scope, size, and level of detail of the ontology under construction. Their success has been demonstrated in a number of applications. After surveying an extensive set of ontology development methodologies [26, 29, 30, 34, 44–46], we can classify them into two large categories [39]: (1) methodologies focusing on building a single ontology for a specific domain of interest and (2) methodologies focusing on the construction of ontology networks.

On the one hand, the methodologies that focus on building a single ontology presently described in the literature, can be further distinguished in:

- Those aiming at building ontologies from scratch, or by reusing pre-existing ontologies, or by using non-ontological resources, according to the resources that are available to developers.
- Collaborative and non-collaborative, according to the degree of participation of the involved ontology engineers, users, knowledge engineers, and domain experts, in the ontology engineering process.
- Application dependent, semi-application dependent, and application independent, according to the degree of dependency of the developed ontology on the final application.
- Manually, semi-automatically, and automatically constructed ontologies, according to the degree of human involvement in the building process.

On the other hand, only one methodology appears in the literature [44] that focuses on building ontology networks. It deals with the definition of different scenarios for building a collection of single interconnected ontologies, related to each other via meta-relations, such as hasPriorVersion, if the ontology to be developed is a new version of an existing one, isExtension, if the ontology to be developed extends another existing ontology, etc. In that sense, it emphasizes on reuse, reengineering and merging of available ontological and non-ontological resources.

By surveying the existing methodologies, one can conclude that there is no methodology that fits all cases. In each case, the selection of the best methodology depends on a set of contextual factors, which includes the number of the participants in the ontology development process, the kind of the ontology to be built (domain, upper level), the application for which the ontology is planned to be used, the participants' knowledge level of a particular domain, or of the ontology field, their cultural biases and skills, the tools that are going to be used, the availability of participants and resources (documents, ontologies, thesauri), etc.

In the case of building ontologies for modeling NGAIE entities, where various stakeholders, with different skills in knowledge engineering are involved in the ontology building process and the entities under description are complex by nature, as is the case with the ATRACO project, the above-mentioned contextual factors are restricted as follows:

- Since each ATRACO entity describes an isolated domain of interest, such as a specific device, domain ontologies are built.
- Only one ontology developer participates in the engineering process, imposing his own interpretation of the domain. For example, a software developer models a UPnP device, while a knowledge engineer models a user profile.
- Non-ontology, or non-domain experts are involved in the ontology building process, since, for example, each manufacturer provides his conceptual model for a specific device he builds.
- Ontological, or non-ontological resources are not easily available, in order to be reused, or reengineered.

Based on these considerations, we propose in Sect. 3.3.1 a methodology for building domain ontologies, which covers the drawbacks of the existing methodologies, while, at the same time, benefiting from their advantages. Our methodology is based on different scenarios, with respect to the skills of the ontology creators and the available resources. In Sect. 3.3.2, we apply this methodology, in order to develop specific resource and other ontologies.

3.2.3 Ontology Alignment Approaches

Ontology alignment can be a key issue in knowledge-based open-ended environments, such NGAIEs, as it permits the discovery and representation of links (lexical, syntactic, semantic, etc.) between pairs of ontologies, while keeping the original ontologies unaltered. By using alignments, a network of interconnected ontologies is created, which contains the overall knowledge about the domain, albeit distributed in several ontologies and alignments. In this way, the management of the global knowledge contained in an NGAIE becomes a problem of managing many small pieces of knowledge, which describe independent resources and their collaboration. The management of local resource ontologies is cost-effective, because it can be done in a distributed manner (each entity owns its ontology and operates autonomously within the NGAIE). Then, resource collaboration, at the knowledge level, can be described using ontology alignments, which essentially are links that connect the local resource ontologies.

More formally, the problem of ontology alignment is described as: Given two ontologies O_1 and O_2, an alignment between them is defined as a set of relations (equivalence, subsumption, disjointness) between pairs of entities (classes, properties, instances), belonging to the original ontologies. The application of ontology alignment to the interaction of entities in NGAIEs, leads to a set of specific problems:

- Since in open systems like NGAIEs, it's impossible to know in advance the properties of the entities that will interact, it is impossible to use a priori (pre-computed) ontology alignments. Thus the alignment process has to be performed at run time, sometimes with (hard) real-time constraints.
- Since the interacting entities may share common purposes, or competencies, an alignment between two ontologies can be discovered in most applications, although the ontologies may differ in granularity (detail of description of the same entities).

Several alignment approaches have been proposed to tackle the problem of discovering semantic correspondences between entities of different ontologies. They mainly focus on the following aspects [14]:

- Lexical comparison, which relies only on the labels on the ontological entities
- Structural comparison, which relies only on the structure of the ontologies

- Instance comparison, which compares the instances of each ontological entity
- Comparison based on "background knowledge source"

Many researchers have investigated the problem of ontology alignment, mostly by proposing several ontology alignment tools and matchers (or alignment algorithms) [11, 15, 16, 22, 28] which exploit various types of information in ontologies, that is, entity labels, taxonomy structures, constraints, and entities' instances. These tools can be classified into two large categories: those that make use of a single matcher in order to calculate similarities between ontology entities and those which use a family of parallel, or sequential matchers in composition. In the latter category, the similarity between two ontology entities is finally computed by a composite method, such as a weighted aggregation of the similarities obtained by each matcher separately. A challenging issue, while applying these methods, consists in deciding whether a single matcher, or a combination of different matchers, performs better and in what cases, that is, for which kind of ontologies in question. Hence, given a specific pair of ontologies to be aligned, one should define a criterion to determine when a special matcher should be used. Based on this consideration and the specification of ontology alignment problems in NGAIEs, we propose in Sect. 3.4, an ontology alignment strategy which includes the calculation, during a pre-alignment step, of two similarity coefficients, which estimate whether the resemblance of the ontologies in question is mainly lexical, or structural. Then, depending on their values, an agent charged with the task of the alignment process, can select the application of suitable matchers, in order to establish correspondences between ontology entities.

3.3 Own Approach to Knowledge Representation and Management in NGAIES

In this section, we discuss the main challenging issues of the ATRACO framework for the development of NGAIE applications, by considering the creation and management of heterogeneous knowledge, which is an inherent feature of NGAIEs. Firstly, we briefly present the fundamental notion of the ATRACO activity sphere, from a knowledge point of view and the problems that arise during its realization.

The ATRACO project aims at conceptualizing an ambient ecology, that is, a space populated by different entities (devices, services, humans, agents), which are interrelated and in interaction with their environment, in order to realize activity spheres. Each activity sphere is considered as a new ATRACO entity, which can be adapted to different conditions (spatial, temporal, user preference related, environmental). For example, an activity sphere serving for the conceptualization and realization of the "reading" activity of a specific user in his/her private room (spatial condition), can be transferred into a hotel by changing this spatial condition and the devices (facilities like book, desk, chair) and services (light, ambient temperature) provided by the specific environment, or it can be transferred into a "reading" activity of another user, by changing the user preference conditions. Moreover, an ATRACO activity sphere, considered as another ATRACO entity, can be interrelated with other ATRACO entities, that is, other activity spheres, in order to describe and realize more complex activities in pervasive computing environments.

The intuition behind this is that once an elementary ATRACO activity sphere is produced, the knowledge associated with it can be reused, in order to produce new activity spheres. Having this in mind, the context of ATRACO is suitable for realizing elementary ATRACO spheres which can then, either be extended, by adding to the sphere, in an appropriate way, an entity of the ambient ecology, or be embedded into another activity sphere, in order to produce a new more complex one.

In order for a specific ATRACO activity sphere to be successfully realized, ontologies, Ontology Managers, and agents are used. These provide information representation, semantic interoperability, and exchange mechanisms, so that the entities/members of an ambient ecology that participate in the specific activity sphere can communicate and collaborate. According to the ATRACO approach, each entity of an ambient ecology (a human, an agent, a device, or a service) maintains locally an ontology, which represents the complete set of knowledge associated with it and which is managed only by the owning entity. As the entities are distinct, their ontologies will certainly be heterogeneous. As they are engineered by different creators and for different purposes, an alignment process is needed for them to come in a mutual agreement. This alignment process aims at semantically relating heterogeneous ontologies, in order for a network of interlinked ontologies to be realized.

Since the building blocks of knowledge of an ATRACO activity sphere are the ontologies and the alignments of the interrelated entities participating in the ambient ecology, we can consider each ATRACO activity sphere as a network of interlinked ontologies, which encodes the necessary knowledge for its realization. In that sense, ATRACO spheres, as networks of semantically linked ontologies, are means of sharing and reuse. Sharing refers to the fact that different ATRACO spheres can make use of the same ATRACO resources/entities, that is two, or more ATRACO spheres may use the same entity of the ambient ecology for servicing different purposes. For example, the light service provided by a specific lamp device can be eventually shared by two distinct ATRACO activity spheres that are realized simultaneously, i.e., the "reading" activity performed by an individual or for the "playing cards" activity performed by a group of persons. Reuse means to build a new ATRACO sphere by assembling already built ATRACO spheres. For example, a "cooking pasta" activity sphere may be realized by embedding a "reading" (a recipe) activity sphere into a "cooking" activity sphere.

Our experimental research has shown that in order to deal with knowledge representation and management of a network of interlinked ATRACO ontologies, one has to resolve the following fundamental issues:

1. How to build complete and consistent domain heterogeneous ontologies, which define concepts closely related to distinct AIEs entities?

2. Which are the ontologies that need to be created, in order to represent the knowledge which is efficient and necessary for the applications in NGAIEs to be carried out successfully?
3. How to strategically guide the alignment process of a pair of heterogeneous ontologies in order to discover their semantic linkage?
4. How to align a pair of heterogeneous ontologies?
5. In what extent and in which order can an initial alignment of a pair of ontologies be further populated with alignments, in order to produce a network of interlinked ontologies?
6. How can, a replacement of a specific ontology within a network of ontologies, be achieved?
7. How to evolve a network of interlinked ontologies, in the case an ontology evolves, undergoes changes, or is replaced by an equivalent or similar one?

The above-mentioned challenging issues refer to the ontology engineering with respect to the nature of NGAIEs, as well as to the application of alignment techniques in networks of ontologies, which seems to be a rather unexplored topic so far. Considering this context, a methodology for domain ontology building has been proposed and following it, various heterogeneous ATRACO ontologies have been created; these are presented in Sects. 3.3.1 and 3.3.2, respectively. Moreover, thorough experiments have been carried out involving the various ATRACO ontologies, in order to confront the third and fourth issues. Considering the third, we propose two ontology resemblance coefficients, presented in Sect. 3.4.1.1, which detect whether their similarity is mainly lexical or structural, in order to select the most appropriate combination schema of alignment algorithms to be used, while for the fourth, we propose an alignment strategy, presented in Sect. 3.4, which involves the use of a "trusted third party," in order for a pair of heterogeneous ontologies to be aligned. The remaining issues are confronted in Sect. 3.5, within the mathematical framework of Category Theory, which, independently of the language used to represent ontologies and alignments, permits the realization of basic operations in networks of ontologies that comprise the ATRACO activity spheres.

3.3.1 Ontology Engineering in ATRACO

Taking into account the considerations of Sect. 3.2.2 about ontology engineering issues that emerge in NGAIEs, we designed a task-based ontology engineering process, in order to build the ATRACO ontologies [39].

Initially, we restrict the kind of the ontology to be built, to a domain ontology, since within the ATRACO project, each entity describes a domain, such as a user profile, or a specific device. Moreover, we restrict the number of participants in the ontology development process, to a single novice ontology developer, since, in the context of ATRACO, each manufacturer, who is not an ontology expert, provides his own personal domain ontological model. We further focus on two parameters, which are related to the above-mentioned restrictions: the knowledge level of a particular domain the ontology developer has (domain, or non-domain expert) and the strategy followed, in order to build the ontology (from scratch, or by reuse); the strategy depends on the availability of resources (documents, ontologies, group of experts). According to the above-mentioned restrictions, the novice ontology developer will have to choose between four scenarios, in order to build his personal domain ontology. In each case, the choice depends on his knowledge level of the particular domain and his access to existing ontologies, related literature, or domain experts. The possible scenarios, as depicted in Fig. 3.1, are:

- 1st scenario: creating a domain ontology from scratch, by a novice ontology developer, who is also domain expert.
- 2nd scenario: creating a domain ontology by semantically reusing existing ontologies, by a novice ontology developer, who is a non-domain expert.
- 3rd scenario: creating a domain ontology from scratch, by a non-domain expert novice ontology developer.
- 4th scenario: creating a domain ontology, by semantically reusing existing ontologies, by a domain expert novice ontology developer.

During the application of each scenario, the ontology developer needs to carry out four mandatory tasks, namely analysis, definition, selection, and evaluation. These tasks are scattered in all phases of the ontology engineering process, i.e., the specification, conceptualization, implementation, and evaluation phase. All These phases appear to be common in the all methodologies available in the literature.

In our approach, we preserve the ordering of the specification, conceptualization, implementation, and evaluation phases, generally adopted by all the methodologies proposed in the literature. Moreover, we further group the activities present in each phase, in four clusters of tasks, each one containing mandatory sub-tasks, according to the selected scenario.

More precisely, concerning these clusters of tasks, the ontology developer needs: (a) To select the available resources, in order to help him conceptualize the domain. These resources include related literature, existing ontologies and a group of experts of the domain under description; (b) To select the appropriate tool and language, in order to implement his conceptualization; (c) To analyze the selected resources, in order to define what is important for the description of the specific domain, through the competency questions; (d) To evaluate the selected resources, as well as the result of his attempt.

The detailed tasks and the sequential sub-tasks that an ontology developer needs to follow, in order to build his domain ontology, according to the selected scenario, are also depicted in Fig. 3.1. Relations between sub-tasks are denoted by appropriate arrows.

The task of analysis includes sub-tasks, such as analysis of the literature related to the domain, analysis of the classes of the selected ontology and analysis of the properties of the selected ontology. These sub-tasks belong to two different phases, the specification and the conceptualization phase, respectively. In this way we denote the temporal sequence in which these sub-tasks must be carried out. The task

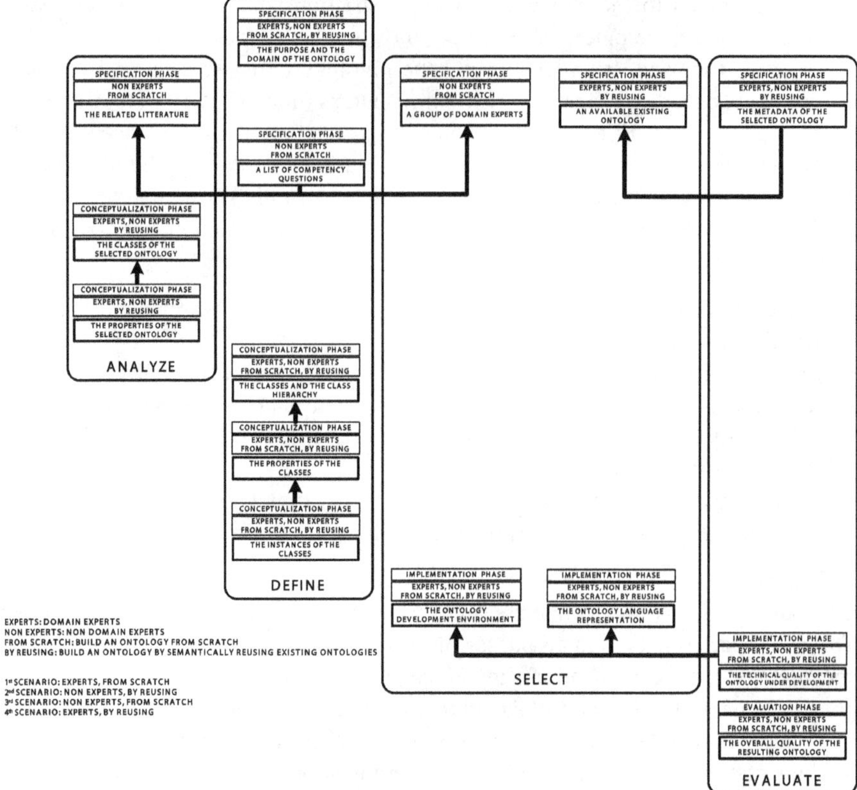

Fig. 3.1 An overview of the possible scenarios in the task-based ontology engineering approach

of definition includes sequential sub-tasks, such as the definition of the purpose and the domain of the ontology, the definition of a list of competency questions, the definition of the classes, the class hierarchy, the properties, and the instances of the ontology.

The selection task includes the selection of a group of experts, the selection of an available ontology, and the selection of the appropriate ontology development environment and the ontology representation language.

The task of evaluation involves sub-tasks that are distributed in the specification phase, in order to evaluate the metadata of the selected ontology, in the implementation phase, in order to evaluate the technical quality of the ontology under construction and in the evaluation phase, in order to evaluate the overall quality of the resulting ontology.

All the above-mentioned sub-tasks included in the four mandatory tasks show a temporal ordering, which is denoted by the phase in which they belong.

Our aim was to create a methodology for building isolated domain ontologies by a single creator, which allows to:

- Define each task to be carried out precisely, that is, to state clearly its purpose, input and output, the participants involved, the right management of the available resources, the time at which its execution is more convenient and the possible ways of executing it
- Be presented in a prescriptive way, in order to facilitate non-ontology experts
- Be general enough, in the sense that it can help any developer to build his personal domain ontologies, by using any ontology development tool

The tasks proposed in this methodology have been followed in order to build the ATRACO ontologies, which are described in details, in the next section. More precisely, the ATRACO ULO was created by following the tasks described in the 1st scenario of the methodology, Device and Service Ontologies (DSOs) were developed by adopting the 2nd and the 4th scenarios, the Privacy Policies Ontology (PPO) was created by following the 3rd scenario, while several UPOs were built by adopting the tasks described in the 2nd scenario of the proposed methodology.

Though cost models exist, that aim at predicting the cost involved in developing an ontology, in terms of effort and duration, it is extremely difficult to evaluate an ontology building methodology, since experimentation involves a multitude of uncontrollable conditions. Moreover, it is unlikely that someone would accept to pay twice for building the same complex ontology with different approaches. Thus, we have not conducted such an evaluation for our ontology building methodology, but tailored it to suit the specific contextual factors met in the ATRACO project, as explained in the preceding analysis. In addition, even though we have not evaluated the ontology building methodology that we followed, we have evaluated its output, i.e., the resulting ontologies, through competency questions.

3.3.2 ATRACO Ontologies

In ATRACO, each sphere entity (user, agent, device, or service) maintains locally an ontology, which describes its features, capabilities, and current state. Each local ontology is managed only by the owning entity. In other words, the local ontology of an entity represents the complete set of knowledge offered about this entity. However, it is expected that these ontologies will be heterogeneous, because they will reflect the conceptualizations of the involved stakeholders (manufacturers, developers, users, experts, etc.). On the other hand, this knowledge must be accessible by all ATRACO entities, in order to support the semantic interoperability among them, so a primary requirement is to define the basic concepts of the ATRACO world model, which are described in the ATRACO ULO.

Nevertheless, the local ontologies do not have to be semantically compatible with the ATRACO ULO; the aim of the project is to align these ontologies by using, when necessary, the ATRACO ULO as a "pivot ontology," or other General Foundational Ontologies (GFOs) as a "third party" background knowledge.

In the context of ATRACO, the following ontologies have been developed:

- The ATRACO ULO (Upper Level Ontology), which serves as a common semantic reference between ATRACO entities. It encodes the basic concepts of ATRACO (i.e., devices, users, agents, services, policies) and their interrelations. Its role is to enhance knowledge and information sharing, between the inherently heterogeneous components of an ambient ecology. Because local ontologies are treated as black boxes (i.e., they can be altered only by the entities they own them), the ATRACO ULO is used by the ATRACO Ontology Manager, in order to optimize the ATRACO Sphere Ontology.
- DSOs (Device and Service Ontologies), which encode, in dissimilar ways, the basic characteristics of each device, as well as the features of the services that they provide. They are maintained by each device.
- POs (Policy Ontologies), which encode entities and rules that describe specific policies, such as user privacy. They are maintained by specific managers in the ambient intelligence (AmI) space.
- UPOs (User Profile Ontologies), which encode user traits and preferences. A user can assume different personas, based on context. They are also maintained by specific managers in the AmI space.

Moreover, GFOs (Generic Foundational Ontologies) are also used for providing background knowledge, whenever this is required.

All these ontologies have been created by following the ontology engineering approach of Sect. 3.3.1, as described above. In particular, OWL-DL was adopted as the ontology representation language, since it offers maximum expressiveness, while retaining computational completeness and decidability.

3.3.2.1 ATRACO ULO

The basic goal of the ATRACO ULO is to provide a shared referent that will enhance knowledge and information sharing among the ATRACO components (users, agents, devices, services), whenever necessary. Thus, it describes the core terms of ATRACO domain model and their interrelations. Some of the main classes of the ATRACO ULO, as depicted in Fig. 3.2, are:

- ContextEntity, which is the root class of the ontology
- Person, which corresponds to all human entities
- Activity, which refers to a person's current activity
- AbstractPlan, which describes what should be done during a specific activity
- Time, which corresponds to the time at which a specific activity is carried out
- Space, which corresponds to the physical space in which a specific person, physical object, or artifact is placed
- Artifact, which represents a set of devices that offers services
- Agent, which represent a set of all agents in the ATRACO context
- Service, which represents a set of services provided by specific devices, or by the network

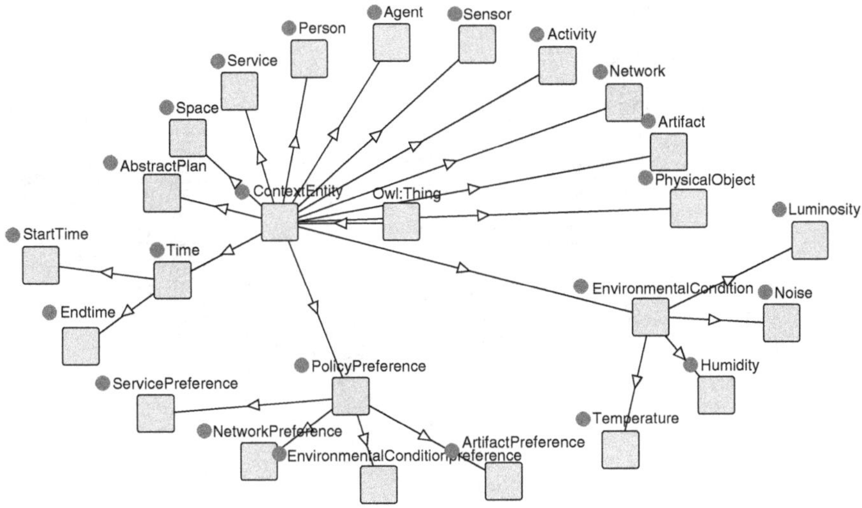

Fig. 3.2 The ATRACO ULO

3.3.2.2 ATRACO DSOs

In the ATRACO scenarios, the following knowledge must be provided: (a) which device offers a specific service, (b) is it permissible, or not, for a user to use a specific service in order to carry out a specific activity, (c) which is the policy of using a specific service, (d) what are the features of a device that are necessary, in order for a user to carry out a specific activity, etc. This kind of knowledge is scattered in the local ontologies of the different autonomous context elements (devices, users) that participate in the execution of an activity sphere; these local ontologies are heterogeneous and may be also inconsistent, or incomplete, as they are developed independently by device manufacturers, software developers, or non-domain experts.

Thus, generally, the ATRACO DSOs have the following characteristics: (a) they are domain ontologies that are short models of the domain under description, (b) they do not have extensive is_a hierarchies, (c) the same concept may be represented at the class level in one ontology and at the instance level in another one, (d) they have complex relations, where classes are connected by a number of different relations, (e) their terminologies, in certain cases, are not identical, even if the ontologies describe the same domain, (f) the modeling principles for them are not well defined and documented. Some examples of the ATRACO DSOs are depicted in Figs. 3.3 and 3.4.

All these ontologies contain information about the features of each device and the services it provides, in a different structure, or in different terminology, as the context of ATRACO states. For example, a Device has DeviceDescription, which contains basic information related to this specific device, such as Device-

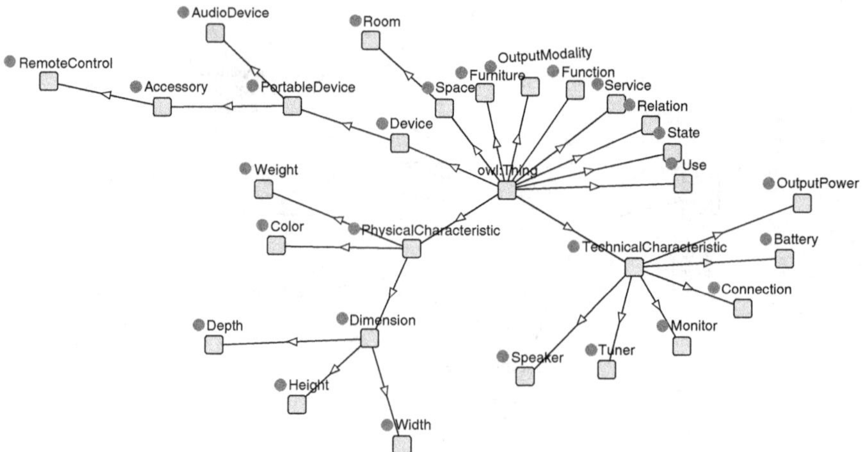

Fig. 3.3 An example of an mp3 player DSO

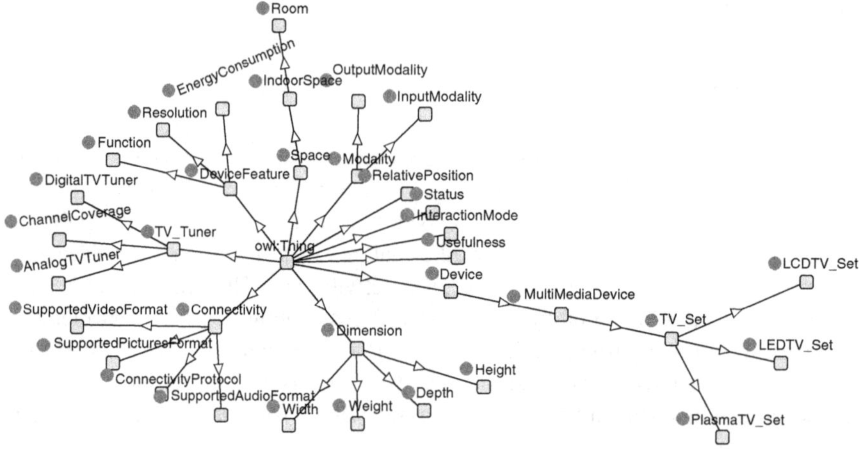

Fig. 3.4 An example of a TV-set DSO

Name, DeviceType, DeviceVendor and its PhysicalCharacteristics such as Shape, Color, etc.

The Device has also a HardwareDescription, which contains the details of its CPUDescription, ConnectionDescription to the network, its MemoryDescription and UIDescription, as well as SoftwareDescription, which contains the details of the operating system of the device. Another important class in DSOs is the class Service, which provides information about the service(s) hosted on the device concerned. An instance of a floor lamp ontology is depicted in Fig. 3.5, where a Lamp with id: Lamp_0234 is described. It has LampGeneralCharacteristics, such as its name, id, etc., LampFunctionalCharacteristics, where the service that it offers is described,

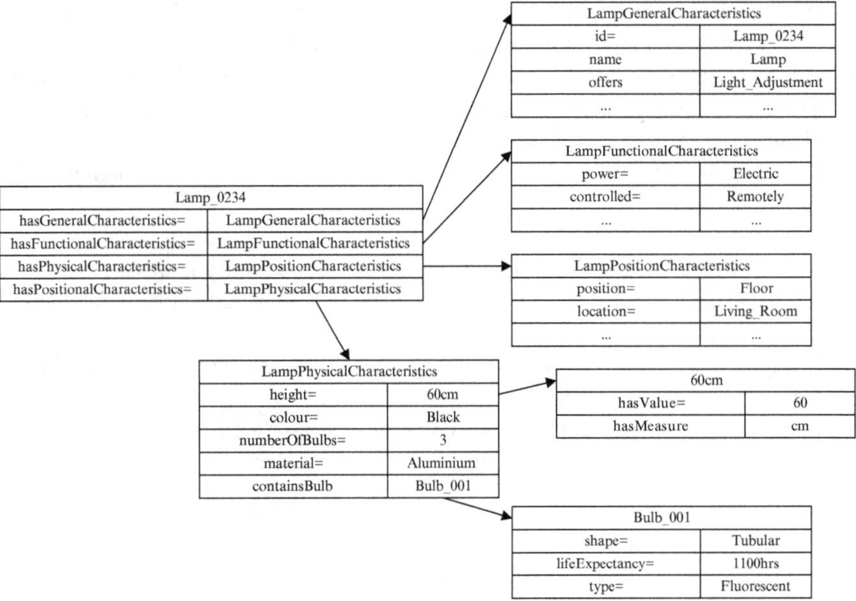

Fig. 3.5 An instance of a floor lamp DSO

LampPositionCharacteristics, such as its location and LampPhysicalCharacteristics, such as height, colour, numberOfBulbs, material, etc. Each Bulb, which is contained in the specific lamp has its own features, such as its type: Fluorescent, shape: Tubular and lifeExpectation.

3.3.2.3 ATRACO PPO

The Privacy Policy Ontology can be used to define privacy rules, which the various sphere components can use while interacting with each other in the context of a specific activity. Note that the privacy policy is bound to the activity supported by the activity sphere, but not to its particular components. It addresses issues such as: What resource type is needed for a service? What is done with this resource? Who has access to a specific resource? Who processes a user's data? In which way are the data transported? In which way are the data stored? Who are the recipients of the data? Are the recipients humans, or machines? Who is the creator of a specific policy? Which are the mechanisms incorporated in a specific policy? Which policies have authentication, encryption, right, or identification as a policy mechanism? What kind of access does a user have to a specific device?

In order for these questions to be answered effectively, the following basic concepts comprise the main classes of the PPO:

- Mechanism: the set of the mechanisms (actions) that can take place in the context we described and have to do mostly with the data of the system
- PolicyMechanisms: all the mechanisms which are required for the implementation of a policy, such as access control (authentication, authorization, and identification)
- Resource: the set of the resources of a ubiquitous computing environment (e.g., agents, devices)
- User: the users in the context
- ID: the identities of the entities of the system. An identity is a list of attribute values of an entity, that allows this entity to be distinguished among other entities within the context
- Policy: all the policies which are defined by the users of the system, according to their needs and preferences
- Time: is a basic concept to describe some actions (e.g., the duration of data storage, or the duration of a policy)

In Fig. 3.6 a graph with all the classes and their subclasses, as they are defined in the ontology, is depicted.

In order to develop the ontology, the entity classes are connected and hence related to each other, through properties. These are called object properties and provide the available capabilities of each class. Similarly, in order to define in detail the properties and characterize each class, data types (data properties) are created. Some of these properties that have been created, in order to describe the holistic ontology concept, are shown below:

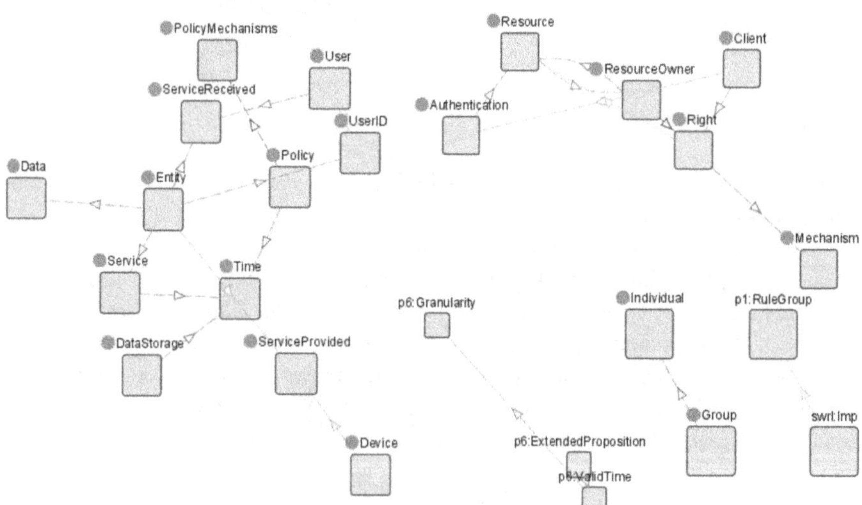

Fig. 3.6 The main classes and their subclasses of the PPO

- hasPolicyMechanism: this property connects the classes Policy and PolicyMechanisms. Indicates that any policy created by a user must have at least one policy mechanism for its proper function
- hasDuration: this property relates the classes DataStorage/Policy/Service, and Time. It provides the duration of the data storage, or a policy, or a particular service
- BankAccount, HomeAddress, MedicalProfile, PhoneNumber, etc.: these properties characterize the class SensitiveData. They indicate which data of a user profile are considered as sensitive
- creatorOfPolicy and policyDescription: these properties characterize the class Policy. They provide information about the user, who has created a concrete policy and the description of this policy, respectively
- identifiesID: this property connects the classes Entity and ID. It provides information about the entity which identifies the identity of a user, or another entity (agent, or device)

The PPO has been enriched with a set of rules for a better understanding of the basic concepts and also for the representation of privacy principles [32]. Some examples of the rules added, are: (a) the user who has an identifiable ID, which is identified by a resource, always has positive authentication on this resource, (b) the owner of a resource has always positive authentication on this resource, etc.

3.3.2.4 ATRACO UPO

In the ATRACO project, each user has a profile which is described in its UPO. Different users may be involved within the same activity, so heterogeneous ontologies are used in order to semantically describe them. In Figs. 3.7 and 3.8, some examples of ontologies describing user profiles within the context of ATRACO are depicted.

The creation of these ontologies, as explained in more details in [42], is driven by a set of competency questions, such as: What are the physical characteristics of a specific user? What is the current activity that a specific user is carrying out? What are the preferences of a specific user, concerning a specific activity? What are the preferences of a specific user concerning a specific service that the user uses, in order to carry out a specific activity? What is the general information (name, age, gender, nationality) of a user? What is the contact information (home address, email address, phone number) of a user? Where is a user situated when carrying out a specific activity? When does a user carry out a specific activity? What is the current location of the user? Which is the temperature that a user prefers? What are the user's interests? Does the user belong to a group? What are the light and illumination levels that the user prefers? What are the media options that the user prefers? What modalities does the user prefer for which devices/tasks (e.g., TV <-> remote controller, MP3 player <-> voice)? Which is the profile of a user, according to a specific activity? Which is a user's permanent profile? Which is the user context that is related to a specific activity? What user related information is contained in a

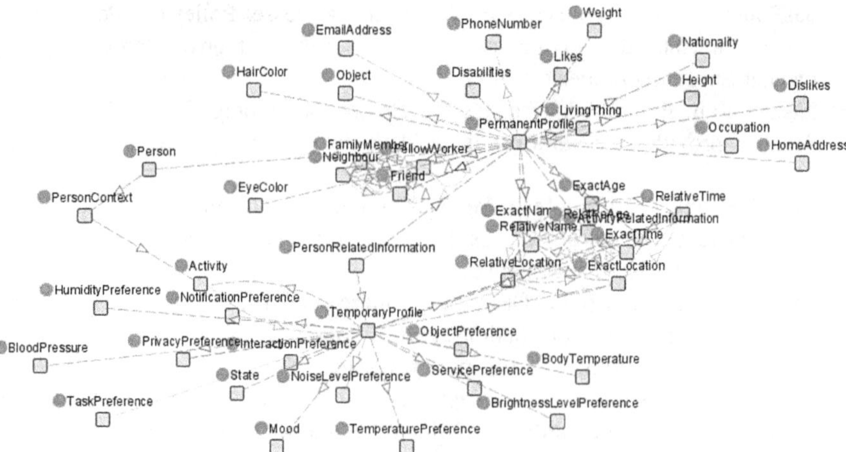

Fig. 3.7 Suki's User Profile Ontology within the ATRACO context

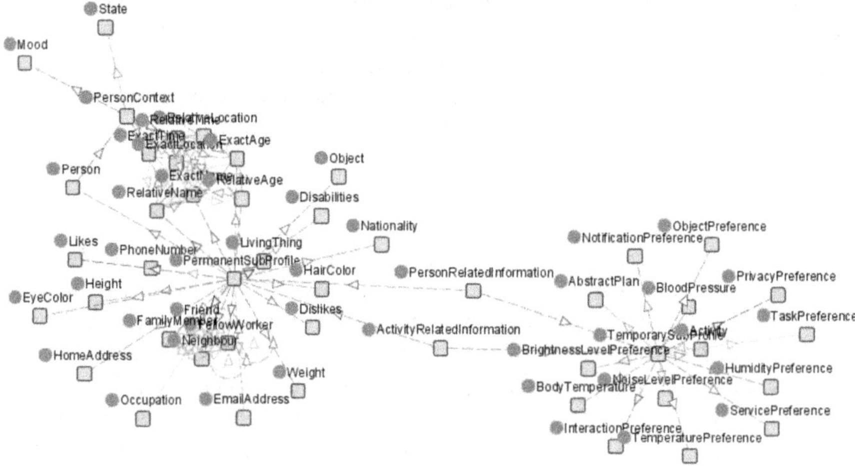

Fig. 3.8 Jona's User Profile Ontology within the ATRACO context

specific person's permanent profile? What activity related information is contained in a specific temporary sub-profile?

Based on these competency questions, many ontologies with different structure and dissimilar labels can be constructed, but they will all return semantically identical responses to the same queries. We have adopted the following structure of a user profile: each User has a PermanentSubProfile and more than one TemporarySubProfile. His PermanentSubProfile contains PersonRelatedInformation, such as GeneralInformation (Gender, Name, Age, Occupation, Nationality); Likes; Dislikes; Disabilities; ContactInformation (PhoneNumber, EmailAddress, HomeAddress); Possessions (LivingThing, Object); SocialInformation (Friend,

FamilyMember, FellowWorker, Neighbor); PhysicalInformation (EyeColor, HairColor, Weight, Height). Each TemporarySubProfile is related to a specific Activity and depends on a PersonContext, which contains the Location of the user, the Time at which the Activity is carried out, the State of the user, which means whether the user is Alone, With_Friends, etc., and the person's Mood. A TemporarySubProfile contains ActivityRelatedInformation, such as the person's BiologicalInformation (BloodPressure, BodyTemperature) and the user's Preferences, such as InteractionPreference, NotificationPreference, TaskPreference, ObjectPreference, ServicePreference, PrivacyPreference, and EnvironmentalConditionPreference (BrightnessLevelPreference, NoiseLevelPreference, TemperaturePreference, HumidityPreference), which are associated with the Activity that a user carries out. For Privacy related issues, there is a distinction in some sensitive personal data, such as Age, Name, Location, and Time. They are distinguished in, for example, ExactAge (e.g., 28 years old) and RelativeAge (e.g., Adult). In that sense, RelativeAge is lessDetailedThan ExactAge.

In order to gain a better understanding of the core structure of this ontology, we exemplify an instance of it, depicted in Fig. 3.9. User: Suki, who has a specific PermanentSubProfile, has the following UserPermanentSubProfile: SukiPermanentSubProfile. He is situated in RelativeLocation: Living_Room, at Time: Evening and he is in State: With_Friends. He is currently carrying out an Activity: Feel_Comfortable. This activity is associated with his TemporarySubProfile: FeelComfortableSubProfile, which contains his TaskPreference: he prefers to Listen_To_Music, his ServicePreference: TemperatureService, his ObjectPreference: he prefers to use the Air_Conditioning and the TV_Set, his EnvironmentalConditionPreference: he prefers Low_Temperature, etc.

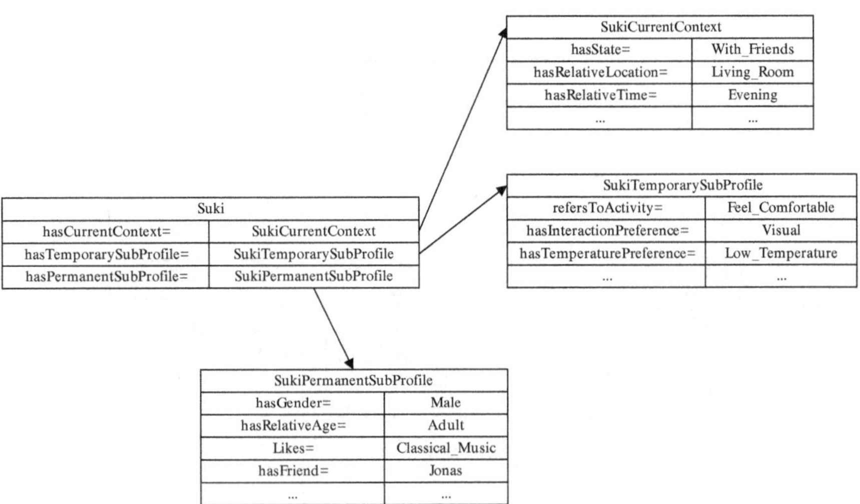

Fig. 3.9 An instance of Suki's profile ontology

3.4 ATRACO Ontology Alignment Strategy

The proposed alignment strategy contains three main steps, as depicted in Fig. 3.10:

(1) Pre-processing step, where we try to make an initial estimation, at the ontology level, of whether the ontologies to be aligned appear to be similar, lexically, or structurally. (2) Processing step, where, based on the previous step, we decide about the proper sequence of execution of appropriate matchers (lexical, structural, instance-based, constraint-based algorithms); and (3) Post-processing step, where the final results are evaluated. The above-mentioned steps are analyzed in the following subsections.

Once ontology alignments are produced and evaluated, they are further used by the ATRACO Ontology Manager, presented in the respective book chapter, for tasks such as ontology merging, data translation, and query answering.

In the ATRACO perspective, an alignment is considered as a meaningful object, that is, an explicit information describing semantic links between different ontologies. Thus, the produced alignments are also shared and reused by different applications within and among NGAIEs.

3.4.1 Pre-Processing Step

In this first step, given two source ontologies, we consider their particularities (of small, or medium size, of limited hierarchy structure, with entity labels being tightly oriented towards the creators' view of the domain under description) and their similarity characteristics (they have similar labels, or structure), as well as

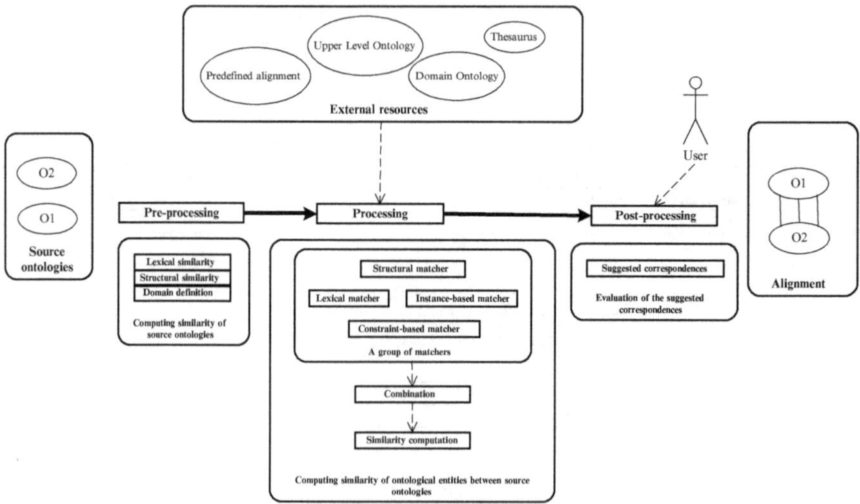

Fig. 3.10 The ATRACO alignment strategy

the description of their domains, in order to decide which family of ontology alignment algorithms, or matchers (lexical, structural) to use, whether to combine matchers sequentially, or in parallel, what kind of external resources to use, in order to enhance the alignment process in case of ontologies that differ greatly either lexically, or structurally, or when no instances and constraints are provided, whether the user should be involved, or not in the alignment process, etc.

More precisely, in order for the ATRACO Ontology Manager to be able to execute successfully each alignment process, during this pre-processing step, it must be aware of the particularities of the source ontologies', as well as of which matchers are suitable for each of the input ontologies. We have designed new two similarity coefficients, which we present in the next subsection, in order to detect the lexical, or structural resemblance of the source ontologies. These coefficients are used, in order to select the suitable family of matchers, as well as their way of combination. Their output values fall into the range [0, 1]. These similarity coefficients measure similarities at an ontology level and are of two kinds: lexical and structural. The former examines the relative structure of the two ontologies, based on the comparison of the lengths of all paths leading from the root of each ontology to each of its leaves. The latter, after discovering concepts with identical labels in both ontologies, considers also the relative proximity of these common concepts, inside each of the ontologies to be aligned.

3.4.1.1 Similarity Coefficients for Detecting Ontologies Resemblance

We detail hereafter the computational aspects of the similarity coefficients used in the pre-processing step of the ATRACO alignment strategy.

Structure Similarity Coefficient

Given two ontologies O_1 and O_2, we define the Structure Similarity Coefficient, denoted by $\sigma(O_1, O_2)$, which is a similarity metric at an ontology level (as opposed to an entity level), with values that range from 0 to 1. The Structure Similarity Coefficient describes the similarity between two ontologies globally (as opposed to local structural similarities between ontology entities), based on their structural resemblance. In order to compute it, one has to follow the constructive procedure described below:

Definition—Structure Similarity Coefficient: Given two ontologies O_1 and O_2, calculate the vectors $\overline{l_1}$, $\overline{l_2}$ having as elements the lengths of all the paths from the root of each ontology, to all its leaves, i.e.,

$\overline{l_1} = [l_{11}, l_{12}, \ldots, l_{1i}, \ldots]$, with l_{1i} = length of the path from the root of ontology O_1 to its ith leaf, $i = 1, 2, \ldots, \#$ leaves of ontology O_1

$\overline{l_2} = \left[l_{21}, l_{22}, \ldots, l_{2j}, \ldots\right]$ with l_{2j} = length of the path from root of ontology O_2 to its jth leaf, $j = 1, 2, \ldots, \#$ leaves of ontology O_2

Let $L = \max\{|\overline{l_1}|, |\overline{l_2}|\}$, with $|\overline{l_i}|$ the dimension of vector $\overline{l_i}$, $i = 1,2$. Create two new vectors $\overline{a}$, $\overline{t}$, by choosing between the vectors $\overline{l_i}$ $i = 1,2$ the one that has the biggest dimension and by completing the other vector with leading zeros. Both vectors $\overline{a}$, $\overline{t}$, have dimension L.

If $|\overline{l_i}| > |\overline{l_j}|$, $i,j \in \{1,2\}$ and $i \neq j$, then $\overline{a} = \overline{l_i}$, $\overline{t} = \left[\overline{0}, \overline{l_j}\right]$, with the dimension of $\overline{0}$ being equal to $L - \min\{|\overline{l_1}|, |\overline{l_2}|\}$.

Compute now a square *LxL* matrix C, with elements $c_{ij} = |a_i - t_j|$, $i,j = 1,2,\ldots,L$. Then, create two new vectors $\overline{r}$ and $\overline{s}$, by appropriately reordering the vectors $\overline{a}$ and $\overline{t}$, as explained hereafter.

Let us consider two sets B and T with cardinalities equal to L and let β_i, τ_i, $i = 1,2,\ldots,L$, denote their respective elements. Consider the bipartite graph having as nodes the elements of the sets B and T and containing all possible edges between respective elements of the two sets. The edge linking β_i, to τ_j $i,j = 1,2,\ldots,L$, has a weight equal to $c_{ij} = |a_i - t_j|$. One can then always find a square matrix X with dimensions *LxL* having elements x_{ij}, $i,j = 1,2,\ldots,L$, such that the following relations hold:

1. $\forall i = 1,2,\ldots,L, \sum\limits_{j=1}^{L} x_{ij} = 1$
2. $\forall j = 1,2,\ldots,L, \sum\limits_{i=1}^{L} x_{ij} = 1$
3. $\forall i,j = 1,2,\ldots,L, x_{ij} \geq 0$
4. $\sum\limits_{i=1}^{L}\sum\limits_{j=1}^{L} c_{ij}x_{ij}$ is minimized

It can be proven that such elements x_{ij}, $i,j = 1,2,\ldots,L$, exist and assume either the value 0, or the value 1. If $x_{ij} = 1$, then the ith element of the reordering $\overline{r}$ is $r_i = a_i$, while the jth element of the reordering $\overline{s}$ is $s_j = t_j$. The structural similarity between the two ontologies is finally calculated as the cosine of the angle between the vectors $\overline{r}$ and $\overline{s}$:

$$\sigma(O_1, O_2) = \frac{\overline{r}.\overline{s}}{||\overline{r}||.||\overline{s}||} = \frac{\sum\limits_{i=1}^{L} r_i s_i}{\sqrt{\sum\limits_{i=1}^{L} {r_i}^2}\sqrt{\sum\limits_{i=1}^{L} {s_i}^2}}.$$

In order to clarify the computation of the Structure Similarity Coefficient, we compute it for the pairs of ontologies of Figs. 3.11 and 3.12. For the ontologies of Fig. 3.11, we compute the Structure Similarity Coefficient as

$$\sigma(O_1, O_2) = \frac{2 \cdot 2 + 1 \cdot 1 + 1 \cdot 0}{\sqrt{2^2 + 1^2 + 1^2}\sqrt{2^2 + 1^2 + 0^2}} = \sqrt{\frac{5}{6}} = 0.9129.$$

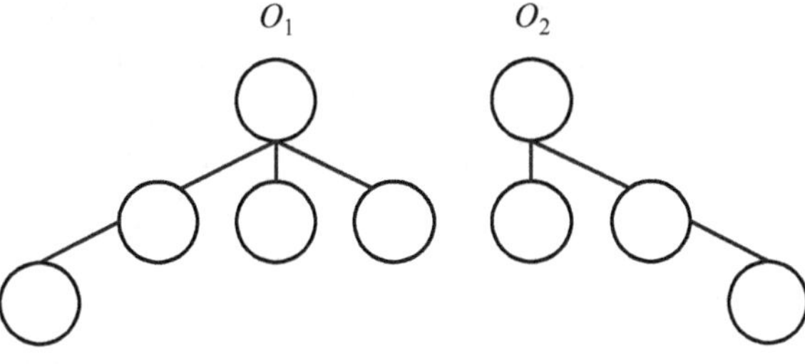

Fig. 3.11 The ontologies of example 1

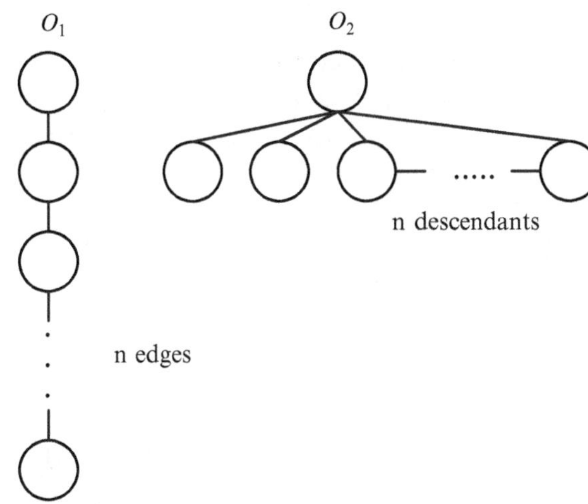

Fig. 3.12 The ontologies of example 2

The Structure Similarity Coefficient depicts accurately the similarity of structure between the two ontologies, which becomes apparent when flipping O_1 horizontally. The Structure Similarity Coefficient for the ontologies of Fig. 3.12 is calculated as

$$\sigma(O_1, O_2) = \frac{n}{\sqrt{n}\sqrt{n^2}} = \frac{1}{\sqrt{n}},$$

that is, $\sigma(O_1, O_2) \to 0$ as $n \to \infty$.

Lexical Similarity Coefficient

We define the Lexical Similarity Coefficient, at an ontology level, with values ranging from 0 to 1. In order to calculate the Lexical Similarity Coefficient, we consider two factors. The first factor is based on the number of concepts/classes having the same label in both ontologies (inter-ontology factor), while the second one takes into account the relative proximity that these common concepts have among them, inside each one of the ontologies (intra-ontology factor).

Definition—Lexical Similarity Coefficient: Given two ontologies O_i and O_j, $i,j = 1,2, i \neq j$, with a number of cc pairs of concepts with the same label, that is, $(\varepsilon_1^{O_i}, \varepsilon_1^{O_j})$, $(\varepsilon_2^{O_i}, \varepsilon_2^{O_j})$,...,$(\varepsilon_k^{O_i}, \varepsilon_k^{O_j})$, $i,j = 1,2, i \neq j$, $k = 1,2,\ldots,cc$, respectively, the Lexical Similarity Coefficient is calculated as:

$$\lambda(O_i, O_j) = \frac{\sum_{k=1}^{cc} [1 - \frac{\left|\delta\left(\varepsilon_k^{O_i}\right) - \delta\left(\varepsilon_k^{O_j}\right)\right|}{\max(\delta\left(\varepsilon_k^{O_i}\right), \delta\left(\varepsilon_k^{O_j}\right))}]}{\max(\#\text{concepts of } O_i, \#\text{concepts of } O_j)},$$

$i,j = 1,2, i \neq j$, where the term $\delta\left(\varepsilon_k^{O_i}\right)$ ranks concept $\varepsilon_k^{O_i}$ of ontology O_i, by taking into account how far, in terms of number of edges, the remaining common concepts $\varepsilon_p^{O_i}, p \neq k$ are from concept $\varepsilon_k^{O_i}$ in ontology O_i and is given by

$$\begin{aligned}
\delta\left(\varepsilon_k^{O_i}\right) = \varphi_k^{O_i} &+ \frac{d\left(\varepsilon_k^{O_i}, n_1\right)}{cc - 1}\left(1 - \varphi_k^{O_i}\right)\rho \\
&+ \frac{d\left(\varepsilon_k^{O_i}, n_2\right)}{cc - 1}\left(1 - \varphi_k^{O_i}\right)\rho\,(1-\rho) + \cdots \\
&+ \frac{d\left(\varepsilon_k^{O_i}, n_m\right)}{cc - 1}\left(1 - \varphi_k^{O_i}\right)\rho(1-\rho)^{m-1} \\
&+ \frac{d\left(\varepsilon_k^{O_i}, O\left(n_{m+1}\right)\right)}{cc - 1}\left(1 - \varphi_k^{O_i}\right)(1-\rho)^{m},
\end{aligned}$$

where

$$\varphi_k^{O_i} = 1 + (\alpha - 1)\text{sgn}\left[(cc-1) - d\left(\varepsilon_k^{O_i}, n_1\right)\right]^{1}$$

with α, $(0 \prec \alpha \prec 1)$ a constant added to the rank of common concept $\varepsilon_k^{O_i}$, due to its lexical similarity to concept $\varepsilon_k^{O_j}, i,j = 1,2, i \neq j$ and where we define:

n_1 to be the 1-neighborhood of concept $\varepsilon_k^{O_i}$, containing all common concepts $\varepsilon_p^{O_i}, p \neq k$, that are within a distance of exactly one edge from $\varepsilon_k^{O_i}$ in O_i,
n_2 to be the 2-neighborhood of concept $\varepsilon_k^{O_i}$, containing all common concepts $\varepsilon_p^{O_i}, p \neq k$, that are within a distance of exactly two edges from $\varepsilon_k^{O_i}$ in O_i,...,
n_m to be the m-neighborhood of concept $\varepsilon_k^{O_i}$, containing all common concepts $\varepsilon_p^{O_i}, p \neq k$, that are within a distance of exactly m edges from $\varepsilon_k^{O_i}$ in O_i,

[1]The signum function is defined as: $\text{sgn}(x) = \begin{cases} -1 & \text{if } x < 0 \\ 0 & \text{if } x = 0 \\ 1 & \text{if } x > 0 \end{cases}$.

$O(n_{m+1})$ to be the remote-neighborhood of concept $\varepsilon_k^{O_i}$, containing all common concepts $\varepsilon_p^{O_i}, p \neq k$, that are within a distance of more than m edges from $\varepsilon_k^{O_i}$ in O_i.

Then, $d\left(\varepsilon_k^{O_i}, n_q\right), q = 1, 2, \ldots, m$, denotes the number of common concepts $\varepsilon_p^{O_i}, p \neq k$, that are within a distance of exactly q edges from $\varepsilon_k^{O_i}$ in O_i and $d\left(\varepsilon_k^{O_i}, O(n_{m+1})\right)$ denotes the number of common concepts $\varepsilon_p^{O_i}, p \neq k$, within a distance of more than m edges from $\varepsilon_k^{O_i}$ in O_i.

$\frac{1}{2} \prec \rho \prec 1$ is a forgetting factor, penalizing more severely the common concepts $\varepsilon_p^{O_i}, p \neq k$ that are more distant from $\varepsilon_k^{O_i}$ in O_i (in more distant neighborhoods).

For clarification reasons, we consider the ontologies of Fig. 3.13 and compute the Lexical Similarity Coefficient of pairs O_1 and O_2 and O_1 and O_3, respectively, by choosing $a = 0.8$ and $\rho = 0.6$. The ontologies O_1, O_2, and O_3 have common labels A, B, and C . Thus, $cc = 3$ and we choose $m = 2$, limiting ourselves to $1-, 2-$neighborhoods n_1, n_2 and $O(n_3)$.

When computing the Lexical Similarity Coefficient between O_1 and O_2, since each common concept distributes in the same way the remaining common concepts in its neighborhoods, in both ontologies, it results that

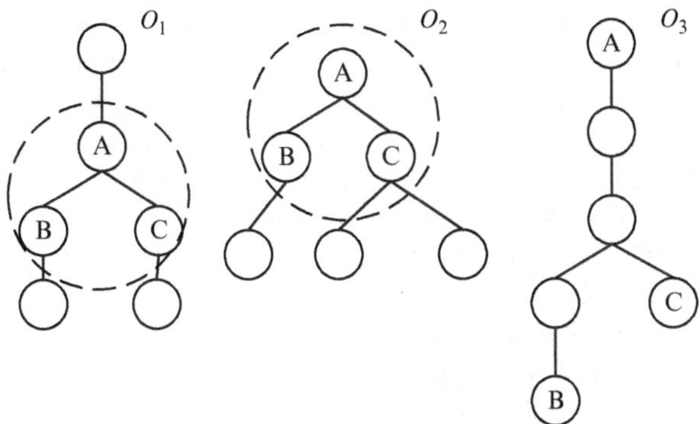

Fig. 3.13 The ontologies of example 3

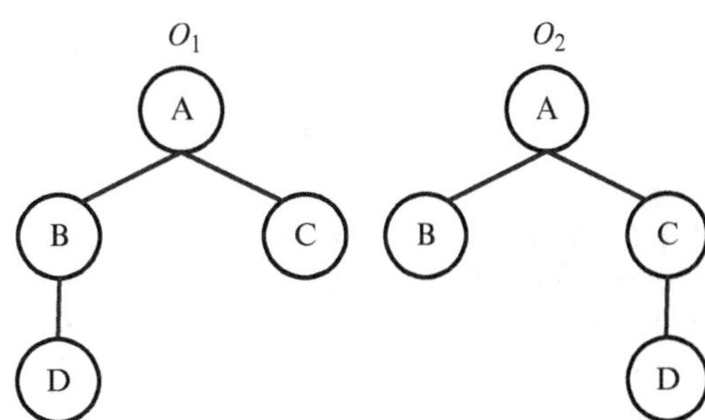

Fig. 3.14 The ontologies of example 4

$$\lambda\left(O_1, O_2\right)=\frac{1+1+1}{6}=0.5.$$

When comparing lexically O_1 to O_3, it is $\delta\left(A^{O_1}\right)$=1,

$$\delta\left(B^{O_1}\right)=\delta\left(C^{O_1}\right)=0.8+\frac{1}{2}\cdot 0.2\cdot 0.6+\frac{1}{2}\cdot 0.2\cdot 0.6\cdot 0.4=0.884$$

and

$$\delta\left(A^{O_3}\right)=\delta\left(B^{O_3}\right)=\delta\left(C^{O_3}\right)=0.8+1\cdot 0.2\cdot 0.4^2=0.832$$

resulting in

$$\lambda\left(O_1, O_3\right)=\frac{(1-0.168)+(1-0.0588)+(1-0.0588)}{6}=0.4524.$$

The difference computed is due to the fact that the interrelations among the common concepts *A*, *B*, and *C* are preserved in ontologies O_1 and O_2, while in the case of the lexical similarity between O_1 and O_3, the value of the coefficient is less than 0.5, depicting the differences in the interrelations among common concepts in these ontologies. The detection of such differences in interrelations among common concepts is essential, since it restricts the problem of polysemy (words that have multiple senses), occurring when comparing ontology entities on the basis of their labels. Indeed, intuitively, groups of common labels in both ontologies, are more probably referring to the same concepts, while distant distinct common labels, may reflect homonyms and thus name different concepts. Another example is depicted in Fig. 3.14, where the ontologies O_1 and O_2 have four common concepts. Here, the common concepts *B* and *C* distribute differently the remaining common concepts (*A*, *C*, and *D* for *B* and *A*, *B*, and *D* for *C*), while *A* and *D* distribute their respective remaining common concepts in the same way, in both ontologies. The result obtained is $\lambda\left(O_1, O_2\right)=0.9832$, less than 1, due to the differences in interrelations between the common concepts in the two ontologies. The exact amount of the difference obtained, can be adjusted by a proper choice for the values of the weighting coefficients a and ρ.

3.4.2 Processing Step

The output of the pre-processing step, i.e., the values of the structure and lexical similarities that indicate whether the source ontologies resemble mostly structurally, or lexically, respectively, guide the processing step of the ATRACO alignment strategy and more precisely, guide the selection and sequence of use of the appropriate matchers. The alignment process includes several matchers and it is organized sequentially, or in parallel, according to the similarity characteristics of the two source ontologies. These matchers calculate similarities between the ontological entities from the different source ontologies. The higher the similarity of the two ontological entities, the more likely they can be aligned. The matchers can implement techniques based on linguistic matching, structural matching, constraint-based approaches, instance-based techniques and approaches that use auxiliary information, or a combination of the above.

During the phase of linguistic matching, multiple algorithms, or matchers based on linguistic matching, such as Edit Distance and Synonym Matcher through the WordNet synonyms thesaurus [14], are executed. These algorithms make use of textual descriptions of ontological entities, such as names, synonyms, and definitions. The similarity measure between entities is based on comparisons of the textual descriptions. The main idea in using such measures is based on the fact that similar entities usually have similar names and descriptions, among different ontologies. More specifically, during this phase we consider establishing correspondences between the entities of the two source ontologies, by using a similarity function and calculating it for the two entities (each belonging to one of the ontologies). If the returned value is greater, or equal to a threshold set up by a domain expert, then the two entities match. Edit Distance, for example, is defined as the number of deletions, insertions, or substitutions required to transform one string into another. The greater the Edit Distance is, the more different the strings are.

The phase of structural matching uses a number of structural matchers (including Descendant's Similarity Inheritance (DSI) and Sibling's Similarity Contribution (SSC) methods [8]), which use the structure of the ontologies, in order to enhance the alignment results. The algorithms used during this phase rely on the intuition that two elements of two distinct models are similar, when their adjacent elements are similar.

In another phase of the alignment strategy, constraint-based approaches are used. In these approaches axioms are used, in order to enhance the previously generated correspondences. For instance, if the range and domain of two relations are the same, this is an indication that a correspondence exists between these relations.

If several instances are available in both source ontologies, then instance-based approaches can be used, in order to define similarities between ontological entities.

Dictionaries, thesauri representing general knowledge, intermediate domain, or upper level ontologies, or predefined alignments, may be also used to enhance the alignment process.

The choice of the appropriate phases (linguistic, structural, constraint-based, and instance-based) and their temporal sequence is driven by the values of the similarity coefficients between ontologies that are calculated during the pre-processing step of the alignment process, according to the following scenarios (see Fig. 3.15):

- If only the label similarity coefficient of the source ontologies is equal, or greater than a threshold H, which is adjusted experimentally, with a value close to 1 (for example, $H = 0.9$), then the alignment process should rely only on lexical matchers. For better results, we can further execute a structural, a constraint-based and an instance-based matcher (if any instances are available) sequentially,

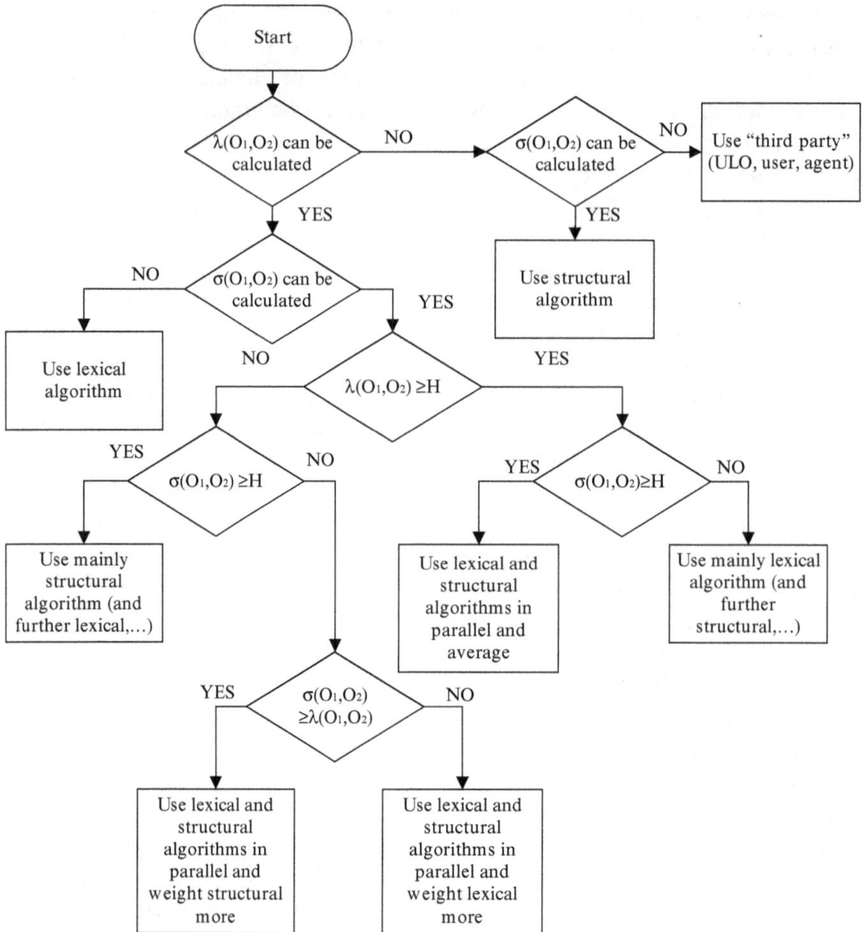

Fig. 3.15 Flowchart indicating the choice of matchers

because, even if some ontological entities have the same label, they may represent totally different concepts. Specifically, starting from a lexical matcher, such as Edit Distance, the entity pairs that score a high similarity value are further compared using a structural matcher, such as one which uses the is_a and part_of hierarchies. Any correspondences between entity pairs with a low similarity value, according to the lexical matcher, are discarded. Then, the pairs with high similarity values are filtered by using constraint-based information. For example, two relations from the source ontologies that have the same range and domain may be equivalent. Following this, an instance-based matcher may be applied, if instances in both ontologies are available. The intuition behind this is that, if two classes have the same set of instances, it is more likely for them to be equivalent.

- If only the structural similarity coefficient of the source ontologies is equal, or higher than H, then the alignment process starts by firstly applying a structural matcher, then a lexical one, and so on, because even if the structure of the two source ontologies is the same, it is difficult to discover the correct correspondences between the ontological entities, due to terminological heterogeneity.
- If the label similarity coefficient of the source ontologies is greater than their structural similarity coefficient, but not high enough (not higher than H), we execute all kinds of matchers in a parallel composition and all the produced similarities are aggregated through a weighted sum, where the similarity value produced by the lexical matcher has the greatest weight.
- If the structural similarity coefficient of the source ontologies is greater than their label similarity coefficient, but not high enough (not higher than H), we execute all kinds of matchers in a parallel composition and all the produced similarities are aggregated through a weighted sum, where the similarity value produced by the structural matcher has the greatest weight.
- If we cannot calculate the similarity coefficients of the source ontologies, the alignment process uses all kinds of matchers in a parallel composition. In this case, the similarity values between candidate pairs of ontological entities are computed, either as the average, or by taking the maximum of all the similarity values that the matchers produce. In the case where no computation of similarity between source ontologies is accomplishable, we additionally try to find out the domain that the two source ontologies describe. Since the meaning (importance and relevance) of a certain ontological entity essentially depends on the specific usage and purpose for which each ontology has been modeled and on different views that creators have on the domain, it is essential to specify the domain that the source ontologies describe, in order to enhance the alignment process by using external resources as a "third party," as we further explain in Sect. 3.4.4.
- If both the label and structural similarity coefficients of the source ontologies are greater than H, then both lexical and structural matchers are executed in parallel and their average value is used as the similarity of the candidate pairs. Then, entity pairs with high similarity values according to the lexical and structural matchers are filtered by using constraint-based information. Following this, an instance-based matcher may be applied, if instances in both ontologies are available.

In the case of a parallel combination of matchers, their similarities are aggregated through computing their average, or by selecting their maximum value, or through a weighted sum, depending on whether we want, for a given alignment, either all matchers to have the same influence, or place emphasis on their highest score, or finally want to give more importance to some of them, respectively. The weighted sum is defined as $\mathrm{sim}(e_1, e_2) = \sum_{k=1}^{n} w_k \cdot \mathrm{sim}_k(e_1, e_2)$, where n is the number of the combined matchers and sim_k and w_k represent the similarity values between the entities e_1 and e_2 belonging to the two source ontologies, and weights, respectively, for the different matchers. It is also required that $w_k \in [0, 1]$ and $\sum_{k=1}^{n} w_k = 1$. If the final similarities obtained after aggregation exceed a threshold, they are selected for the resulting alignment.

3.4.3 Post-Processing Step

In this step, the produced alignment should be evaluated and further refined, before being stored in RDF\XML format by the Alignment Module of the ATRACO Ontology Manager. In practice, this can be done in three ways: (1) assessing individual correspondences, (2) comparing the alignment to a reference alignment, or (3) evaluating the application that uses the alignment. In this step, the user is involved in the alignment process, in order to decide on the final alignment as described in the following subsection. He can accept, or reject some of the produced correspondences between ontological entities from the two source ontologies. The user plays the role of the checker, in order for misalignments to be avoided and the role of a helper that enhances the ontology alignment process, by refining suggested correspondences, or adding new matching pairs.

3.4.4 Using a "Trusted Third Party" Within the Alignment Strategy

A "trusted third party" is used within the ATRACO alignment strategy in two ways. One First, in order to obtain an indirect alignment, when the direct alignment of a pair of non-overlapping ontologies is impossible and the other in order to evaluate the alignments produced. Considering the first way during the processing step of the ATRACO alignment strategy, in order to result in correct correspondences, in the case where the input ontologies describe unrelated domains, the alignment process can rely on ontologies sharing a similar context with the input ontologies. A few approaches [1, 2, 36] have already considered the use of external knowledge, as a way to obtain semantic matching between dissimilar ontologies. Considering the second way a "trusted third party" is used to evaluate the produced alignments during the post-processing step of the ATRACO alignment strategy.

A "trusted third party" can be of different types; it depends on the role that it plays within the alignment strategy:

- Use of external ontological resources: as we are in the specific domain of NGAIEs, the missing information during the processing step of the alignment strategy, in case the ontologies to be aligned are very dissimilar, is acquired from existing ontologies sharing a similar context with the aligned ones, or by using WordNet as an external resource, in order to make use of the synonyms that this resource provides.

- Manual intervention: at the time being, the automatic alignment of ontologies generates, in most cases, incomplete, or incorrect, or no correspondences at all—it depends on the type of input ontologies (that is, complete, or not, consistent, or not, overlapping, low overlapping, or non-overlapping ontologies). Therefore, we prefer semi-automatic alignment, where the Alignment Module which implements the ATRACO alignment strategy suggests correspondences between concepts of the two ontologies to be aligned and the user, either discards, or follows these suggestions.
- Use of agents: another approach that stands between the two approaches mentioned above, is to employ the services of agents that specialize in ontology alignment. These agents can (a) refer to an internal knowledge base of semantic mappings, (b) access public knowledge resources (as suggested in the first item), (c) infer new valid semantic alignments, and (d) interact with people to resolve unknown, or ambiguous situations (as suggested in the second approach).

3.5 Using Category Theory as the Formalization Framework of the Network of ATRACO Ontologies

In the ubiquitous computing environment of the ATRACO project, the coexistence of intrinsically different models of local knowledge (ontologies), makes the exchange of information difficult. This exchange of information sought can be achieved through the ontology alignment process. Some approaches followed in the literature for the formalization of ontologies and their operations, consist in using the information flow theory [23, 25], Goguen's work on institution theory [27] and Category Theory. Category Theory offers several ways in order to combine and integrate objects and has been used as a mechanism to formalize ontology matching, by providing operations to compose and decompose ontologies (alignment, merging, integration, mapping) [6, 21, 24, 48]. Since the basic objects in Category Theory are the relationships between ontological specifications and not the internal structure of a single knowledge representation, it permits to view global operations involving ontologies (like alignment, merging, matching), independently from the languages used to express their local entities and from the techniques used. In the framework of the ATRACO project we have adopted Category Theory as the most appropriate formalization framework. This is because the chosen formalism:

- Focuses on relationships (categorical morphisms, functors, and natural transformations) and not on entities (categorical objects, and categories)
- Allows the coexistence of heterogeneous entities, since it provides the ability to define several categories, according to the kinds of entities to be described (category of ontologies, category of alignments, category of networks of interlinked ontologies), which can be related by the definition of special morphisms (categorical functors)

- Offers a set of categorical constructors for creating new categories, by using predefined ones
- Provides a means for the combination of categorical objects (colimits can be used to compose them and limits to decompose them), and for the combination of categorical functors (natural transformations)
- Provides a multi-level study of its categorical notions, by defining three interrelated levels (the level of categories, the level of functors, and the level of natural transformations).

3.5.1 Categorically Formalizing Ontologies and Alignments

In the following, we review the main results of the categorical formalization of ontology operations.

The concepts of an ontology are structured in a taxonomy (hierarchy of concepts related by the subsumption relation) and can also be related by more complex non taxonomic relations, like the mereologic ones. Moreover, correspondences can be established between entities (concepts, properties, instances) belonging to two distinct ontologies (this operation is called matching), resulting in a set of correspondences between them, called an alignment. Correspondences between entities belonging to different ontologies can be restricted to equivalence only, or could be binary relations of a wider nature, like disjointness, generalization/specification, etc. On the other hand, a category consists of a collection of objects and of a collection of morphisms (or arrows, or maps), each morphism being a relation from a domain object to a codomain object. Each category is equipped with an associative composition operator defined for any composable pair of morphisms (the codomain of one of the morphisms is the domain of the other one) and unique identity morphisms acting as the units of the composition operator. By restricting ourselves to subsumption relations between concepts of the same ontology and to equivalence relations between concepts belonging to different ontologies and in order to have a categorical view of ontologies [6], ontologies are considered as category objects and the morphisms of the category are functions f between a domain and a codomain ontology, mapping concepts, or relations of the domain ontology to respective concepts, or relations of the codomain ontology. The morphisms are such that they preserve any relations between entities in the domain ontology, for example, if c_1 is related to c_2 through the relation r in the domain ontology, then $f(c_1)$ is related to $f(c_2)$ in the codomain ontology by the relation $f(r)$, establishing thus a graph homomorphism between the two ontologies, if the two ontologies are seen as graphs.

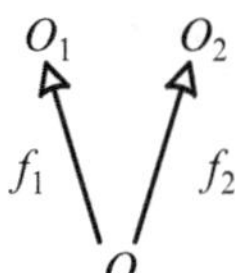

Fig. 3.16 V-alignment

The alignment between two ontologies O_1 and O_2, is the task of establishing binary relations between the entities of the two ontologies. In a categorical perspective, an alignment is not expressed as a direct relation between entities belonging to the two ontologies. Instead, each binary relation between entities belonging to O_1 and O_2, respectively, can be decomposed into a pair of mappings from a common intermediate source ontology, O, to the ontologies O_1 and O_2 [24]. The mappings from O to O_1 and from O to O_2, specify how the concepts and relations of O are understood in O_1 and O_2, respectively. This structure, comprising the ontologies O_1 and O_2 to be aligned, the intermediate ontology O and the morphisms f_1,f_2, is called (due to its shape) a V-alignment (see Fig. 3.16); it is a span, in the Category Theory terminology.

The operation of integrating two aligned ontologies into a single one is called merging and can be accomplished with V-alignments. The ontology resulting from the unification process of merging, embodies the semantic differences of the two ontologies and collapses the semantic intersection between them. Merging of aligned ontologies can be described, in the Category Theory formalization, in terms of a Category Theory construct, called pushout, which is a special case of another construct, called colimit. The pushout involves three objects, the three categories O_1, O_2, and O and the two morphisms of the alignment diagram. The pushout is a new object, a new ontology O' in our case, together with morphisms f_1',f_2', such that

$$f_1' \circ f_1 = f_2' \circ f_2.$$

The commutativity of the pushout diagram in Fig. 3.17, means that components of O_1 and O_2 that are images of the same component in O (that is, the semantic intersection of O_1 and O_2), are collapsed in the resulting ontology O' (mapped to the same entity), which is exactly the definition of the merging operation. That is, the pushout ontology realizes the merging of O_1 and O_2. Moreover, for any other object (ontology) O'' for which the commutativity holds, i.e., for which

$$f_1'' \circ f_1 = f_2'' \circ f_2$$

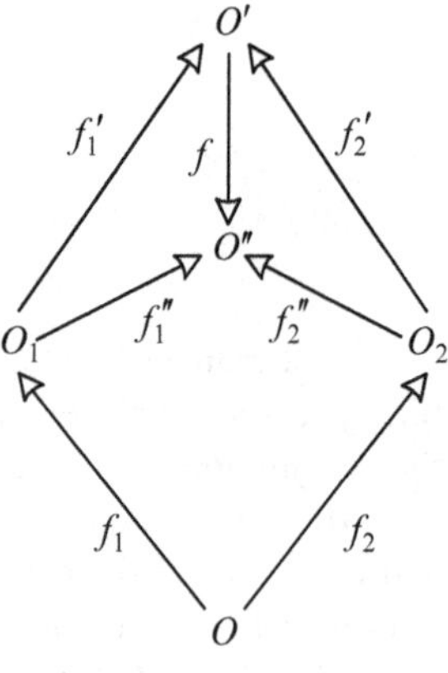

Fig. 3.17 Merging through the pushout construct

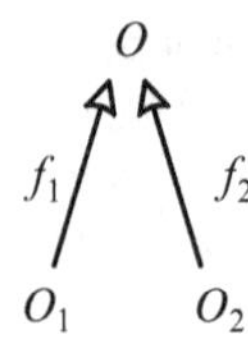

Fig. 3.18 Λ-alignment

there exists a unique morphism f such that

$$f \circ f_1' = f_1''$$
$$f \circ f_2' = f_2''$$

that is, the pushout O' is the most compact ontology that can embody the union of O_1, O_2 and possibly comprises collapsed components (i.e., embodies the semantic differences and collapses the semantic intersection).

In Category Theory, dual concepts arise from the process of reversing all the morphisms in a diagram. Thus, the dual concept of pushout is a construct called pullback, which is a particular case of another construct called limit (dual of colimit). The pullback is used in order to formalize the matching operation, by which similarities between ontologies are detected. We start with what is called a Λ-alignment, of the form depicted in Fig. 3.18.

Here, O_1 and O_2 are the ontologies to be matched and O is an intermediate ontology that guides the matching. The pullback is a new ontology O', together with morphisms f_1', f_2', such that

$$f_1 \circ f_1' = f_2 \circ f_2'$$

that is, the pullback O'embodies all information of O_1 and O_2 that is semantically equivalent.

The commutativity of the pullback diagram in Fig. 3.19, means that components of O_1 and O_2 that have the same image in O (i.e., are semantically equivalent),

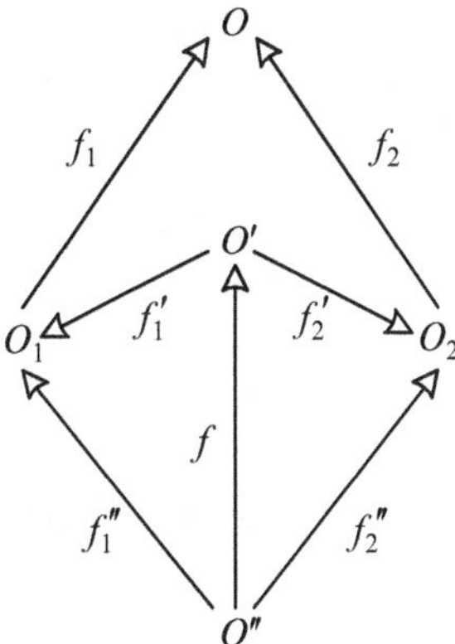

Fig. 3.19 Matching through the pullback construct

are images of the same component in O', which is exactly the definition of the matching operation. Thus, the pullback operation realizes the matching of O_1 and O_2. Moreover, for any other object (ontology) O'' for which the commutativity holds, i.e.,

$$f_1 \circ f_1'' = f_2 \circ f_2''$$

there exists a unique morphism f, such that

$$f_1' \circ f = f_1''$$
$$f_2' \circ f = f_2''$$

that is, the pullback O' is the biggest ontology that includes all the semantic intersection of O_1 and O_2.

3.5.2 *Categorical Formalization of Activity Spheres*

In the ATRACO environment, the activity spheres conceptualize an ambient ecology populated by different entities, such as devices, services, humans, or agents, interrelated and in interaction with their environment. Once an activity sphere is produced, the knowledge associated with it can be reused, in order to produce new activity spheres. To this end, activity spheres can be extended by the proper addition of ambient ecology entities, or can be embedded into other activity spheres, to produce more complex ones. Thus, activity spheres undergo changes, each time they adapt to different conditions, or each time a reconfiguration of their structure occurs, due to entities entering, or leaving the sphere. This dynamics can be captured by the alignment composition operation. Indeed, if we have an alignment between ontologies O_1 and O_2 and an alignment between ontologies O_2 and O_3,

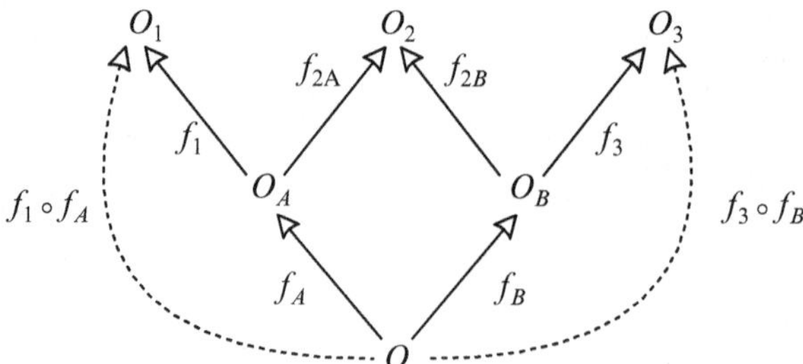

Fig. 3.20 Composition of alignments

we can compose them and obtain an alignment between ontologies O_1 and O_3, thus depicting relations holding between the entities of ontologies O_1 and O_3.

The operation of alignment composition can be formulated in the categorical framework [48], as the composition of spans in Category Theory, through the use of the pullback construct. In Fig. 3.20, if (O_A, f_1, f_{2A}) is a V-alignment expressing the alignment between ontologies O_1 and O_2 and (O_B, f_{2B}, f_3) is a V-alignment expressing the alignment between ontologies O_2 and O_3, then the ontology O together with the morphisms $f_1 \circ f_A$ and $f_3 \circ f_B$ constitute the composition of the two V-alignments sought, where O, together with the morphisms f_A, f_B is the pullback of the Λ-alignment of O_2 (O_2 with f_{2A}, f_{2B}).

In the ATRACO perspective, an activity sphere is a network of already aligned ontologies. In order to define an activity sphere under a categorical perspective, we define a category having ontologies as objects. There exists a morphism between two ontologies, objects of this category, if and only if there exists a V-alignment between them. Thus, composable morphisms in this category, reflect to the composition of V-alignments, which becomes then the main operation needed in the ATRACO perspective. More precisely, as depicted in Fig. 3.21, whenever an entity represented by an ontology joins an already established activity sphere of entities represented by a network of already aligned ontologies, it suffices to align it to a single anchor ontology, already participating in the activity sphere. The anchor alignment produced, is then composed to already established alignments involving the anchor ontology, producing a batch of composition-generated alignments that remain, even if later on the anchor ontology leaves the activity sphere.

In cases where subsumption, or more elaborate relations, between concepts belonging to different ontologies is to be expressed, since these relations cannot be represented with the vocabulary of any of the two ontologies, it is externalized in an additional new ontology (called bridge ontology), as a bridge axiom. The following diagram depicts the situation, with the original ontologies O_1 and O_2 containing the concepts related via subsumption and the bridge ontology B containing the bridge axioms. The fact that there exist concepts of the ontologies O_1 and O_2 occurring within the bridge ontology, is represented by the two V-alignments between the bridge ontology and the ontologies O_1 and O_2. Thus, the so-called W-alignment is defined (see Fig. 3.22).

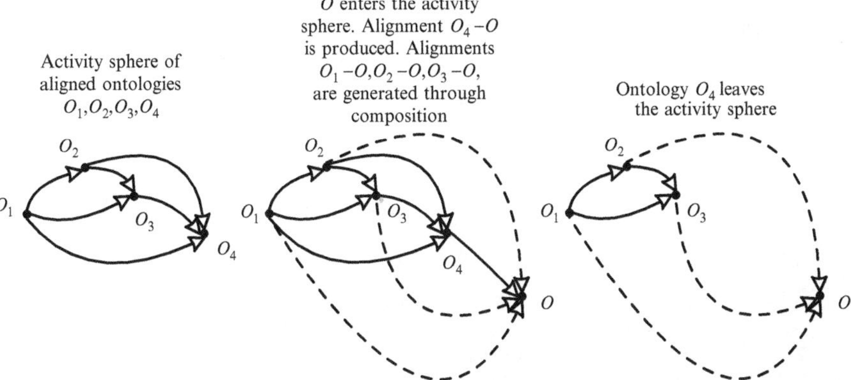

Fig. 3.21 Ontologies entering or leaving an ATRACO activity sphere

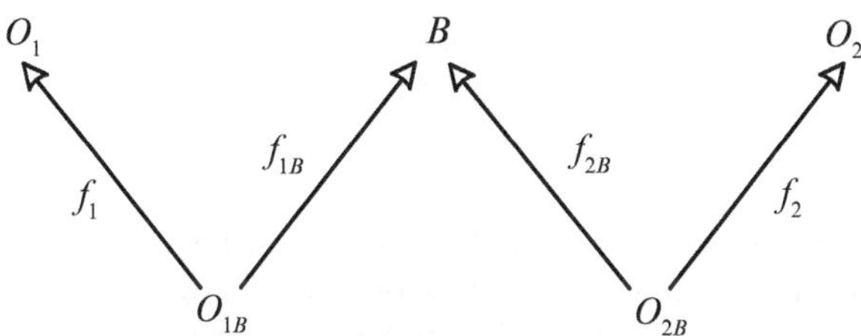

Fig. 3.22 W-alignment

The merging operation in this case, is defined as the colimit of the alignment diagram in Fig. 3.23 and is computed by successive pushouts [48].

In a similar way, one can compose W-alignments. If a W-alignment exists between ontologies O_1 and O_2 with bridge ontology B_1 and if also a W-alignment exists between ontologies O_2 and O_3 with bridge ontology B_2, by composing the two W-alignments, it results that a W-alignment exists between ontologies O_1 and O_3, with bridge ontology B, which is obtained if the merging operation is applied to the bridge ontologies B_1 and B_2. The problem of this approach, consists in incorporating in the new bridge ontology B bridge axioms from the ontologies B_1, B_2, and O_2, that might be irrelative to O_1 and O_3.

Another solution to the problem of more elaborate relationships (subsumption, strict inclusion, strict containment, disjointness, overlapping with partial disjointness, temporal relations), between entities belonging to different ontologies, is to enhance the category of ontologies with more elaborate morphisms that denote the relationship that holds between the syntactic entities of the two ontologies (subsumption, strict inclusion, etc.) [13, 48]. In this case, when applying the composition operation, if an entity in ontology O_2 has an elaborate relation to entities in the ontologies O_1 and O_3, there is some kind of relation between the two entities in O_1 and O_3. The latter relation depends strongly on the former one. For example, if an

Fig. 3.23 Merging with W-alignments

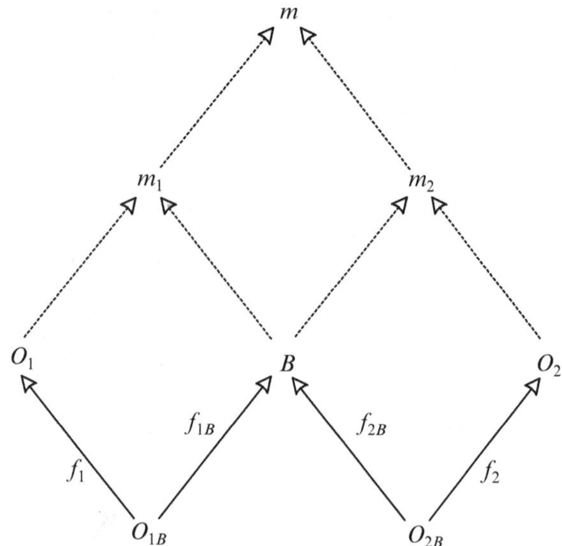

entity in O_1 is related to an entity in O_2 with strict inclusion and the same entity in O_2 is related to an entity in O_3 with strict containment, then the entity in O_1 can be related to the entity in O_3 by either of the following relationships: equivalence, strict inclusion, strict containment, disjointness, overlapping with partial disjointness. More specifically, in [13], the formalism of algebras of binary relations has been proposed, in order to solve the problem of expressing the relations between entities of different ontologies in a general way, support alignment composition and deduce new alignments from existing ones, via composition tables. In order to capture the dynamic nature of an activity sphere in the case of more general relations between ontological entities, the question remains to construct the appropriate category that captures the structure.

3.5.3 *Entailments in NGAIEs*

An NGAIE, in the ATRACO perspective, is composed of a pleiad of activity spheres. Each activity sphere can be considered as a network of aligned ontologies. In this setting, assertions holding locally in standalone ontologies, induce relations between entities belonging to remote ontologies, via paths of consecutive ontologies and alignments forming a chain in the network, or provide new relations holding between local entities of a standalone ontology, through paths forming cycles in the network [17]. We assume that entities e_i and e_j can be related through a relation r assuming values from a set of base relations (we write $e_i r e_j$). We further assume that relations can be composed through an associative composition operator, $\circ$,

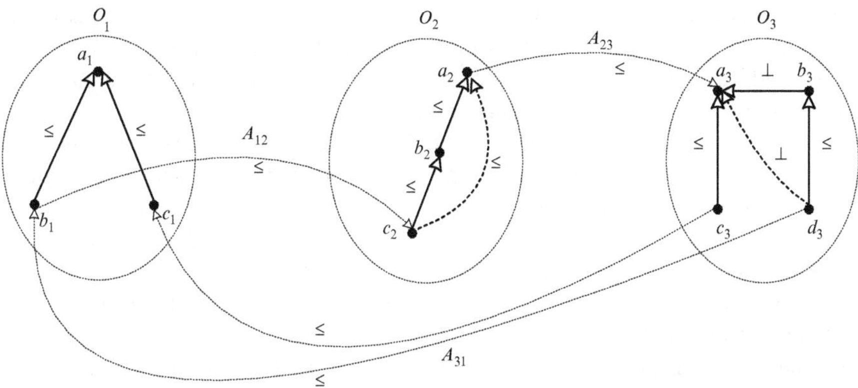

Fig. 3.24 Detecting inconsistencies in networks of aligned ontologies

that permits alignment reuse, i.e., if $e_i r e_j$ and $e_j r' e_k$, then $e_i(r \circ r')e_k$. Let us, for example, consider the network of aligned ontologies of Fig. 3.24 adapted from [13], where, we have an alignment A_{12} between ontologies O_1 and O_2, an alignment A_{23} between ontologies O_2 and O_3, and an alignment A_{31} between ontologies O_3 and O_1. In this network, new relations can be deduced, for example, by relating ontological entities belonging to remote ontologies O_1 and O_3 through the particular path $O_1 - A_{12} - O_2 - A_{23} - O_3$ of ontologies and alignments in the network, or by relating ontological entities belonging to the same ontology, but through a particular path forming a cycle starting and ending at this specific ontology. This is, for example, the case for the cycle $O_3 - A_{31} - O_1 - A_{12} - O_2 - A_{23} - O_3$, where the relation $d_3 \leqslant a_3$ is revealed by relation composition along the path, while simultaneously $d_3 \perp a_3$ results locally in ontology O_3, making apparent consistency problems emerging from the network induced relations.

It is thus crucial, on the one hand, to rigidly characterize the different types of entailments that may appear in such settings, by the interaction of the constituent ontologies and to define, on the other hand, suitable operators for propagating local knowledge through the network, in order to detect logical contradictions, or redundancies.

With regard to the above-mentioned critical issues, we provide a characterization of locally and globally induced entailments in a network of aligned ontologies, by propagating local knowledge through the network. This contributes to (a) detecting inconsistencies, that is, conceptual errors so as to revise the knowledge emerging from the whole network, (b) composing alignments in order to support transitive knowledge propagation among interlinked ontologies, and (c) reasoning on ontologies which are independently conceived and related by means of alignments. The main principles that guide this formalization are: (a) it should be based on a natural algebra of binary relations (i.e., equipped with an associative composition relation) in order to characterize entity relations within a network of aligned ontologies, (b) it should be able to represent both ontologies and alignments in a suitable way, in order to guarantee the composition of relations and define appropriate mathematical structures for propagating relations in a network of aligned ontologies, and (c) it should define composition operators, in order to combine entailments along chains of ontologies and alignments in a network.

In this perspective, in order to characterize possible entailments in a network of aligned ontologies, we start by representing each ontology by a labeled multigraph (quiver), where ontology entities are represented by vertices and a certain relation holding between two ontology entities is represented by an edge connecting the respective nodes of the quiver and labeled by this relation [41]. Since in the ontology framework successive relations can be composed giving rise to new relations holding between entities not directly related, we then consider the free category associated with the quiver representing the ontology. In this category, the objects are the vertices of the quiver (i.e., the ontology entities) and the morphisms of the free category are lists of consecutive edges and vertices of the quiver, forming paths of successive relations that can be composed. These compositions are performed under the rules of the associative algebra of binary relations, where we require the entity relations to belong to. Thus, in the free category, we make apparent all possible paths of composable relations, as incoming arrows to the objects-ontology entities. Finally, we build what we call the representing category of the ontology, where the lists of composable successive relations forming a path in the free category, have now been replaced by the result of the composition of the successive relations, calculated on the basis of the composition operator defined in the algebra of binary relations (see Fig. 3.25).

An analogous representation is used for alignments between network ontologies, where the main difference is that alignment correspondences involve entities belonging to different ontologies.

Moreover, in order to represent all the compositions of relations inside a standalone ontology and to be able to compose alignments between ontologies in a network (i.e., be able to further propagate relations in a network of ontologies via alignments), based on the afore-mentioned representation of an ontology (or an alignment) with the free category of its quiver and the respective representing

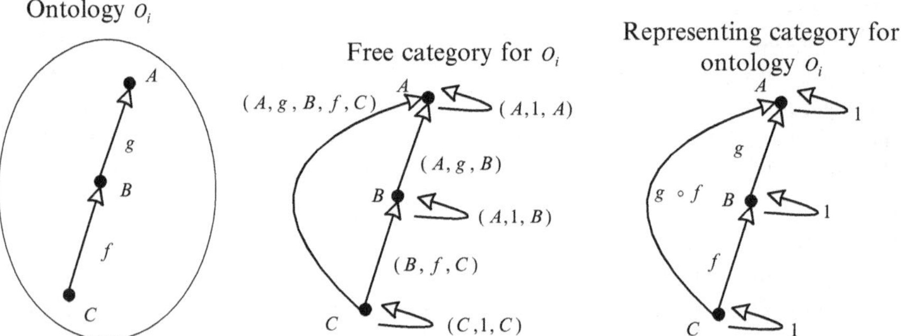

Fig. 3.25 Categorical representation of an ontology

category, we define for each ontology and each alignment a so-called propagation matrix P, where we express the composition of relations along all paths connecting source entities that admit incoming relations from the network, and target entities that further propagate these relations in the network. Thus, for example, for an intermediate ontology in a chain of consecutive ontologies and alignments in the network, the source entities of the ontology are those involved in the preceding alignment and the target entities are those involved in the succeeding alignment. The columns of a propagation matrix are related to the contravariant representable functors of objects in the representing category.

For the example of Fig. 3.24 and concerning the cycle $O_3 - A_{31} - O_1 - A_{12} - O_2 - A_{23} - O_3$ of consecutive ontologies and alignments in the network, the source entities (denoted vertically), the target entities (denoted horizontally), and the propagation matrices for the ontologies and alignments involved, are depicted hereafter:

$$\underset{O_3}{\begin{array}{c|c} & c_3\ d_3 \\ \hline a_3 & \\ b_3 & \\ c_3 & \\ d_3 & \end{array}\begin{bmatrix} 0\ 0 \\ 0\ 0 \\ 1\ 0 \\ 0\ 1 \end{bmatrix}} \quad \underset{A_{31}}{\begin{array}{c|c} & b_1\ c_1 \\ \hline c_3 & \\ d_3 & \end{array}\begin{bmatrix} 0 & \leqslant \\ \leqslant & 0 \end{bmatrix}} \quad \underset{O_1}{\begin{array}{c|c} & b_1 \\ \hline b_1 & \\ c_1 & \end{array}\begin{bmatrix} 1 \\ 0 \end{bmatrix}} \quad \underset{A_{12}}{\begin{array}{c|c} & c_2 \\ \hline b_1 & \end{array}[\leqslant]} \quad \underset{O_2}{\begin{array}{c|c} & a_2 \\ \hline c_2 & \end{array}[\leqslant]} \quad \underset{A_{23}}{\begin{array}{c|c} & a_3 \\ \hline a_2 & \end{array}[\leqslant]} \quad \underset{O_3}{\begin{array}{c|c} & a_3\ b_3\ c_3\ d_3 \\ \hline a_3 & \end{array}[1\ 0\ 0\ 0]}$$

These propagation matrices can be composed by suitably defining sequential and parallel composition operators, in order to obtain the propagation of relations, through alignments, in a network of aligned ontologies. These operators are defined as follows [41]:

- Sequential composition operator: We denote by $P = P^i * P^{ij}$ the sequential composition of the propagation matrices $P^i = \left[r^i(m,l)\right]$ and $P^{ij} = \left[r^{ij}(l,n)\right]$, which represent an ontology O_i and the succeeding alignment A_{ij} with ontology O_j, respectively. We define the operator of sequential composition as $P = P^i * P^{ij} = [r(m,n)]$, where $r(m,n) = \bigcap_l r^{ij}(l,n) \circ r^i(m,l)$ (i.e., compositions of relations are considered conjunctively).
- Parallel composition operator: If $R = [r(m,n)]$ and $S = [s(m,n)]$ are the sequential composition operators evaluated across two different paths, having the same source and target ontologies, then the parallel composition operator is evaluated as $P = R + S = [p(m,n)]$, where $p(m,n) = r(m,n) \cap s(m,n)$.

The categorical representation adopted for an ontology or an alignment can be extended in a straightforward manner in the case of a network of aligned ontologies. Initially, an ontology network is represented by a labeled quiver, where now labeled vertices represent the ontologies and edges between vertices represent alignments between the respective ontologies. The representing category of the network, that is constructed based on the free category of its quiver, has morphisms that represent

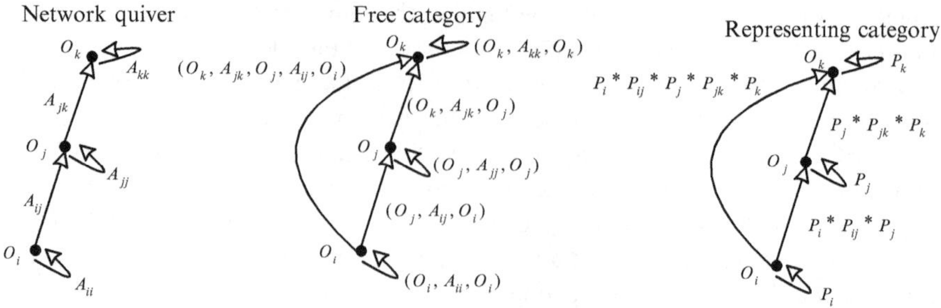

Fig. 3.26 Categorical representation of a network of aligned ontologies

an alignment A_{ij} between ontologies O_i and O_j as the sequential composition $P = P^i * P^{ij} * P^j$ of their respective propagation matrices. The composition of morphisms in the representing category of the network is equivalent to the composition of alignments (see Fig. 3.26).

The proposed categorical representation is advantageous, because (a) it increases expressiveness (entities belonging to two network ontologies can now be related with any possible relation belonging to a set of available composable relations and entities belonging to the same ontology can be related through more elaborate relationships), and (b) it makes apparent new relations emerging from compositions of existing ones, over the network.

3.6 Conclusion

The knowledge management framework developed in the context of the ATRACO project, was presented in this chapter. We mainly focused on the following axes of research:

- An ontological representation, in order to cope with the dynamic nature and heterogeneity of NGAIEs components, realized in the context of ATRACO with the ATRACO ULO and a set of DSOs, Policy Ontologies, UPOs, and GFOs.
- A task-based ontology engineering process, for the specification, conceptualization, implementation, and evaluation of the ATRACO ontologies.
- A three-step ontology alignment strategy, guided by two similarity coefficients, in order to detect the lexical, or structural resemblance of the source ontologies.
- A formalization framework, based on Category Theory, permitting to confront the dynamic nature of ATRACO activity spheres, the constituent networks of interlinked alignments of the ATRACO project.

The presented knowledge management framework provides us with theoretical and practical tools so as to overcome the barrier of heterogeneity intrinsic in the ATRACO world, due to the different conceptualizations of its actors.

3.7 Further Readings

The reader interested in learning more about the principles and techniques of ontological engineering in any context, is encouraged to consult [3, 18, 43]. To learn more about tools, languages, matchers, and approaches on ontology alignment, [10, 14] are the most recent extensive references. Concerning Category Theory, [33] provides a presentation of the basic constructions and terminology of Category Theory and illustrates applications of Category Theory to programming language design, semantics and the solution of recursive domain equations, while in [5] the material covered includes the standard core of categories, functors, natural transformations, equivalence, limits and colimits, functor categories, representables, Yoneda's lemma, adjoints and monads.

References

1. Aleksovski, Z., ten Kate, W., van Harmelen, F.: Exploiting the structure of background knowledge used in ontology matching. In: Proceedings of the 1st International Workshop on Ontology Matching (OM-2006) Collocated with the 5th International Semantic web Conference (ISWC-2006), Athens, Georgia, USA, November 5, pp. 13–14 (2006)
2. Aleksovski, Z., Klein, M., ten Katen, W., van Harmelen, F.: Matching unstructured vocabularies using a background ontology. In: Proceedings of the 15th International Conference on Knowledge Engineering and Knowledge Management (EKAW'06). Lecture Notes in Artificial Intelligence. Springer, Berlin (2006)
3. Allemang, D., Hendler, J.: Semantic Web for the Working Ontologist: Effective Modeling in RDFS and OWL. Morgan Kaufmann, San Francisco (2008)
4. Anaya, F., Vazquez, J.: Semantic technologies and techniques for interoperable information. In: First International Workshop on Semantic Interoperability for Smart Spaces (SISS), in Conjunction with IEEE ISCC2010 (2010)
5. Awodey, S.: Category Theory. Oxford University Press, Oxford (2010)
6. Cafezeiro, I., Haeusler, E.: Semantic interoperability via category theory. In: 26th International Conference on Conceptual Modelling - ER Auckland, pp. 197–202 (2007)
7. Chen, H.: An intelligent broker architecture for pervasive context-aware systems. Ph.D. thesis, University of Maryland, Baltimore (2004)
8. Cruz, I., Antonelli, F., Stroe, C., Keles, U., Maduko, A.: Using agreement maker to align ontologies for oaei 2009 overview, results, and outlook. In: Fourth International Semantic Web Conference (2009)
9. Davies, J., Suder, R., Warren, P.: Semantic Web Technologies: Trends and Research in Ontology-Based Systems. Wiley, New York (2008)
10. Ehrig, M.: Ontology Alignment: Bridging the Semantic Gap. Springer, New York (2007)

11. Ehrig, M., Sure, Y.: FOAM—framework for ontology alignment and mapping results of the ontology alignment evaluation initiative. In: Integrating Ontologies Workshop Proceedings, p. 72 (2005)
12. Ejigu, D., Scuturici, M., Brunie, L.: Coca: a collaborative context-aware service platform for pervasive computing. In: Fourth IEEE International Conference on Information Technology, pp. 297–302 (2007)
13. Euzenat, J.: Algebras of ontology alignment relations. In: International Semantic Web Conference, pp. 387–402 (2008)
14. Euzenat, J., Shvaiko, P.: Ontology Matching. Springer, Berlin/Heidelberg (2007)
15. Euzenat, J., Loup, D., Touzani, M., Valtchev, P.: Ontology alignment with ola. In: Third International Semantic Web Conference (2004)
16. Euzenat, J., Le Bach, T., Barrasa, J., Bouquet, P., De Bo, J., Dieng-Kuntz, R., Ehrig, M., Hauswirth, M., Jarrar, M., Lara, R.: State of the art on ontology alignment. Deliverable 2.2. 3. Tech. Rep., IST Knowledge Web NoE (2004)
17. Fionda, V., Pirro, G.: Semantic flow networks: Semantic interoperability in networks of ontologies. In: Proceedings of the Joint International Semantic Technology Conference. Lecture Notes in Computer Science, vol. 7185, pp. 64–79. Springer, Berlin/Heidelberg (2012)
18. Gómez-Pérez, A., Fernández-López, M., Corcho, O.: Ontological Engineering: With Examples from the Areas of Knowledge Management, e-Commerce and the Semantic Web. Springer, Berlin (2004)
19. Gruber, T.: Towards principles for the design of ontologies used for knowledge sharing. Int. J. Hum. Comput. **43**, 907–928 (1995)
20. Heinroth, T., Kameas, A., Pruvost, G., Seremeti, L., Bellik, Y., Minker, W.: Human-computer interaction in next generation ambient intelligent environments. Intell. Decis. Technol. Spec. Issue Knowl.-Based Environ. Serv. Hum. Comput. Interact. **5**(1), 31–46 (2011)
21. Hitzler, P., Krotzsch, M., Ehrig, M., Sure, Y.: What is ontology merging?: a category theoretic perspective using pushouts. In: First International Workshop on Concepts and Ontologies: Theory, Practice and Applications, pp. 104–107 (2005)
22. Hu, W., Qu, Y., Cheng, G.: Matching large ontologies: a divide-and-conquer approach. Data Knowl. Eng. **67**(1), 140–160 (2008)
23. Kalfoglou, Y., Schorlemmer, M.: Information-flow-based ontology mapping. In: Ont the Move to meaningful Internet Systems 2002: CoopIS, DOA and ODBASE: Confederated International Conferences, pp. 1132–1151. Springer, Berlin (2002)
24. Kalfoglou, Y., Schorlemmer, M.: Ontology mapping: the state of the art. Knowl. Eng. Rev. **18**(01), 1–31 (2003)
25. Kent, R.: The iff foundation for conceptual knowledge organization. Cat. Classif. Q. **37**, 187–203 (2000)
26. Kotis, K., Vouros, G., Alonso, J.: Hcome: tool-supported methodology for collaboratively devising living ontologies. In: Workshop on Semantic Web and Databases (2004)
27. Kutz, O., Mossakowski, T., Codescu, M.: Shapes of alignment: construction, combination, and computation. In: Workshop on Ontologies: Reasoning and Modularity, WORM-08, ESWC (2008)
28. Li, J., Tang, J., Li, Y., Luo, Q.: RiMOM: a dynamic multistrategy ontology alignment framework. IEEE Trans. Knowl. Data Eng. **21**, 1218–1232 (2008)
29. Lopez, M., Gómez-Pérez, A., Sierra, J., Sierra, A.: Building a chemical ontology using methontology and the ontology design environment. IEEE Intell. Syst. Appl. **14**(1), 37–46 (2002)
30. Nicola, A., Navigli, R., Missikoff, M.: Building an eProcurement ontology with upon methodology. In: The 15th e-Challenge Conference (2005)
31. Noy, N., McGuinness, D.: Ontology development 101: a guide to creating your first ontology. Tech. Rep., Stanford Knowledge Systems Laboratory (2001)
32. Panagiotopoulos, I., Seremeti, L., Kameas, A., Zorkadis, V.: Proact: an ontology-based model of privacy policies in ambient intelligence environments. In: Proceedings of 14th Panhellenic Conference on Informatics (PCI), pp. 124–129 (2010)

33. Pierce, B.: Basic Category Theory for Computer Scientists. The MIT Press, Cambridge (1991)
34. Pinto, H., Staab, S., Tempich, C.: Diligent: towards a fine-grained methodology for distributed loosely-controlled and evolving engineering of ontologies. In: The 16th European Conference on Artificial Intelligence (2004)
35. Román, M., Hess, C., Cerqueira, R., Ranganathan, A., Campbell, R., Nahrstedt, K.: A middleware infrastructure for active spaces. IEEE Pervasive Comput. **1**(4), 74–83 (2002)
36. Sabou, M., d'Aquin, M., Motta, E.: Using the semantic web as background knowledge for ontology mapping. In: Proceedings of the 1st International Workshop on Ontology Matching (OM-2006) Collocated with the 5th International Semantic web Conference (ISWC-2006), Athens, Georgia, USA, November 5, pp. 1–12 (2006)
37. Seghrouchni, A., Breitman, K., Sabouret, N., Endler, M., Charif, Y., Briot, J.: Ambient intelligence applications: introducing the campus framework. In: 13th IEEE International Conference on Engineering of Complex Computer Systems (2008)
38. Serafini, L., Tamilin, A.: Drago: distributed reasoning architecture for the semantic web. In: The Semantic Web: Research and Applications, pp. 361–376. Springer, Berlin (2005)
39. Seremeti, L., Kameas, A.: A task-based ontology engineering approach for novice ontology developers. In: Fourth Balkan Conference in Informatics (2009)
40. Seremeti, L., Kameas, A.: Tools for ontology engineering and management. In: Theory and Applications of Ontology: Computer Applications, pp. 131–154. Springer, New York (2010)
41. Seremeti, L., Kameas, A.: Composable relations induced in networks of aligned ontologies: a category theoretic approach. Axiomathes (2014). doi:10.1007/s10516-014-9242-y
42. Seremeti, L., Kameas, A., Panagiotopoulos, I.: An alignable user profile ontology for ambient intelligence environments. In: Proceedings of 7th International Conference on Intelligent Environments, IE'11 (2011)
43. Sharman, R., Kishore, R., Ramesh, R.: Ontologies: A Handbook of Principles, Concepts and Applications in Information Systems. Springer, Berlin (2007)
44. Suarez-Figueroa, M.: Neon development process and ontology life cycle - deliverable d5.3. Tech. Rep., NeOn Project (European Commission's Sixth Framework Programme Under Grant Number IST-2005-027595) (2007). D5.3
45. Sure, Y.: Methodology, tools and case studies for ontology based knowledge management. Ph.D. thesis, University of Karlsruhe (2003)
46. Uschold, M., King, M.: Towards a methodology for building ontologies. In: Workshop on Basic Ontological Issues in Knowledge Sharing (1995)
47. Wang, X., Dong, J., Chin, C., Hettiarachchi, S., Zhang, D.: Semantic space: an infrastructure for smart spaces. IEEE Pervasive Comput. **3**(3), 32–39 (2004)
48. Zimmermann, A., Krotzsch, M., Euzenat, J., Hitzler, P.: Formalizing ontology alignment and its operations with category theory. In: FOIS'06 (2006)

Chapter 4
Privacy and Trust in Ambient Intelligent Environments

Bastian Könings, Florian Schaub, and Michael Weber

Abstract Privacy and trust are critical factors for the acceptance and success of next generation ambient intelligent environments. Those environments often act autonomously to support a user's activity based on context information gathered from ubiquitous sensors. The autonomous nature, their accessibility to large amounts of personal information, and the fact that actuators and sensors are invisibly embedded in such environments, raise several privacy issues for users. Those issues need to be addressed by adequate mechanisms for privacy protection and trust establishment. In this chapter, we provide an overview of existing privacy enhancing technologies (PETs) in the area of ambient intelligent environments and present novel adaptive privacy mechanisms as used in the ATRACO architecture and in an ambient calendar system. Further, we will discuss how computational trust mechanisms and social trust aspects can be utilized to support privacy protection and the establishment of trust between system components and between the system and users. After describing the integration of these mechanisms in the overall system architecture of ATRACO, we conclude by giving an outlook on future directions in this area.

4.1 Introduction

The field of *Ambient Intelligence* (AmI), or ubiquitous computing, has been the focus of much research in the last decades. New technological solutions emerging from this research area offer great opportunities for a large number of new applications ranging from assisted or daily living systems, entertainment systems, to wearable computing and intelligent transportation systems. Such systems will be invisibly embedded into our everyday environments through a pervasive transparent

---

B. Könings (✉) • M. Weber
Institute of Media Informatics, Ulm University, 89081 Ulm, Germany
e-mail: bastian.koenings@uni-ulm.de; michael.weber@uni-ulm.de

F. Schaub
School of Computer Science, Carnegie Mellon University, Pittsburgh, PA 15213, USA
e-mail: fschaub@cmu.edu

© Springer International Publishing Switzerland 2016
S. Ultes et al. (eds.), *Next Generation Intelligent Environments*,
DOI 10.1007/978-3-319-23452-6_4

infrastructure consisting of a multitude of sensors, actuators, processors, and networks. The interplay of those components allows the system to support, interact with, and adapt to individuals in a seamless and unobtrusive way.

However, in order to provide such a high degree of flexibility, support, and adaptation, AmI systems require a large amount of information, such as real-time data gathered from ubiquitous sensors, personal user information, and the ability to intervene in the user's physical environment. These requirements raise several issues regarding privacy and trust which need to be addressed in order to satisfy user needs and acceptance. Solutions to achieve these goals include *Privacy Enhancing Technologies* (PETs) and mechanisms for trust establishment. PETs are required to cope with changing privacy preferences of users and the dynamic nature of AmI systems. Thus, privacy enforcement strategies must respect individuals' privacy preferences and support context-based privacy adaptation. The AmI system itself should be trustworthy to the user and the user should be able to control the system in any privacy relevant situation.

In this chapter, we first discuss different definitions and forms of privacy in the context of AmI. We then introduce the state of the art in PETs with focus on AmI and propose two approaches for adaptive privacy mechanisms as used in ATRACO and in an ambient calendar system. We provide an overview of trust definitions and mechanisms of trust establishment before discussing our approach towards trust management in ATRACO. The chapter closes with a discussion of lessons learned and research challenges for future work on privacy and trust in AmI.

4.2 Privacy in Ambient Intelligent Environments

AmI environments have been the focus of much research in the last decades [5, 19, 73]. While at the beginning of this research field only a small number of research focused on privacy and trust issues [11, 45], we can see more efforts towards enhancing privacy and trust in AmI today. We give an overview of existing work in privacy protection mechanisms in general, and for AmI environments in particular, after discussing common definitions for privacy.

4.2.1 Definitions of Privacy

As there is no universally accepted definition of privacy, the formulation of its meaning is one of the most intractable problems in privacy research. Besides hundreds of existing definitions one of the oldest and most cited still remains influential: Samuel Warren and Louis Brandeis's 1890 declaration of privacy as the "right of an individual to be let alone" [72]. This traditional way of understanding privacy can also be described as "a state in which one is not observed or disturbed by others" [70].

However, most of today's common definitions and understandings of privacy in the context of information technology are based on Westin's definition from 1967. Westin defined privacy as "the claim of individuals, groups or institutions to determine for themselves when, how, and to what extent information about them is communicated to others" [74]. This definition is focused on information collection, dissemination, and use, which is often referred to as *information privacy*.

Philosopher Sissela Bok combines both informational and physical privacy aspects in her definition of privacy [12] as "the condition of being protected from unwanted access by others—either physical access, personal information, or attention." Robert Ellis Smith [65] also covers both these privacy aspects in his definition of privacy as "the desire by each of us for physical space where we can be free of interruption, intrusion, embarrassment, or accountability and the attempt to control the time and manner of disclosures of personal information about ourselves."

AmI systems have both physical and information privacy implications, which need to be considered together in the design of AmI systems.

4.2.2 *Privacy Classification*

In the context of AmI, we distinguish between two forms of privacy, namely *information privacy* and *territorial privacy* as depicted in Fig. 4.1. Information privacy refers to the protection of personal information (or personal data), whereas territorial privacy refers to the protection of personal spaces (or territories). Some privacy classifications [24] also list *communication privacy* and *bodily privacy* as separate privacy categories, whereas we consider them to be subcategories of information privacy and territorial privacy, respectively.

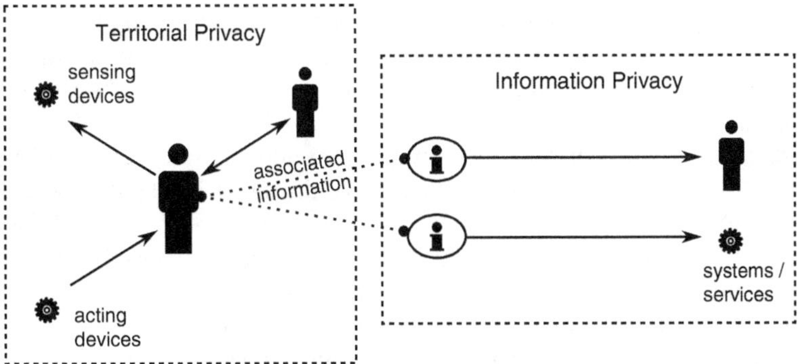

Fig. 4.1 Privacy classification in AmI: information privacy and territorial privacy

4.2.2.1 Information Privacy

Information privacy refers to the definition of Westin [74] given in Sect. 4.2.1. Thus, persons want to control "when, how, and to what extent information about them is communicated to others." In the context of AmI environments, this means that any kind of sensitive information needs to be protected from unintentionally leaving the borders of the surrounding AmI environment and from unauthorized access by other entities. Thus, personal information should remain at any time under full control of the user. Such information includes *personal data* (e.g., pictures, e-mails, web history), *personal interests* (e.g., age, size, weight), *personal interests* (e.g., favorite music, movies, fashion), or *personal behavior* (e.g. doing what, when, how often).

Personal information can either be *static* (e.g., emails, pictures) or *dynamic* (e.g., location, activity) and can be collected *directly* (e.g., database access, physically sensed) or *indirectly* (e.g., analytically derived, compositionally derived).

4.2.2.2 Territorial Privacy

While protecting information privacy privacy may be sufficient in the context of most common IT domains, it is only one aspect of privacy protection in the context of AmI systems. The pervasive nature of such systems equipped with multiple sensors, actuators and computational devices, constitutes an additional facet of privacy protection, that we call *territorial privacy*. Territorial privacy aims to satisfy a more traditional expectation of privacy such as being in "a state in which one is not observed or disturbed by others" [70].

In an AmI environment a user might be continuously observed by cameras, microphones, or other sensors that are present in the user's *physical territory* which is delimited by physical boundaries, e.g., the walls of an apartment. Through pervasive communication technologies the data gathered by those sensors can easily be forwarded to other *virtual entities* beyond physical boundaries; for example, a remote entity receiving a live video stream of a user. Therefore, it is required that the user is able to control whether or not an entity is allowed to observe him in a certain way.

Further, the control of how other entities are allowed to disturb a user in his private space is an important aspect of territorial privacy. Disturbances can occur in the user's physical territory by undesired outputs of visual or acoustic signals or by undesired interactions or automations. These disturbances can also be triggered by virtual entities outside the user's physical territory.

The goal of territorial privacy is to control all physical or virtual entities which are present in the user's *virtual extended territory* in order to mitigate undesired *observations* and *disturbances*, and to exclude undesired entities from the *private territory*. Based on these concepts and terminologies we have proposed a model for territorial privacy, which represents entities in a privacy graph with privacy implications modeled as *observation channels* and *disturbance channels*.

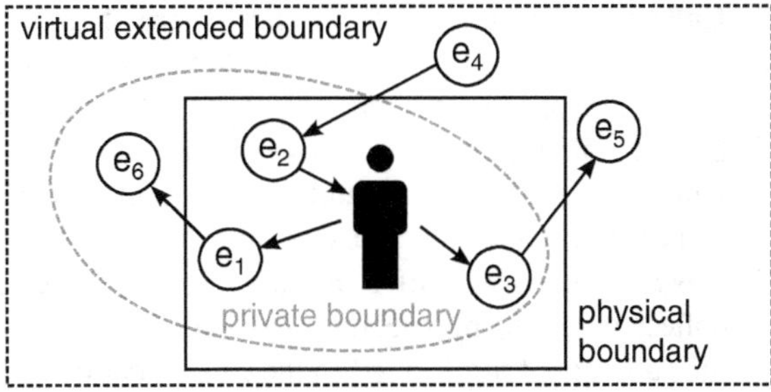

Fig. 4.2 A territorial privacy boundary with desired and undesired observers and disturbers

An example of a person and its physical, virtual extended, and private territory is depicted in Fig. 4.2. Desired entities are separated from undesired ones by the private boundary. A detailed discussion and formalization of this approach is given in [37–39].

4.2.3 *Mechanisms for Privacy Protection*

In general, privacy protection mechanisms can be divided into the four main categories of *regulatory strategies*, *policy matching*, *prevention and control*, and *detection*. Mechanisms belonging to the last three categories are often referred to as PETs. Even though some of the PETs discussed below were not primarily designed for AmI environments, they provide important approaches which can be further extended to create new AmI privacy solutions.

4.2.3.1 Regulatory Strategies

Regulatory strategies refer to governmental rules on the use of personal information or respective self-regulation efforts by industry. Because individual privacy preferences often differ in several ways, it is important to find general principles for privacy protection that fit the most common requirements. Those principles can be used to expand voluntary agreements or regulations enforced by law, but can also serve as important input for the design process of AmI systems and the subsequent treatment of personal information in such systems.

The most popular principles relating to information privacy are listed in the *Organization for Economic Cooperation and Development* (OECD) guidelines on the protection of privacy and transborder flows of personal data [54]. The guidelines state eight principles for the protection of privacy together with numerous other and similar sets of privacy protecting rules. The four core principles are:

1. **Collection Limitation Principle:** There should be limits to the collection of personal data and any such data should be obtained by lawful and fair means and, where appropriate, with the knowledge or consent of the data subject.
2. **Data Quality Principle:** Personal data should be relevant to the purposes for which they are to be used and, to the extent necessary for those purposes, should be accurate, complete, and kept up-to-date.
3. **Purpose Specification Principle:** The purposes for which personal data are collected should be specified not later than at the time of data collection and the subsequent use limited to the fulfillment of those purposes or such others as are not incompatible with those purposes and as are specified on each occasion of change of purpose.
4. **Use Limitation Principle:** Personal data should not be disclosed, made available or otherwise used for purposes other than those specified in accordance with the Purpose Specification principle except:

   (a) with the consent of the data subject; or
   (b) by the authority of law.

These four principles contain the essence of privacy protection whereas the remaining four describe procedural aspects. The fundamental role of these principles is reflected by the fact that they were adopted in many international privacy regulations, including the EU Data Protection Directive 95/46/EC [25] and the Canadian PIPEDA law [14].

However, the OECD guidelines, as well as similar existing regulations, are often in conflict with common AmI characteristics. For instance, the principles of limitation and purpose specification of data collection are conflicting with the active, pervasive, and continuous collection of data in AmI environments. PETs which try to enforce existing guidelines are therefore often a trade-off between privacy, and benefit or usability of the AmI system.

4.2.3.2 Policy Matching

Policy matching techniques try to minimize privacy risks by specifying and matching user privacy policies with involved service provider policies. The most common policy languages of that kind are the *Platform for Privacy Preferences* (P3P) [20], the *Enterprise Privacy Authorization Language* (EPAL) [4], and the *eXtensible Access Control Markup Language* (XACML) [28], which are XML-based languages to support the definition and enforcement of privacy policies and obligations. The main drawback of privacy policy systems is that they generally cannot enforce privacy and instead rely on trustworthiness and regulatory pressures to ensure policy compliant behavior.

4.2.3.3 Prevention and Control

Prevention and control mechanisms seek to prevent misuse of personal information by adapting information in several dimensions or by avoiding privacy-critical operations or access. Such mechanisms include *pseudonymization* [17], *obfuscation* [75], or *access control* [52].

Pseudonymization and obfuscation techniques gained popularity in the domain of location privacy, which was one of the first privacy concerns raised by ubiquitous computing [73]. As mobile devices with integrated GPS-receivers are becoming more and more ubiquitous, location privacy has attracted considerable research in the last decade [22, 41]. Beresford and Stajano [7] employ mix zones and anonymity sets to analyze location privacy. They propose to use changing pseudonyms instead of unique identifiers to prevent tracking of users. Gruteser and Grunwald [30] applied the concept of k-anonymity to location privacy. Instead of pseudonymously reporting exact location information, a person reports a region containing $k-1$ other persons. This way k-anonymity provides privacy protection by guaranteeing that each released record of an individual cannot be distinguished from records of at least $k-1$ other individuals. The larger the value of k, the greater the implied privacy since no individual can be identified through linking attacks with probability exceeding $1/k$. Duckham and Kulik [21] propose the use of obfuscation, that is decreasing the accuracy of exchanged personal information to the minimum needed by a service provider in order to provide the required service. A simple solution is giving a more general location, e.g., only the street or city instead of detailed coordinates.

Also several architectures for privacy protection and control in ubiquitous environments have been proposed in the last decades. Hong and Landay [31] proposed Confab, a privacy-aware architecture for ubiquitous computing. Confab uses the concept of information spaces by Jiang et al. [33] in combination with in- and out-filters to manage the flow of context information about a person. Langheinrich [42] discussed how privacy guidelines, similar to the OECD guidelines, can be adapted to ubiquitous computing scenarios. He pointed out six principles, which he followed in the design of the privacy awareness system PawS [43], in order to address third party data collection in smart environments.

A major drawback of most approaches is that they provide rather static solutions, e.g., access control mechanisms based on policy matching or preventive data minimization techniques that do not respect individual user preferences. However, privacy preferences are highly dynamic [48] and can depend on several contextual factors which are shaped by individual, social, and cultural expectations and norms [53]. A first approach of a PET respecting context factors was proposed by Moncrieff et al. [51] for a smart home safety system. They distinguish between spatial context (i.e., the current location), social context (i.e., interactions between persons or third parties), hazardous context (i.e., abnormal interactions with hazardous devices), and temporal context (i.e., abnormal periods of inactivity). The current context is mapped to appropriate granularity levels of sensor data to control privacy. The mapping process is further influenced by the purpose of data access (e.g., diagnostic aid by a doctor), the trust level in the entity accessing the data, and the currently assumed thread-benefit trade-off.

4.2.3.4 Detection

Detection mechanisms try to identify and penalize privacy violators. A system for privacy violation detection continuously monitors access to personal data and detects misuse or abnormal behavior.

PRIVDAM [8] is a data mining-based architecture for privacy violation detection and monitoring. The system attempts to detect possible privacy violations based on network characteristics and seeks to prevent them in the future.

An et al. [3] propose to use Bayesian network-based methods to detect insider privacy intrusion in database systems. The advantage of Bayesian networks is that they can effectively deal with uncertainties involved in the activities of database operations.

Ortmann et al. [55] discuss the idea of modeling information flow graphs to identify directly and indirectly available data in a pervasive system and link them to the sources of the raw input data. The graphs can be used to determine which information or data sources need to be disabled to hide forbidden information and adapt the modeled system to given privacy requirements.

4.2.4 Privacy Requirements

Most of the discussed PETs in Sect. 4.2.3 only focus on a small part of privacy protection in the context of AmI environments and in most cases address only information privacy aspects. As a consequence, solutions are required which are widely applicable by combining policy matching techniques with adequate policy enforcement mechanisms. In addition, a comprehensive privacy solution needs to respect and address territorial privacy aspects as well.

Another fundamental requirement for privacy in AmI environments is the consideration of contextual information, e.g., location, activity, or presence of other persons. Furthermore, users prefer to rely on context instead of setting access control policies on specific content [64]. This aspect must be taken into consideration when developing policies and enforcement techniques as it leads to the need of dynamic adaptation of privacy policies or enforcement. In addition, privacy protection needs to respect given user preferences and trust in involved systems components or other persons.

Previous studies [6, 31] have shown that the deployment of adequate feedback and control mechanisms is another essential requirement for the user acceptance of AmI systems. This fact needs to be respected in the design process of privacy solutions as well.

Taking all previous aspects into consideration we identified the following requirements for privacy protection mechanisms in AmI:

- Privacy protection should ensure information privacy and territorial privacy.
- Privacy protection should respect and depend on users' privacy preferences.

- Privacy protection should adapt to contextual changes.
- Privacy protection should respect trust in involved systems and persons.
- Privacy protection should support users in gaining awareness about potential privacy implications.

4.2.5 Privacy Management in ATRACO

The ATRACO approach aims to realize the concepts of an AmI system by addressing heterogeneity of artifacts, system transparency, discovery and management of various artifacts, and autonomous behavior of learning agents. ATRACO uses a *Service-Oriented Architecture* (SOA) at the resource level to support numerous devices and sensors, and at the system level to support AmI applications. Several agents complement the SOA infrastructure by providing high level adaptation to a user's task. ATRACO agents support adaptive planning, task realization, and enhanced human computer interaction. Ontologies are used to address the semantic heterogeneity that arises in AmI environments. The instantiation of agents to support a specific user task is based on the concept of *Activity Spheres* (AS). An AS is the utilization of knowledge, services and other resources required to realize an individual user goal within ATRACO. A more detailed discussion of the ATRACO core concepts and architecture is given in Sect. 1.3.

Our basic approach to privacy enforcement in ATRACO is based on policy matching and access control mechanisms that respect context factors in order to allow dynamic adaptation. All privacy relevant functionalities and components are encapsulated within the *Privacy Manager* (PM). The integration of the PM within the overall architecture and its interaction with other system components is discussed in Sect. 4.2.5.2.

4.2.5.1 Privacy Policy Ontologies

In ATRACO, a *Privacy Policy Ontology* (PPO) describes how, under which conditions, and in which context an entity is allowed to handle personal information or is allowed to participate in a user's activity. In the case of information privacy, policies need to be specified on the user's side as privacy preferences, and on the receiver's side as privacy obligations in order to allow a matching between the two parties. The policies of receivers (e.g., persons, web services, but also system components) can either be specified as policies in PPO representation or in a different policy language, like P3P [20]. For the latter case, a policy wrapper translates such policy languages into PPO representation.

We distinguish between *Information Privacy Policies* (IPPs) and *Territorial Privacy Policies* (TPPs). The key attributes of an IPP are *information item* (the information to apply the policy to), *purpose* (what is allowed to be done with the

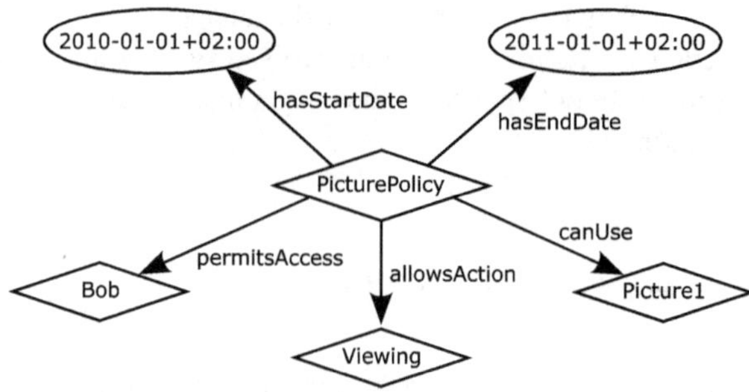

Fig. 4.3 Example of a privacy policy ontology

information), *recipient* (who is allowed to use the information), *retention* (how long can the information be used/stored), and *context* (further access constraints).

A TPP consists of the key attributes *territory* (the space to apply the policy to, explicitly identified by a location, or implicitly by an activity), *participant* (the entity which is allowed to observe or disturb the user), *observation type* (the allowed observation channel, declared by sensor types), *disturbance type* (the allowed type of intervention), and *context* (further access constraints).

For both IPP and TPP a default *deny-all* rule is assumed. Therefore, any access to information or territories must be explicitly granted by either an IPP or TPP, respectively.

In the ATRACO prototype we implemented policies as part of the *User Profile Ontology* (UPO). Each policy is an individual of one of the two class types: `InformationPrivacyPolicy` or `TerritorialPrivacyPolicy`. Policies of the first type may permit access to specific private information, indicated by the object property `canUse`, for different entities via the `permitsAccess` object property. If access to private information should be granted only for specific actions, a policy can specify the `allowsAction` object property. In addition, time constraints can be specified with the data properties `hasStartDate` and `hasEndDate` to limit the usage of private information to a predefined time interval. Figure 4.3 shows a simplified example of an IPP, granting Bob access to view a picture for a time period of 1 year.

4.2.5.2 Privacy Manager

In ATRACO, all privacy components are encapsulated in the Privacy Manager which consists of the *Policy Processing Engine* (PPE), the *Information Privacy Controller* (IPC), the *Territorial Privacy Controller* (TPC), and the *Feedback and Control Component* (FCC). Figure 4.4 provides an overview of the main components of the ATRACO privacy architecture and its interactions with the *Sphere Manager* (SM), *Ontology Manager* (OM), *Interaction Agents* (IAs), and *Trust Manager* (TM).

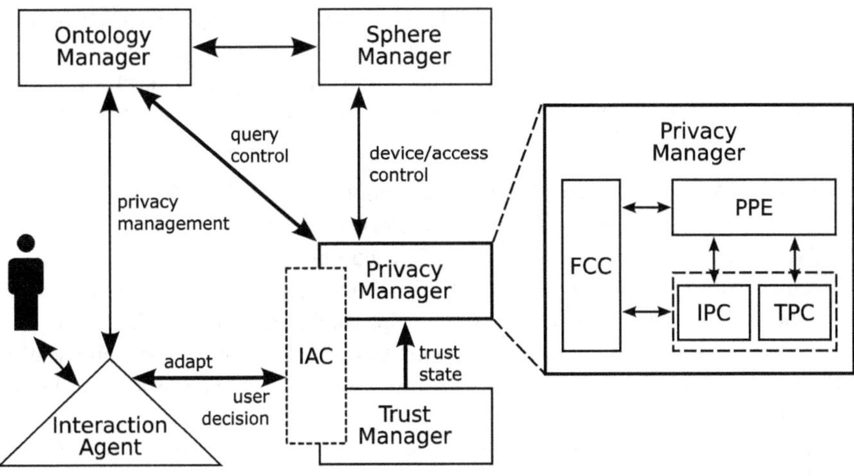

Fig. 4.4 Integration of the Privacy Manager

The SM is responsible for initializing or dissolving an activity sphere (AS) with its associated resources for a specific user goal. The SM acts as an event service to other components of the AS. It monitors the execution of the task workflow and adapts the composition of resources in case of any conflicts, such as failing devices or other exceptions. For a detailed description of the SM see Sect. 1.3.

The OM is responsible for managing the sphere ontology and responds to ontology queries of other components. Ontologies provide a central knowledge base for properties, state and context information, as well as user preferences. The most privacy relevant ontology is the UPO, which contains user preferences, user properties, and other personal-related information. The management of ontologies is discussed in Sect. 3.3.

Several IAs offer multimodal services for user interaction. In order to provide smooth interaction, IAs use distributed multimodal interaction widgets which adapt to the user's context. Therefore, the user interaction can be distributed among available modalities and devices at runtime. For more information about the ATRACO IAs see Sect. 6.3. The interaction between PM and IA is managed by the *Interaction Agent Controller* (IAC) which is part of the PM as well as of the TM, and provides shared functionalities for controlling and event handling of the IA. The TM will be discussed in detail in Sect. 4.3.4.2.

The Privacy Manager serves as a privacy interface for other components in the ATRACO architecture. Whenever a privacy relevant event occurs, such as a query of personal information, a territory access request, or context change, the PM requests related policies and contextual information from the OM. The PPE uses this information to perform policy reasoning and provides the results to the corresponding privacy controller, which will perform further privacy enhancing mechanisms based on the results. If any conflict occurs during this reasoning process, the PPE can trigger the IA via the FCC and IAC in order to ask the user for a conflict resolution, such as modifying policies or adding policy exceptions. Furthermore, the FCC provides a user interface for controlling and modifying privacy settings that are stored as part of the UPO. The interplay and role of privacy controllers (IPC and TPC) is described below, along with a discussion of potential privacy affecting events.

Queries of Personal Information: The main knowledge base for personal information is the UPO. In order to ensure information privacy, the OM triggers the PM to validate queries on ontologies, in particular on the UPO. The privacy validation will be handled by the PPE, involving the requester ID, associated policies, and current context. Based on the validation result the IPC will grant or deny access to the query results. Optionally, the IPC can perform obfuscation and pseudonymization techniques to further enhance privacy or react to policy conflicts.

A special case occurs whenever IAs request personal information. The IPC then checks if any other persons are located inside the user's sphere by querying the OM. If this is the case, the IPC uses the IA's ontology to determine the presentation modality for that information and whether it is accessible by other persons. If the presentation is accessible by other persons, for example a large screen display, the PPE checks the related policies, as described above. On a deny decision, the IPC may either retain the personal information or adapt the interaction modalities of the IA via the IAC to ensure that unauthorized access to personal information is prevented.

Territory Access Requests: The TPC performs territorial access control in order to protect territorial privacy whenever external entities attempt to join the user's sphere. Access requests are forwarded by the SM. An access request may ask either for an observation or disturbance. Observation access is requested when an entity attempts to receive sensed user data, for example via a camera or microphone. A disturbance access is requested whenever an entity attempts to pass the physical borders (e.g., enters a room) or wants to actively intervene in the user's territory, for example, by acoustic or visual output, or by initiating interactions. Access decisions are based on the associated policies, which are validated by the PPE. In addition, access decisions can take into consideration the current trust state of a device which is provided by the TM. This will be discussed further in Sect. 4.3.4.2.

Contextual Changes: Contextual changes, such as an activity change, a location change or a newly arrived person, are reflected as events by the SM. Whenever such an event occurs, the IPC and TPC perform appropriate privacy measures. In the case of a new person, the IPC checks whether personal information is currently presented by any interaction modality the new person would have access to. This information is obtained either from the interaction ontology or directly from the interaction agent. If so, the IPC may again adapt the interaction modalities via the IAC based on the policy reasoning results of the PPE, as described above. In case of activity or location changes, the TPC checks if any device or remote entity is currently observing or disturbing the user in his new territory. This process is analogous to the IPC's operation, in that the PPE validates the associated privacy policies. Based on the results, the TPC will exclude the undesired observing and disturbing entities by switching off relevant sensors or active devices, respectively.

4.2.5.3 Prototype Evaluation

A prototype of the overall ATRACO architecture has been deployed and evaluated at the University of Essex in its iSpace laboratory, a real world testbed for home technology. A first evaluation of the ATRACO privacy components was performed as part of a user test with nine participants during the project's second year. Next, we briefly describe the privacy relevant scenario and results.

Scenario

Alice arrives at home and sits down on the sofa in her livingroom. The ATRACO system asks Alice if she wants to look at holiday pictures on the TV. Alice accepts and the slideshow starts (see Fig. 4.5a). After a while her roommate Bob enters the room. His presence is automatically detected by the ATRACO system. The PM starts a privacy reasoning process about Alice's privacy preferences which may result in the system's decision that Bob should not see some of the pictures. If the access is denied only to some of the pictures, the PM will seamlessly exclude those pictures from the slideshow. If the complete slideshow is private, it is stopped by the PM as shown in Fig. 4.5b. In this case, Alice can manually resume the slideshow to implicitly grant Bob access to the pictures (see Fig. 4.5c). Similar privacy adaptation has been employed in a computer-centric office scenario. If Bob approaches Alice while she is doing sensitive office work on her computer (e.g., online banking), the PM can dynamically hide the sensitive parts on her screen.

Fig. 4.5 The ATRACO Privacy Manager (PM) dynamically adapts the presentation of personal information based on present persons and predefined user preferences. (**a**) A user is watching a private slideshow. (**b**) When the roommate enters the room, the slideshow is stopped as he should not have access. (**c**) The user can manually grant access to the slideshow

User Reactions

The service of automatically hiding private information in presence of other persons was generally accepted by users of the system. However, the approach of explicitly hiding sensitive information (e.g., the complete slideshow instead of some pictures) revealed the issue that direct adaptation makes Bob aware of Alice's privacy preferences with respect to him. This could potentially cause social tensions.

In general system interactions, users were concerned about the ability to control the system's autonomous processes. In situations where the user felt out of control, reactions to the system were mostly negative. Especially in situations when the system initiated undesired interactions that were perceived as an invasion of personal space. For example, users perceived the system as invasive when it asked them whether they would like to listen to music or look at photos as soon as they arrived at home. In some cases, participants were comfortable with the system initiating interactions, such as when the system announced a guest's arrival at the front door. Perceptions of control may also be influenced by the channel of interaction. Voice interactions appeared to imply the existence of a separate social presence which could lead to a feeling that the personal space has been invaded. Also mood changes could have an impact on the way a user would prefer to interact with the system.

These results show the technical feasibility of adaptive privacy mechanisms for protecting sensitive information. They also highlight that privacy concerns of participants in AmI environments often involve territorial privacy aspects, such as the fear of losing control about the physical environment, the perception of privacy invasion, or particular forms of interaction. Furthermore, the social implications of making others aware of one's own privacy preferences by dynamic adaptations must be considered in order to avoid social conflicts.

4.2.6 Context-Adaptive Privacy in Ambient Calendar Displays

Encouraged by our experiences with ATRACO's Privacy Manager and components, we further investigated the role of context in privacy decision making and reasoning. Next, we outline PriCal, an ambient calendar display system that implements resulting context-adaptive privacy mechanisms. We further present the results of a deployment study of the PriCal system.

4.2.6.1 PriCal Architecture

PriCal [60] is an ambient calendar display system designed for a multi-office setting. PriCal displays are positioned in individual offices. These displays show the calendars of present persons. Hereby, the visibility of a user's calendar entries is dynamically adapted to other present persons and the user's privacy preferences

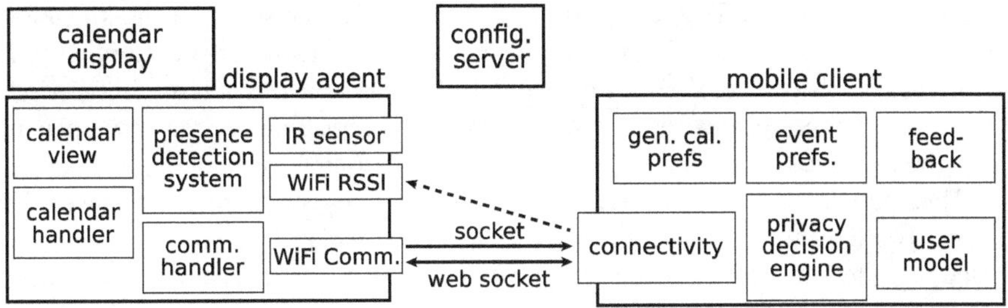

Fig. 4.6 The PriCal architecture consists of a *display agent* that manages display content and a *mobile client* that performs privacy reasoning

for those persons. While ATRACO provided privacy reasoning centrally through the Privacy Manager, PriCal moves privacy reasoning to a privacy decision engine running on a user's smartphone, which has access to the user's calendar data. Figure 4.6 outlines PriCal's system architecture.

When a registered user enters a PriCal-equipped office, the office's PriCal display detects their presence and initiates a connection to the mobile client running on the user's smartphone over WiFi. The mobile client establishes a WebSocket connection with the display over which it receives the list of currently present persons and respective updates whenever someone enters or leaves the office. This information is used by the mobile client's privacy decision engine to perform reasoning. Privacy reasoning is hereby based on a lean user model that combines rule-based and case-based privacy preferences. The reasoning process results in calendar entries adapted to the user's individual privacy preferences for the other present persons. More specifically, calendar entries may be shown in full, only as busy, or be completely hidden. The display agent receives the already adapted calendar entries and renders them on the display.

For example, a user's personal events may be displayed when the user is alone in an office but may be completely hidden when other persons are present; during a meeting with a colleague, events concerning projects that person is not involved in may only be shown as busy to still provide a full overview of one's schedule in order to facilitate scheduling of follow-up meetings.

As noted in the previous section, user feedback concerning the ATRACO system revealed that context-adaptive privacy changes also require privacy protection, because other persons may get upset if they realize that information is being hidden from them or that the system adapts because of their presence. In PriCal, we addressed this issue by introducing a *hide then reveal* paradigm for dynamic privacy adaptations. Present persons are being detected in a two-step process. Light barriers in the doorways trigger immediately when a person enters or leaves the room, which causes the display to be cleared. As the person has been detected but not yet identified, mobile clients connected to the display agent receive an update that an *unknown person* has been detected, which causes the mobile clients to perform privacy reasoning for the changed situation and send accordingly adapted calendar events to the display. Triggering the light barriers further starts an identification process which employs WiFi received signal strength measurements to identify a registered user's device. If a user can be detected this way in a certain time period, the display agent updates its present person list by replacing the unknown person with the identified user. This triggers subsequent reasoning by the mobile clients.

Thus, the calendar display is cleared every time a person enters or leaves the room. The advantage of this approach is that the display behavior is always the same for entering persons. When the person enters, the display is empty and is subsequently populated by adapted calendar entries received from the mobile clients. The *hide then reveal* paradigm provides a certain level of plausible deniability to other users, because newly arriving persons cannot distinguish whether events were adapted or just reloaded because of their arrival.

The two-step presence detection approach has further privacy advantages. The light barriers can detect persons that are not registered users of the system and list them as unknown persons. Thus, users can create privacy preferences specifically for unknown users.

Another issue revealed in the ATRACO evaluation was a perceived lack of control concerning system actions. We addressed this issue proactively in the design of PriCal. PriCal's mobile client enables users to manage their privacy preferences on multiple levels, allows them to rate adaptation actions in order to refine the privacy decision engine, and also offers direct controls to remove currently shown information from the display. The PriCal system is described in more detail in [59, 60].

4.2.6.2 Deployment Study Results

The PriCal system has been evaluated in a deployment study in which participants used ambient calendar displays with their own calendars as part of their typical work routines over the course of 3 weeks [60].

Seven PriCal displays were deployed in the Institute of Media Informatics at Ulm University in the fall of 2013. Ten participants (nine male, one female) used the PriCal system for 3 weeks. All but one participant had a PriCal display in their office, as many of them shared an office with a colleague. Participants either received an Android phone with the PriCal mobile client pre-installed or installed the mobile client on their own phone. Three interviews were conducted with each participant during the study to learn about their experiences with PriCal. Here, we discuss the overall results of the study, a detailed analysis can be found in [59, 60].

The deployment study with the PriCal system showed that physical placement of ambient displays and components of AmI systems in general may have implications for users' territorial privacy. For instance, the orientation and viewing angle of displays determines how easily other persons can view a display's content and whether privacy adaptation of content will be noticed. The PriCal displays were placed inside offices facing away from the door. This placement enabled users inside the office to easily view the ambient displays, while displays were shielded from views of passers-by and entering persons.

Context-based privacy adaptation was perceived as usable and useful by our study participants, but the interviews also highlighted the importance of robust context detection and adaptation. For instance, unreliable detection of present persons may lead to decreased trust in the system, which may drive users towards using privacy settings more conservative than desired in order to protect their information in cases where detection errors occur.

Study participants were comfortable with the provided privacy controls and opportunities to dynamically express privacy preferences. They particularly valued the ability to specify preferences for unknown, unregistered persons. This highlights the importance of providing flexible adaptation and considering uncertainty and unknown context features in the design of AmI privacy mechanisms.

Next, we discuss the role of trust in the design of AmI systems which also serves as a prerequisite for the development and trustworthy deployment of the previously discussed privacy protection mechanisms.

4.3 Trust in Ambient Intelligent Environments

Because a major goal of AmI is the seamless and invisible integration of technology in our everyday environments, trust in these technologies becomes a significant determinant for its acceptance. In this section, we will briefly discuss the different definitions and forms of trust in the context of AmI and provide an overview of technical mechanisms for trust establishment. Based on the different forms of technical and social trust we discuss the integration of trust management in ATRACO.

4.3.1 Definitions of Trust

Although trust is a common concept, on which we rely in our everyday life, it is relatively difficult to define concisely. Like privacy, the term trust has a huge definitional diversity, which often leads to more confusion than to a better understanding of the term. As a consequence, some researchers argue that their "purposes may be better served ... if they focus on specific components of trust rather than the generalized case" [50].

Nevertheless, there is a set of trust properties, which are commonly recognized among most trust researchers in the context of information technologies. They agree that trust is *subjective*, *asymmetric*, *context- and situation-dependent*, *dynamic and non-monotonic*, and typically *not transitive* [1, 16, 49, 71].

Some of these properties are reflected in one of the most adopted trust definitions formulated by sociologist Diego Gambetta [26]: "trust is a particular level of the subjective probability with which an agent assesses that another agent or group of agents will perform a particular action, both before he can monitor such action ... and in a context in which it affects his own action."

Depending on the environment in which trust is specified, it can be considered as a composition of different attributes like *reliability*, *dependability*, *honesty*, *truthfulness*, *security*, *competence*, and *timeliness*. Utilizing these attributes, Grandison and Sloman [29] define trust as "the firm belief in the competence of an entity to act dependably, securely, and reliably within a specified context."

Chopra and Wallace [18] tried to find a common denominator of various definitions and argued that an integrated definition of trust consists of at least three elements. A *trustee* to whom trust is directed, *confidence* that the trust will be upheld, and a *willingness* to act on that confidence. Based on these elements they define trust as "the willingness to rely on a specific other, based on confidence that one's trust will lead to positive outcomes."

Further, trustworthiness of an entity can be interpreted as a level of trust, that could be established over time by monitored information. Therefore, "trust can be seen as a complex predictor of the entity's future behavior based on past evidence" [62].

Although the given definitions do not explicitly use the term risk in their formulation, the inclusion of risk is implied by most definitions in terms of uncertainty and negative outcomes if trust assessment fails. Thus, most researchers agree that trust is inherently related to risk but often fail to understand the precise relation between the two notions [47, 66].

4.3.2 Trust Classification

In order to realize a trustworthy AmI system, different kinds of trust need to be considered. We distinguish between *technical trust* in system components and devices, and *social trust* in other persons (see Fig. 4.7).

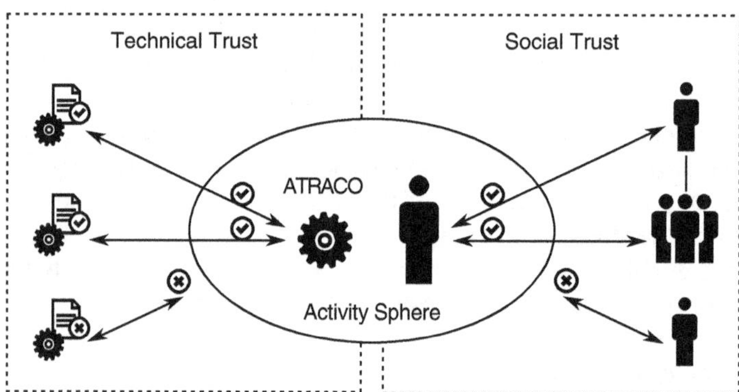

Fig. 4.7 Trust classification in ATRACO: technical trust and social trust

4.3.2.1 Technical Trust

An AmI system consists of several interacting software and hardware components of different types. Technical trust refers to trust in these components with respect to their intended functionality, reliability, and safety. These aspects are particularly important in AmI environments with high flexibility and adaptivity of involved components. This form of trust can be partially established by credential-based trust mechanisms and Public Key Infrastructures (PKIs). For example, a component can provide a digitally signed certificate issued by the manufacturer's Certificate Authority (CA). If the manufacturer CA is in a valid PKI path to a trusted root CA, the device will be trusted in the current AmI environment. The inclusion of manufacturers in a trusted PKI might depend on several factors, such as quality guidelines or performance and reliability assessments. Despite certain short comings of the PKI system [23], this approach allows an initial binary trust assignment, which can support the decision whether a component should be used in a given context.

Further, the use of digital signed certificates allows verification of several component properties, which could be utilized for trust establishment. Thus, a certificate might not only certify the device manufacturer, but also the place of origin, application fields, or other device specific properties. This way different credentials can be rated to achieve a more fine granular trust level assessment.

To further enhance trust assessment in terms of dynamic establishment and adaptations, we propose to combine credential-based mechanisms with reputation-based mechanisms, as depicted in Fig. 4.8. For example, a new component of an unknown manufacturer will be classified as untrusted when it initially appears in an AmI environment. The operations and interactions of the component will be continuously monitored to dynamically assess the device's proper functioning and reliability. The results of this monitoring phase will be fed to a central database in order to collaboratively collect and receive monitored information about a specific component. On the one hand, the creation of such a reputation database will ease the trust assessment of components, which are not certified by a trusted CA. On the other hand, trust in properly certified components could further be refined, for example, decreased if the reputation database reports unreliable behavior for the given component.

4.3.2.2 Social Trust

Social trust refers to the way human beings perceive and understand the notion of trust in social interactions and social relationships. As AmI environments are not always single-user oriented, social trust needs to be respected whenever a system component is involved in multi-user situations. In this context, we focus on the establishment of trust groups, which represent a group of individuals with the same level of assigned trust. The concept of trust groups addresses the problem of defining specific trust levels, which are typically difficult to determine computationally [44].

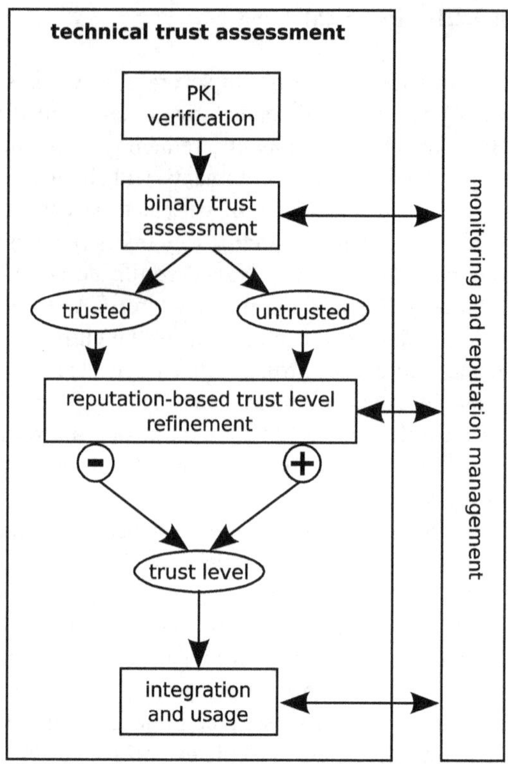

Fig. 4.8 Technical trust assessment process combining credential-based and reputation-based trust establishment

Approaches for computing such specific levels of trust need to map real life trust to a given computational metric. The problem is to find metrics that are flexible enough to model the dynamic and context dependent trust perception of real persons. Furthermore, maintaining such trust models for other persons would require frequent monitoring and assessment of their behavior, which would constitute an infringement of their privacy. Trust groups avoid the need of specifying those metrics and specific trust levels by providing an adequate abstraction layer for user-oriented trust management. Thus, these groups act as surrogates for individual inter-personal trust.

We propose to have several predefined trust groups, which could be further refined and rearranged by a user. Those groups refer to common social groups, namely *family*, *friends*, and *colleagues*. Users could be offered the ability to create further groups that reflect their social groups with different trust associations. By default a trust group will not have a specific trust level assigned. Trust will be implied by creating specific access policies either for personal information that utilize trust for privacy protection, or for control instances of a personal AmI environment. For example, a home AmI system which controls the heating or light levels should be accessible only by family members.

4.3.3 Mechanisms for Trust Establishment

As introduced in Sect. 4.3.1, the concept of trust may be used in interactions for estimating the future behavior of an interacting entity. For instance, whether that entity will respect a user's privacy preferences and specifications. Trust allows people to deal with uncertainty and assess risk in uncertain interactions or situations.

In social interactions, trust can be established based on personal assessments, experience or reputation. In the context of AmI environments and ICT systems in general, the establishment of trust is often based on the concept of *computational trust*. Existing mechanisms and models for computational trust can be classified into *credential-based* and *reputation-based* mechanisms [13, 40, 58].

4.3.3.1 Credential-Based Trust

Credential-based computational trust refers to cryptographic solutions for establishing trust by obtaining and verifying credentials of an entity. These credentials are usually issued in form of digital certificates by a trusted third party or CA. This approach is commonly used to authenticate identities or memberships in Internet applications, often based on PKIs [27, 57]. A PKI is an infrastructure for public key management and for issuing digital certificates. The main part of a PKI is a hierarchically arrangement of digitally signed certificates for assuring bindings of public keys or attributes to specific identities.

Based on this notion of trust, Blaze et al. [9] introduced the term *trust management*, which involves the formulation of policies and credentials, determining satisfaction of credentials to policies and deferring trust to third parties. They proposed a simple language for specifying trusted actions and trust relationships, which was used in their trust management system *PolicyMaker* [10]. Li et al. [46] combined the strengths of role-based access control with similar trust management systems in their *RT framework* to support delegation of role definition and role activations.

Lund et al. [47] view trust management as a particular form of risk management with the same weaknesses but higher complexity and dynamics. In order to account for evolution in risk and trust assessment, they distinguish between three main scenarios: *maintenance*, *before-after*, and *continuous-evolution*. They focus on understanding the impact of subjective trust on factual risks in each of these scenarios.

4.3.3.2 Reputation-Based Trust

Reputation-based trust establishment uses the history of an entity's past behavior or recommendations and experiences about this entity provided by other entities to compute a certain trust level [34]. Well-known applications for reputation-based systems are electronic markets like eBay[1] or Amazon,[2] where buyers rate the reliability of sellers and merchants. The main challenge here is to find adequate trust metrics and models to support this kind of computation. One of the first attempts at formalizing such trust metrics was made by Marsh [49]. His model was based on linear equations and was further extended by Abdul-Rahman and Hailes [1] to address trust in virtual communities.

Other approaches to compute trust and reputation values are based on the use of probability theory [61], fuzzy logic [15], or the use of entropy [63]. Similar models have been adopted by distributed mechanisms for computing reputation values in peer-to-peer networks [2, 36, 68, 76].

4.3.4 *Trust Management in ATRACO*

Trust in ATRACO is comprised of technical trust and social trust, as described in Sect. 4.3.2. This section discusses the main requirements for building these two types of trust into an AmI environment and then describes how this is achieved in the context of ATRACO by the Trust Manager component.

4.3.4.1 Trust Requirements

The high dynamics in the availability of devices as well as their autonomous nature make the establishment of technical trust for those devices indispensable. A user relies on the technical functionalities of devices in the environment and must therefore have the assurance that the devices are trustworthy. In ATRACO, we refer to a device as trustworthy in terms of reliability, compatibility, intended functioning, and functioning without unwanted or even malicious behavior. As a first step towards such trustworthiness, all available devices need to be authenticated when they connect to an ATRACO activity sphere. In addition, the technical reliability and compatibility with the ATRACO environment needs to be proven and certified by an ATRACO certificate authority which issues a digital certificate to a valid device.

As certification allows only binary trust assessment, the process should be extended by reputation mechanisms. This allows for finer trust assessment, on the one hand, and the inclusion and establishment of trust for devices not certified by an ATRACO CA, on the other hand. The most recent trust state of a device needs to be respected by ATRACO when devices are used in an activity sphere. The trust state should further influence device selection when more than one device of a specific category is available. Assessed trust states of devices should be visible to an activity sphere's user in an intuitive way. Further, the user should be able to influence

[1] http://www.ebay.com.

[2] http://www.amazon.com.

selection or usage of devices when trust states are not available or insufficient for automated selection.

In addition to technical trust considerations, we need to consider social trust assessment in ATRACO whenever an activity sphere includes more than one person. This is essential to ensure that the owner of an activity sphere retains complete control over it. Thus, access to user interfaces which can influence the state of an activity sphere need to be restricted to authorized, trusted users. To achieve such access control, a user needs to be supported in managing social trust groups and access control policies in an intuitive manner.

In summary, the following basic design requirements for realizing technical trust and social trust in ATRACO were defined:

- Device binding and usage must depend on technical trust decisions.
- Technical trust should be provided by digital certificates and signatures.
- Monitoring and reputation mechanisms should be integrated to refine binary trust assignments.
- A user should be aware of the trust states of active and involved components.
- Social trust must be respected in multi-user scenarios.
- A user should be able to manage and control access policies and trust groups.

4.3.4.2 Trust Manager

The Trust Manager (TM) is the central component for controlling and managing trust decisions in ATRACO. The TM consists of the *PKI Verifier*, the *Technical Trust Controller* (TTC), the *Social Trust Controller* (STC), the *Monitoring and Reputation Component* (MRC), and a *FCC*. Figure 4.9 provides an overview of the TM integration in the overall ATRACO architecture and its interactions with the Sphere Manager (SM), Ontology Manager (OM), Interaction Agent (IA), and Privacy Manager (PM). The interaction between TM and IA is managed by the shared IAC. The interplay of these different components will be discussed for technical trust decisions and social trust decisions, respectively.

Technical Trust Decisions

Technical trust decisions are made by the TM whenever a new activity sphere starts or a new component appears in an already running activity sphere. Upon the initialization phase of an activity sphere, the SM requests a trust verification for relevant components from the TM. Based on the computed trust states, the SM decides which of the components can be used to create the new activity sphere. Thus, an activity sphere will always be created with the most trusted components. The SM will request the same trust verification if a new component appears in an already existing activity sphere.

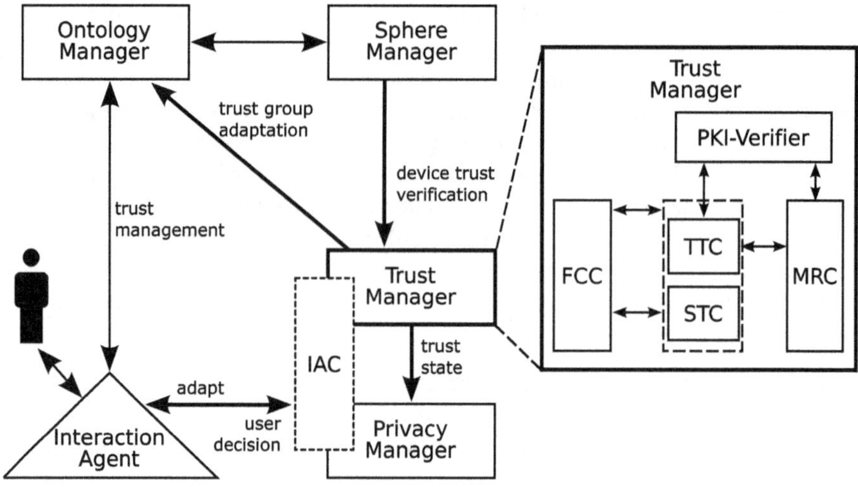

Fig. 4.9 Integration of the Trust Manager

The verification requests of the SM are processed by the TTC. First, the PKI-Verifier verifies the provided certificate of the given component. A certificate confirms the validity of several properties of the component, such as unique identifier, manufacturer, type, or serial number. In this way, the component is uniquely identifiable and can get assigned a reputation-based trust level by the MRC, even if the certificate was signed by a CA, which is untrusted or unknown in the context of ATRACO. However, to prevent Sybil attacks, the assignment of reputation-based trust must be made by trusted ATRACO components.

The MRC has a local database of monitored components, which stores statistical information needed to infer reputation metrics. This information includes the number of failures or incorrect behavior, as well as the duration of reliable functioning. The database can be further synchronized with an online repository to enable collaborative reputation gathering.

In the case of a new device which is both rated as untrusted by the PKI-Verifier and unable to get a trust level assigned by the MRC due to insufficient reputation information, the TTC will use the FCC to involve the user. The user may decide to use and monitor the new component in his current activity sphere, or may decide to deny the usage of the device. The FCC will receive this user decision from the IAC which in turn will trigger the IA.

The current trust state of a device can further support access control decisions of the PM for protecting territorial privacy.

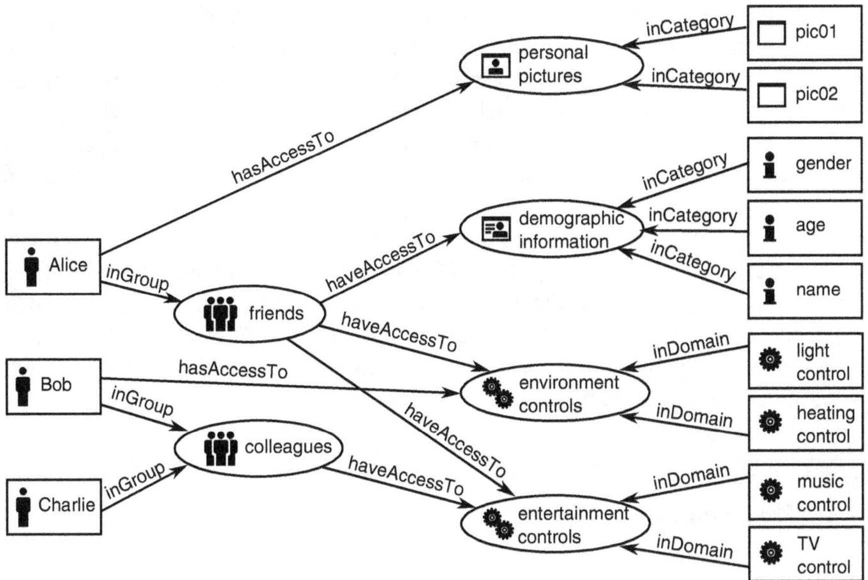

Fig. 4.10 Representation of trust groups and associated access control policies

Social Trust Decisions

A decision about social trust is needed whenever multi-user situations occur during the lifetime of an activity sphere. These decisions primarily support privacy protection and mechanisms for access control of user interfaces that offer control over activity sphere components. As mentioned in Sect. 4.3.2, we adopted the concept of trust groups to support the user in managing social trust preferences and, based on that, to create appropriate access policies. Figure 4.10 shows a possible arrangement of trust groups and access control policies, which are stored and provided as ontologies by the OM. In order to ease the management of access control policies, controls are organized in different control domains, such as *environment controls* or *entertainment controls*. In a similar way, sensitive information items can be organized in different categories, for instance *demographic information* or *personal pictures*.

The main task of the STC is locking or unlocking controls depending on social trust settings. For instance, if Charlie (as depicted in Fig. 4.10) is in physical range of the heating control, the STC will lock it because Charlie is not granted access to environment controls. For this scenario to work, the STC receives the location events of persons who are present from the SM. In addition, the IA registers each user interface belonging to a specific control domain via the IAC with the TM. Thus the STC always knows the location of a control and can lock or unlock it depending on each person's location.

Another task of the STC is to find new members of existing trust groups or to suggest new trust groups based on explicit user interactions. For instance, in Fig. 4.10, Bob does not have access to personal information. If the user of the activity sphere is viewing personal information on a screen when Bob enters the room, the information will be protected by the PM and thus disappear from the screen. However, if the user explicitly makes the information reappear in Bob's presence, the PM will create a new policy which grants Bob access to the information with the context constraint that the user is present. Whenever a new policy is created in such a way, the STC compares access policies of individuals with those of trust groups. If the policies match to a specific threshold, the STC will suggest adding the person to that group. In our example, Bob's individual access policies will match those of the trust group *friends*. As a consequence, the STC will suggest making Bob a member of the group friends.

4.4 Discussion

In this chapter, we have discussed privacy and trust in the context of ambient intelligent environments. While we introduced both terms, their definitions and involved mechanisms separately, we will now briefly elaborate on the relation between privacy and trust. After providing a discussion of the lessons learned we will close this chapter by outlining some of the major research challenges in this domain.

4.4.1 The Relation Between Privacy and Trust

Many researchers agree that the two concepts of privacy and trust cannot be addressed without respecting each other and often even depend on each other [32, 59].

From the privacy point of view, trust in entities can be utilized as a policy restriction on which privacy decisions are based. For example, a person might want to reveal his mobile number only to a trusted entity or to an entity of a certain level of trust. In this context, trust is needed to deal with the risk that privacy preferences will not be respected by the other entity. For example, a person registering at a web shop needs to trust that the shop owner will not forward any sensitive information to third parties. Indeed, the privacy decision of sharing information with someone else often overlaps with the intention and decision to trust that person or entity with the shared information [59]. Jutla and Bodorik [35] propose a model for online privacy that considers institutionalized trust, trust intentions, trust beliefs, and a predisposition to trust as factors influencing users' online behavior with regard to privacy.

On the other hand, the verified use of well-integrated privacy mechanisms can support trust establishment. If a person knows that an online shop applies adequate privacy mechanisms to protect sensitive user data, his trust in this shop will be higher than without these guaranteed privacy protection mechanisms. However, AmI systems that offer few to no privacy control mechanisms require a higher level of initial trust from the user without necessarily warranting it [59].

Some argue that privacy and trust are conflicting [62] and need to be balanced by ensuring minimal trade-in of privacy for the required trust. This perspective implies that the establishment of trust in an entity always comes with a degradation of its privacy, which might be true if trust is solely established by gathering information about an entity, as it is the case in most reputation-based systems. Some research exists that tries to find solutions for this problem. Pavlov et al. [56] present three protocols based on probabilistic schemes that allow partial privacy for feedback providers in decentralized reputation systems. Steinbrecher [69] proposed privacy-respecting design options for centralized reputation systems. However, if trust establishment between two parties is based on assessing former interactions between those parties, privacy is not necessarily affected.

4.4.2 Lessons Learned

While information privacy has been the dominating aspect of privacy research in recent years, next generation AmI environments will raise new demand for the consideration of the more traditional aspect of territorial privacy. The preliminary results of our user study with the ATRACO prototype have further motivated the need for privacy mechanisms that respect privacy implications of disturbances which could be caused by undesired interactions and automations.

The dynamic nature of users' privacy preferences and their dependability on contextual factors raise the need for privacy solutions which could dynamically adapt to the given situation. We showed that this goal can be achieved in ATRACO via reasoning about privacy preferences and context, which both are represented in the form of ontologies. We investigated these aspects further in the PriCal system. PriCal highlighted the utility of context-adaptive privacy mechanisms, that dynamic adaptations can be designed in a way that protects a user's privacy adaptation preferences from others, as well as the importance of considering the privacy implications of physical configurations of AmI systems.

Similar to privacy, different dimensions of trust need to be respected in AmI. We classified these dimensions into technical trust and social trust. The combination of credential-based and reputation-based trust establishment mechanisms facilitates dynamic trust level assessment for technical components. Social trust can be addressed by the concept of trust groups with support of implicit adaptation and creation of group arrangements.

The proposed privacy and trust concepts were integrated in ATRACO in the form of the Privacy Manager and Trust Manager components, respectively. The interplay of these components with the overall ATRACO architecture allows the creation of trustworthy and privacy preserving activity spheres in ambient intelligence environments.

4.4.3 Research Challenges

Privacy and trust aspects in AmI still hold several research challenges and open issues, which need to be solved in order to achieve a comprehensive protection of privacy and trust establishment in AmI environments. Our conceptual solution proposes the protection of territorial privacy by disabling observers and disturbers, e.g., an active camera. However, even if this is possible in a user-controlled environment such as the home, it drastically increases the complexity of the AmI system and raises new problems. For example, a system which depends on active cameras to support a user's activity will fail if cameras are suddenly disabled. Solutions need to be found that address this issue by allowing a dynamic trade-off between privacy and functionality. Further, in environments which are not under full control of the user, e.g., in public environments, disabling of devices might not be possible at all. Therefore, trustworthy privacy mechanisms need to be found that a user can rely on or even provide user empowerment to control privacy as far as possible. The establishment of trust in those mechanisms will be a key requirement for their success.

The dynamic creation and adaptation of privacy policies and trust is another important area for future work. In ATRACO, policies and trust groups can be adapted by explicit user interactions. However, an implicit adaptation depending on several contextual information and constraints can further improve the reliability of adaptation and provide a more fine-grained access control. The PriCal system demonstrated a first feasible approach to dynamically adapt privacy preferences based on learning from user interactions and similar situations with case-based reasoning. Future work needs to investigate how this approach can be applied to more complex scenarios with higher dynamics in privacy preferences, a larger number of users, and a larger number of different privacy implications.

As AmI environments are becoming more and more reality in our everyday lives, it is important to consider privacy and trust issues from the very beginning. If we have to trade our privacy for the benefit we get out of those environments, their successful establishment will likely fail.

4.5 Further Readings

Marc Langheinrich presents a good overview of privacy issues in ubiquitous computing and existing approaches for protecting privacy in his chapter in the book *Ubiquitous Computing Fundamentals* [45].

A comprehensive discussion of privacy problems in the context of legislation and a taxonomy of privacy is provided by Daniel J. Solove in his book *Understanding Privacy* [67].

Privacy International, the Electronic Privacy Information Center (EPIC) and the Center for Media and Communications Studies (CMCS) provide an overview of international privacy development in different states in their annual report *Privacy and Human Rights*.[3]

Christiano Castelfranchi and Rino Falcone provide an extensive overview of socio-cognitive and computational trust models in their book *Trust Theory* [16].

References

1. Abdul-Rahman, A., Hailes, S.: Supporting trust in virtual communities. In: Proceedings of the 33rd Annual Hawaii International Conference on System Sciences, vol. 1, 9 pp. (2000). doi:10.1109/HICSS.2000.926814
2. Aberer, K., Despotovic, Z.: Managing trust in a peer-2-peer information system. In: Proceedings of the 10th International Conference on Information and Knowledge Management, pp. 310–317. ACM, Atlanta, Georgia (2001). doi:10.1145/502585.502638. http://www.portal.acm.org/citation.cfm?id=502638
3. An, X., Jutla, D., Cercone, N.: A Bayesian network approach to detecting privacy intrusion. In: Proceedings of the 2006 IEEE/WIC/ACM International Conference on Web Intelligence and Intelligent Agent Technology, pp. 73–76. IEEE Computer Society (2006)
4. Ashley, P., Hada, S., Karjoth, G., Powers, C., Schunter, M.: Enterprise privacy authorization language (EPAL 1.2). Tech. Rep., W3C. http://www.w3.org/Submission/2003/SUBM-EPAL-20031110/ (2003)
5. Aztiria, A., Izaguirre, A., Augusto, J.C.: Learning patterns in ambient intelligence environments: a survey. Artif. Intell. Rev. **34**(1), 35–51 (2010). doi:10.1007/s10462-010-9160-3
6. Bellotti, V., Sellen, A.: Design for privacy in ubiquitous computing environments. In: Proceedings of the third conference on European Conference on Computer-Supported Cooperative Work, pp. 77–92. Kluwer Academic Publishers, Milan (1993)
7. Beresford, A., Stajano, F.: Location privacy in pervasive computing. IEEE Pervasive Comput. **2**(1), 46–55 (2003). doi:10.1109/MPRV.2003.1186725
8. Bhattacharya, J., Dass, R., Kapoor, V., Gupta, S.: Utilizing network features for privacy violation detection. In: Proceedings of the 1st International Conference on Communication System Software and Middleware, pp. 1–10 (2006). doi:10.1109/COMSWA.2006.1665184
9. Blaze, M., Feigenbaum, J., Lacy, J.: Decentralized trust management. In: Proceedings of the IEEE Symposium on Security and Privacy, pp. 164–173. IEEE (1996). doi:10.1109/SECPRI.1996.502679
10. Blaze, M., Feigenbaum, J., Strauss, M.: Compliance checking in the PolicyMaker trust management system. In: Financial Cryptography. Lecture Notes in Computer Science, vol. 1465, pp. 1439–1456. Springer, Berlin (1998)
11. Bohn, J., Coroama, V., Langheinrich, M., Mattern, F., Rohs, M.: Social, economic, and ethical implications of ambient intelligence and ubiquitous computing. In: Ambient Intelligence, pp. 5–29. Springer, Berlin (2005)
12. Bok, S.: Secrets: On the Ethics of Concealment and Revelation. Vintage, New York (1989)
13. Bonatti, P., Duma, C., Olmedilla, D., Shahmehri, N.: An integration of reputation-based and policy-based trust management. In: Proceedings of the Semantic Web Policy Workshop, Galway, Ireland (2005)

[3]http://www.privacyinternational.org/.

Used icons in graphics from http://www.picol.organdhttp://www.pictoico.org.

14. Canadian Law: Personal information protection and electronic documents act (PIPEDA). http://www.laws.justice.gc.ca/eng/P-8.6/ (2000)
15. Carbo, J., Molina, J.M., Davila, J.: Trust management through fuzzy reputation. Int. J. Coop. Inf. Syst. **12**(1), 135–155 (2003)
16. Castelfranchi, C., Falcone, R.: Trust Theory: A Socio-Cognitive and Computational Model, 1st edn. Wiley, New York (2010)
17. Chaum, D.L.: Untraceable electronic mail, return addresses, and digital pseudonyms. Commun. ACM **24**(2), 84–90 (1981). doi:10.1145/358549.358563
18. Chopra, K., Wallace, W.: Trust in electronic environments. In: Proceedings of the 36th Annual Hawaii International Conference on System Sciences, 10 pp. (2003). doi:10.1109/HICSS.2003.1174902
19. Cook, D.J., Augusto, J.C., Jakkula, V.R.: Ambient intelligence: technologies, applications, and opportunities. Pervasive Mob. Comput. **5**(4), 277–298 (2009)
20. Cranor, L., Dobbs, B., Egelman, S., Hogben, G., Humphrey, J., Langheinrich, M.: The platform for privacy preferences 1.1 (P3P1.1) specification. Tech. Rep., W3C. http://www.w3.org/TR/2006/NOTE-P3P11-20061113/ (2006)
21. Duckham, M., Kulik, L.: A formal model of obfuscation and negotiation for location privacy. In: Pervasive Computing, pp. 152–170. Springer, Berlin (2005)
22. Duckham, M., Kulik, L.: Location privacy and location-aware computing. In: Dynamic & Mobile GIS: Investigating Change in Space and Time, pp. 34–51. CRC Press, New York (2006)
23. Ellison, C., Schneier, B.: Ten risks of PKI: what you're not being told about public key infrastructure. Comput. Secur. J. **16**(1), 1–7 (2000). http://www.reocities.com/CapeCanaveral/orbit/7457/Downloads/pki-risks.pdf
24. EPIC, Privacy International: Privacy and Human Rights Report 2006: An International Survey of Privacy Laws and Developments, 2006 edn. Electronic Privacy Information Center (2007)
25. EU: 95/46/EC-Data protection directive. Official Journal of the European Communities (1995) http://eur-lex.europa.eu/legal-content/EN/TXT/?uri=CELEX:31995L0046
26. Gambetta, D.: Can we trust? In: Trust: Making and Breaking Cooperative Relations, Electronic edition, pp. 213–237. Blackwell Publishers, Department of Sociology, University of Oxford, Oxford (2000)
27. Gerck, E.: Overview of certification systems: X. 509, PKIX, CA, PGP & SKIP. The Bell **1**(3), 8 (2000)
28. Godik, S., Anderson, A., Parducci, B., Humenn, P., Vajjhala, S.: OASIS eXtensible access control 2 markup language (XACML) 3. Tech. Rep., OASIS (2002)
29. Grandison, T., Sloman, M.: A survey of trust in internet applications. IEEE Commun. Surv. Tutorials **3**(4), 2–16 (2000)
30. Gruteser, M., Grunwald, D.: Anonymous usage of location-based services through spatial and temporal cloaking. In: Proceedings of the 1st International Conference on Mobile Systems, Applications and Services, pp. 31–42. ACM, San Francisco, CA (2003)
31. Hong, J.I., Landay, J.A.: An architecture for privacy-sensitive ubiquitous computing. In: Proceedings of the 2nd International Conference on Mobile Systems, Applications, and Services, pp. 177–189. ACM, Boston (2004)
32. Iachello, G., Hong, J.: End-user privacy in human-computer interaction. Found. Trends Hum. Comput. Interact. **1**(1), 1–137 (2007). doi:10.1561/1100000004
33. Jiang, X., Hong, J., Landay, J.: Approximate information flows: socially-based modeling of privacy in ubiquitous computing. In: Proceedings of the 4th International Conference on Ubiquitous Computing, pp. 176–193. Springer, Berlin (2002)
34. Jusang, A., Ismail, R., Boyd, C.: A survey of trust and reputation systems for online service provision. Decis. Support Syst. **43**(2), 618–644 (2007)
35. Jutla, D., Bodorik, P.: Sociotechnical architecture for online privacy. IEEE Secur. Priv. **3**(2), 29–39 (2005). doi:10.1109/MSP.2005.50
36. Kamvar, S.D., Schlosser, M.T., Garcia-Molina, H.: Eigenrep: reputation management in p2p networks. In: 12th International World Wide Web Conference, Budapest, Hungary (2003)

37. B. Könings. User-centered awareness and control of privacy in Ubiquitous Computing. Doctoral dissertation. Faculty of Computer Science and Engineering, Ulm University, (2015). http://vts.uni-ulm.de/doc.asp?id=9603
38. Könings, B., Schaub, F.: Territorial privacy in ubiquitous computing. In: Proceedings of the 8th International Conference on Wireless On-Demand Network Systems and Services, pp. 104–108. IEEE (2011). doi:10.1109/WONS.2011.5720177
39. Könings, B., Schaub, F., Kargl, F., Weber, M.: Towards territorial privacy in smart environments. In: Intelligent Information Privacy Management Symposium. Stanford University, Stanford (2010)
40. Krukow, K., Nielsen, M., Sassone, V.: Trust models in ubiquitous computing. Philos. Trans. R. Soc. A **366**, 3781–3793 (2008)
41. Krumm, J.: A survey of computational location privacy. Pers. Ubiquit. Comput. **13**(6), 391–399 (2008). doi:10.1007/s00779-008-0212-5
42. Langheinrich, M.: Privacy by design: principles of privacy-aware ubiquitous systems. In: Proceedings of the 3rd International Conference on Ubiquitous Computing, pp. 273–291. Springer, London (2001)
43. Langheinrich, M.: A privacy awareness system for ubiquitous computing environments. In: Proceedings of the 4th International Conference on Ubiquitous Computing, pp. 237–245. Springer, London (2002)
44. Langheinrich, M.: When trust does not compute – the role of trust in ubiquitous computing. In: Workshop on Privacy, UbiComp '03, pp. 1–8 (2003)
45. Langheinrich, M.: Privacy in ubiquitous computing. In: Krumm, J. (ed.) Ubiquitous Computing Fundamentals, 1st edn. Chapman & Hall/CRC, New York (2009)
46. Li, N., Mitchell, J., Winsborough, W.: Design of a role-based trust-management framework. In: 2002 IEEE Symposium on Security and Privacy, 2002. Proceedings, pp. 114–130 (2002). doi:10.1109/SECPRI.2002.1004366
47. Lund, M.S., Solhaug, B., Stolen, K.: Evolution in relation to risk and trust management. Computer **43**(5), 49–55 (2010)
48. Margulis, S.T.: On the status and contribution of Westin's and Altman's theories of privacy. J. Soc. Issues **59**(2), 411–429 (2003). doi:10.1111/1540-4560.00071
49. Marsh, S.: Formalising trust as a computational concept. Ph.D. thesis, University of Stirling (1994)
50. McKnight, D.H., Chervany, N.: Trust and distrust definitions: one bite at a time. In: Trust in Cyber-Societies, pp. 27–54. Springer, Berlin (2001)
51. Moncrieff, S., Venkatesh, S., West, G.: Dynamic privacy assessment in a smart house environment using multimodal sensing. ACM Trans. Multimed. Comput. Commun. Appl. **5**(2), 1–29 (2008). doi:10.1145/1413862.1413863
52. Ni, Q., Bertino, E., Lobo, J., Calo, S.B.: Privacy-aware role-based access control. IEEE Secur. Priv. **7**(4), 35–43 (2009). doi:10.1109/MSP.2009.102
53. Nissenbaum, H.: Privacy in Context: Technology, Policy, and the Integrity of Social Life. Stanford University Press, Stanford (2009)
54. OECD: Guidelines on the Protection of Privacy and Transborder Flows of Personal Data. OECD Publishing, Paris (1980)
55. Ortmann, S., Langendörfer, P., Maaser, M.: Enhancing privacy by applying information flow modelling in pervasive systems. In: Proceedings of the OTM 2007 Workshops on the Move to Meaningful Internet Systems, pp. 794–803. Springer, Berlin (2007). doi:10.1007/978-3-540-76890-6_5
56. Pavlov, E., Rosenschein, J.S., Topol, Z.: Supporting privacy in decentralized additive reputation systems. In: Proceedings of the 2nd International Conference on Trust Management, pp. 108–119. Springer, Berlin (2004). doi:10.1007/978-3-540-24747-0_9
57. Perlman, R.: An overview of PKI trust models. IEEE Netw. **13**(6), 38–43 (1999)
58. Sabater, J., Sierra, C.: Review on computational trust and reputation models. Artif. Intell. Rev. **24**(1), 33–60 (2005). http://www.springerlink.com/content/rw03811201223550/

59. Schaub, F.: Dynamic privacy adaptation in ubiquitous computing. Ph.D. thesis, Faculty of Engineering and Computer Science, University of Ulm (2014)
60. Schaub, F., Könings, B., Lang, P., Wiedersheim, B., Winkler, C., Weber, M.: PriCal: context-adaptive privacy in ambient calendar displays. In: Proceedings of the International Joint Conference on Pervasive and Ubiquitous Computing. ACM (2014). doi:10.1145/2632048.2632087
61. Schillo, M., Funk, P., Rovatsos, M.: Using trust for detecting deceitful agents in artificial societies. Appl. Artif. Intell. **14**(8), 825–848 (2000)
62. Seigneur, J., Jensen, C.D.: Trading privacy for trust. In: Jensen, C., Poslad, S., Dimitrakos, T. (eds.) Trust Management. Lecture Notes in Computer Science, vol. 2995, pp. 93–107. Springer, Berlin/Heidelberg (2004)
63. Sierra, C., Debenham, J.: An information-based model for trust. In: Proceedings of the 4th International joint Conference on Autonomous Agents and Multiagent Systems, pp. 497–504. ACM, New York, NY (2005)
64. Smetters, D.K., Good, N.: How users use access control. In: Proceedings of the 5th Symposium on Usable Privacy and Security, pp. 1–12. ACM, Mountain View, CA (2009)
65. Smith, R.E.: Ben Franklin's web site: privacy and curiosity from Plymouth Rock to the internet. Priv. J. (2004) ISBN: 9780930072148
66. Solhaug, B., Elgesem, D., Stolen, K.: Why trust is not proportional to risk. In: Proceedings of the 2nd International Conference on Availability, Reliability and Security, pp. 11–18 (2007). doi:10.1109/ARES.2007.161
67. Solove, D.J.: Understanding Privacy. Harvard University Press, Cambridge (2008)
68. Song, S., Hwang, K., Zhou, R., Kwok, Y.K.: Trusted P2P transactions with fuzzy reputation aggregation. IEEE Internet Comput. **9**(6), 24–34 (2005)
69. Steinbrecher, S.: Design options for privacy-respecting reputation systems within centralised internet communities. Secur. Priv. Dyn. Environ. **201**, 123–134 (2006)
70. The Oxford English Dictionary: "Privacy" Definition, 2nd edn. Oxford University Press, Oxford (2005)
71. Tschannen-Moran, M., Hoy, W.K.: A multidisciplinary analysis of the nature, meaning, and measurement of trust. Rev. Educ. Res. **70**(4), 547–593 (2000). doi:10.3102/00346543070004547
72. Warren, S., Brandeis, L.: Right to privacy. Harv. Law Rev. **4**, 193–220 (1890)
73. Weiser, M.: Some computer science issues in ubiquitous computing. Commun. ACM **36**(7), 75–84 (1993). doi:10.1145/159544.159617
74. Westin, A.F.: Privacy and Freedom. Atheneum, New York (1967)
75. Wishart, R., Henricksen, K., Indulska, J.: Context privacy and obfuscation supported by dynamic context source discovery and processing in a context management system. In: Proceedings of the 4th International Conference on Ubiquitous Intelligence and Computing, Hong Kong, vol. 4611, pp. 929–940 (2007)
76. Xiong, L., Liu, L.: PeerTrust: supporting reputation-based trust for peer-to-peer electronic communities. IEEE Trans. Knowl. Data Eng. **16**(7), 843–857 (2004)

Chapter 5
Novel Approaches to Artefact Adaptation in Ambient Intelligent Environments

Aysenur Bilgin, Hani Hagras, and Christian Wagner

Abstract The paper presents advanced novel approaches to develop strategies that will allow the artefacts to adapt to the uncertainties associated with the changes in the artefacts characteristics, context as well as changes in the user(s) preferences regarding these artefacts in Ambient Intelligent Environments (AIEs). This work is within the framework of an EU funded project entitled ATRACO (Adaptive and Trusted Ambient Ecologies) which aims to contribute to the realization of trusted ambient ecologies in AIEs.

5.1 Introduction

Adaptation is a relationship between a system and its environment where change is provoked to facilitate the survival of the system in the environment. Biological systems exhibit different types of adaptation so as to regulate themselves and change their structure as they interact with the environment.

The dynamic and ad-hoc nature of Ambient Intelligent Environments (AIEs) requires the environment to adapt to changing operating conditions as well as changing user preferences and behaviours while enabling more efficient and effective operation and avoiding any system failure. Thus, there is a need to provide robust mechanisms that will allow handling the varying and unpredictable conditions associated with the dynamic environment and user preferences. In other words, there is a need to provide autonomous intelligent adaptive techniques which should be able to create models that can be evolved and adapted online in a life-long learning mode over short- and long-term intervals. These models need to be transparent and easy to be interpreted by an ordinary user. Besides, these intelligent systems should

A. Bilgin (✉) • H. Hagras • C. Wagner
The Computational Intelligence Centre, School of Computer Science and Electronic Engineering, University of Essex, Wivenhoe Park, Colchester CO43SQ, UK
e-mail: abilgin@essex.ac.uk; hani@essex.ac.uk; christian.wagner@nottingham.ac.uk

© Springer International Publishing Switzerland 2016

S. Ultes et al. (eds.), *Next Generation Intelligent Environments*,
DOI 10.1007/978-3-319-23452-6_5

allow controlling the environment on the user behalf and to his/her satisfaction. In addition, the intelligent approaches should allow for real-time data mining of the user data and create on-the-fly adaptive models of the user preferences while effectively operating on the embedded hardware platforms present in the everyday environments which have limited memory and processor capabilities.

This chapter will present novel adaptation strategies that will allow the artefacts to adapt to the uncertainties associated with the changes in the artefacts characteristics, context as well as changes in the user(s) preferences regarding these artefacts in AIEs.

Section 5.2 presents an overview of the previous work in adaptation in AIEs. Section 5.3 gives the ATRACO project approach to Artefact Adaptation in AIEs. Section 5.4 provides an overview of the employed Artefact Adaptation approaches while Sect. 5.5 provides an overview on the Artefact Adaptation design in AIEs. Section 5.6 presents the experiments and results. In Sect. 5.7, we introduce a novel user-centred approach for adaptive modelling in AIEs. Finally, Sect. 5.8 presents the conclusions and future work.

5.2 Previous Work in Artefact Adaptation in AIEs

Several works have targeted the topic of adaptation in Ambient Intelligent Environments (AIEs) where Davidsson [9] has developed a multi-agent system (MAS) that monitors and controls an office building in order to meet user preferences while conserving energy.

The Oxygen at MIT [19] centres around two rooms containing cameras, microphones, an X-10 controlled lighting system and a multitude of computer vision and speech understanding systems that help the system interpret what people are saying, where they are saying it and what interactions and activities are taking place. The system responds accordingly using speech synthesis when spoken to. The vision system is able to intelligently train itself for a particular environment in less then 5 min using projectors displaying a simulation of someone performing the training. The agents operate in a rather independent intelligence level where each sensor resides in a particular place and uses various local resources.

The goals of the Aware Home [16] agents are to investigate what kind of services can be built on top of an environment that is aware of the activities of its occupants. Besides building models of human behaviour to aid computers in decision making, it aims to support older individuals to "age in place" by helping them to maintain a good diet, take medication when required, notify family members of their well-being and provide support with everyday tasks. The large number of sensors deployed in the Aware Home range from trip sensors (that detect when a door has been opened or closed) and motion detectors, to higher-fidelity sensors, e.g. embedded microphones and cameras. These sensors, all centrally connected to a computer, are mostly used to determine the current activities, facial expressions, locations and gestures of occupants. The sensors and actuators are fixed and dedicated to performing specific tasks only [1].

The Adaptive Home, a.k.a. the Neural Network Home, explored the "learning user's habits" aspects of an AIE. It aimed to predict the occupancy of rooms, hot water usage and the likelihood that a zone is entered in the next few seconds using trained feed-forward neural networks. The context information in the project was again mainly comprised of location, but additional state information from rooms like the status of lights or the temperature set by inhabitants were used. Although learning and prediction were done via feed-forward multi-layer perceptrons with the known limitations, it showed that the prediction of user locations can help to save resources and support users by learning their behaviour and automating simple tasks [7].

The UT-AGENT at Ajou University in Korea focuses on the use of intelligent agents for ubiquitous smart home environments. An intelligent agent model for AIEs is proposed in which the agents must learn the user preferences, which are represented by user profiles, in order to assist them. The proposed UT-AGENT uses case-based reasoning to acquire knowledge about user actions that are worth recording to determine their preferences and a Bayesian Network as an inference tool to model the relationships between them. The UT-AGENT maintains the status of every device present in the home and activates them according to user preference. It generates a sequence of the expected user's query and simultaneously activates the devices with preset preference settings [17].

The MavHome smart home project focuses on the creation of an environment that acts as an intelligent agent, perceiving the state of the home through sensors and acting upon the environment through device controllers [27]. The environment is represented using a Hierarchical Hidden Markov Model and a reinforcement learning algorithm is employed to predict the environmental preferences based on sensors within the environment. Desired actions are proposed for the control of lights within the environment primarily based on motion detection sensors and if the actions are within the bounds of acceptable safety and security policies, they are invoked within the environment. The agent aims to maximize the comfort and productivity of its inhabitants while minimizing the energy consumption.

In the University of Essex, the Incremental Synchronous Learning (ISL) approach was developed for embedded agents to realize Ambient Intelligence (AmI) in ubiquitous computing environments [14]. In [12], the AOFIS system was developed to enhance the capabilities of the ISL agents by including mechanisms to extend the original agent-based approach to use type-2 fuzzy systems [15]. The type-2 agent generated type-2 Fuzzy Logic Controllers (FLCs) in which type-2 membership functions (MFs) with fixed uncertainties directly modelled and handled the long-term environmental uncertainties. Type-2 embedded agents facilitate short-term online adaptation of the rules while being robust to long-term environmental changes. The agent can incrementally adapt the type-2 fuzzy rules and MFs in a life-long learning mode, accommodating for the accumulated long-term uncertainties arising from seasonal changes in the environment and the associated changing user behaviour over extended periods of time [15].

Most of the above systems did not manage to allow the artefacts to adapt to the uncertainties associated with the changes in the artefact characteristics, context as well as changes in the user(s) preferences regarding these artefacts in Ambient Intelligent Environments (AIEs). Hence, this has motivated us to focus on developing advanced adaptive modelling techniques to handle such uncertainties present in AIEs.

5.3 ATRACO Approach to Artefact Adaptation in AIEs

To our knowledge, no other work has targeted the Artefact Adaptation (AA) in AIEs where AA deals with developing strategies that allow the artefacts to adapt to the uncertainties associated with the changes in the artefact characteristics, context as well as changes in the user(s) preferences regarding these artefacts.

AIEs face huge amount of uncertainties which can be categorized into environmental uncertainties and users' uncertainties. The environmental uncertainties can be due to:

- The change of environmental factors (such as the external light level, temperature, time of day) over a long period of time due to seasonal variations.
- The change of the sensors and actuators outputs due to the noise from various sources. In addition, the sensors and actuators can be affected by the conditions of observation (i.e. their characteristics can be changed by the environmental conditions such as wind, sunshine, humidity, rain, etc.).
- Uncertainties associated with the change in the context and operation conditions.
- Wear and tear which can change sensor and actuator characteristics.

Thus, the environmental uncertainties are associated with the change in artefact characteristics and change in the artefact context.

The user uncertainties can be classified as follows:

- Intra-user uncertainties which are exhibited when a user decision for the same problem varies over time and according to the user location and activity. This variability is due to the fact that the human behaviour and preferences are dynamic and they depend on the user's context, mood, and activity as well as the weather conditions and time of year. For the same user, the same words can mean different things on different occasions. For instance the values associated with a term such as "warm" in reference to temperature can vary as follows: depending on the season (for example, from winter to summer), context (in the Arctic 15 °C might be considered "high", while in the Caribbean it would be considered "low"), depending on the user activity within a certain room and depending on the room within the user home and many other factors.
- Inter-user uncertainties which are exhibited when a group of users occupying the same space differ in their decisions in a particular situation. This is because users have different needs and experiences based on factors such as age, sex, profession, etc. For instance, the users might disagree on issues such as how "warm" a room should be on any given day.

Thus, the user(s) uncertainties are associated with the changes and variations in the user(s) preferences regarding the artefacts and their operation.

Therefore, it is crucial to employ adequate methods to handle the above uncertainties to enable the system to produce the desired behaviour to perform a given task. In addition, there is a need to produce models of the users' particular behaviours that are transparent and that can be adapted over long time duration and thus enabling the control of the users' environments on their behalf. Fuzzy Logic Systems (FLSs) are credited with being adequate methodologies for designing robust systems that are able to deliver a satisfactory performance when contending with the uncertainty, noise and imprecision attributed to real-world settings [10]. In addition, FLSs provide a method to construct controller algorithms in a user-friendly way closer to human thinking and perception by using linguistic labels and linguistically interpretable rules. Hence, FLSs can satisfy one of the important requirements in AmI systems by generating transparent models that can be easily interpreted and analysed by the end users. Moreover, FLSs provide flexible representations which can easily be adapted due to the ability of fuzzy rules to approximate independent local models for mapping a set of inputs to a set of outputs. As a result, FLSs have been used in AIEs spaces as in [10, 15, 20].

5.4 An Overview of the Employed Artefact Adaptation in ATRACO

There are many frameworks for dealing with uncertainty in decision-making including those based on probability and probabilistic (Bayesian) reasoning. However, in our experience, fuzzy methods have proven to be more effective than other methods when applied in AmI spaces [10, 15, 20]. This is because the fuzzy methods provide a framework using linguistic labels and linguistically interpretable rules which are very important when dealing with human users.

Recently, type-2 FLSs, with the ability to model second order uncertainties, have shown a good capability of managing high levels of uncertainty. Type-2 FLSs have consistently provided an enhanced performance compared to their type-1 counterparts in real-world applications [8, 15]. A type-2 fuzzy set is characterized by a fuzzy membership function, i.e. the membership value (or membership grade) for each element of this set is a fuzzy set in [0,1], unlike a type-1 fuzzy set where the membership grade is a crisp number in [0,1] [18]. There are two variants of type-2 fuzzy sets—interval type-2 fuzzy sets (IT2 FS) and general type-2 fuzzy sets (GT2 FS). In an IT2 FS, the membership grades are within an interval in [0,1] and the third dimension value equals unity (1). On the other hand, in a GT2 FS, the membership grade of the third dimension can be any function in [0,1]. Most applications to date use IT2 FS due to its simplicity. It has been shown that IT2 FS-based type-2 FLSs can handle the environmental uncertainties and the uncertainties associated with a single user in a single room environment and that type-2 FLSs can outperform their

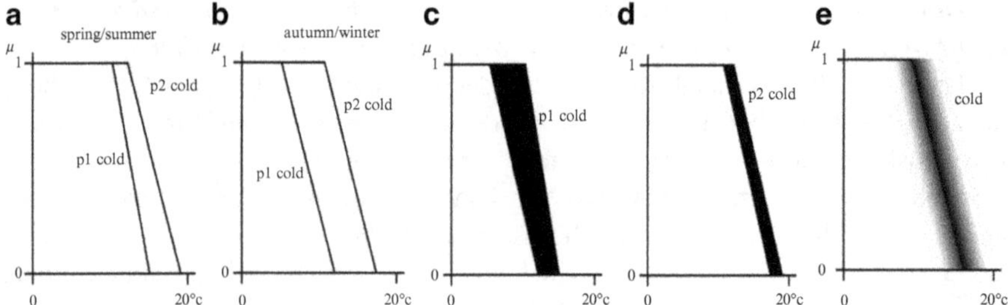

Fig. 5.1 Fuzzy sets modelling the linguistic label cold. (**a**) Type-1 fuzzy systems to depict the differences between two users (p1 and p2) for the concept of "cold" in spring/summer (**b**) Type-1 fuzzy systems to depict the differences between two users (p1 and p2) for the concept of "cold" in autumn/winter periods (**c**) Interval type-2 fuzzy system for the concept of "cold" throughout the year for user p1 (**d**) Interval type-2 fuzzy system for the concept of "cold" throughout the year for user p2 (**e**) General type-2 fuzzy system for the concept of "cold" where the different grey levels correspond to different membership levels in the third dimension

type-1 counterparts [15]. However, no work has tried to approach the challenging area of developing AIEs that can handle the environmental uncertainties as well as the intra- and inter-user uncertainties.

Consequently, in ATRACO, we employ type-2 fuzzy logic to model the uncertainties in AIEs. Consider an example of a central heating system for a living space occupied by two users. In such a situation, each user's concept of "cold" has to be modelled throughout the year. There will be seasonal variations affecting each user's idea of what is "cold", which poses an example of intra-user uncertainty. Each individual user will have his notion of what temperatures constitute "cold", which poses an example of inter-user uncertainty. Modelling either of these uncertainties can be accomplished using a number of existing techniques. The novel challenge with this is to model and cope with the uncertainties created by the relationship between the dynamic interactions of multiple users, each having individual preferences that change over time. For example, Fig. 5.1a, b show the use of type-1 fuzzy systems to depict the differences between two users (p1 and p2) for the concept of "cold" in spring/summer and autumn/winter periods. Figure 5.1c, d show how interval type-2 fuzzy sets might model each user's concept of "cold" throughout the year. Figure 5.1e shows how a general type-2 fuzzy set might encompass both the inter-user and intra-user uncertainties about what "cold" is by employing the third dimension, where the different grey levels correspond to different membership levels in the third dimension.

However, till recently, general type-2 fuzzy systems were perceived to be computationally expensive that they do not have any real applications within real-time systems such as AIEs. However, since the start of ATRACO, University of Essex has developed novel theoretical models which allowed the realization of general type-2 fuzzy logic on real-time embedded systems [2, 6, 21–24].

Each task in ATRACO is associated with a series of devices which are related to this task; for example, in terms of "adjusting the light to comfortable levels", this could be a series of lights, light sensors, dimmer switches, etc. AA is achieved by

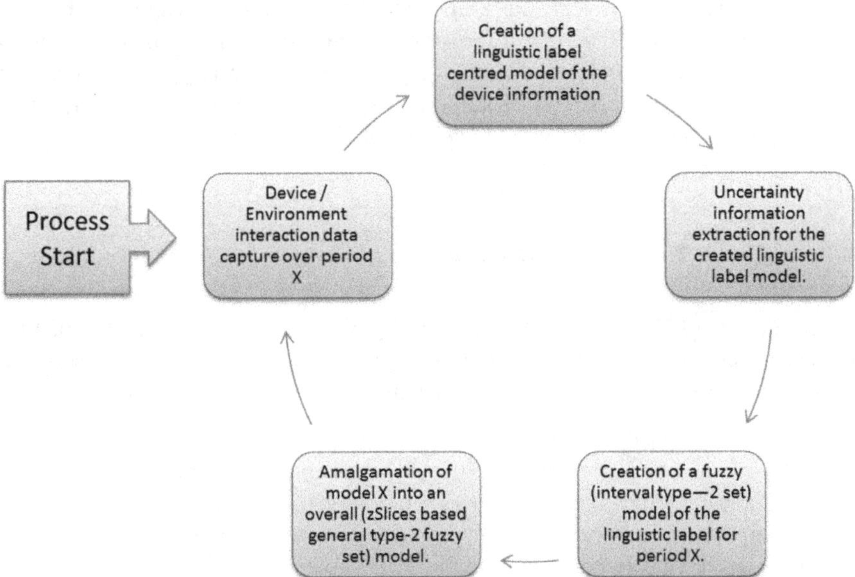

Fig. 5.2 Artefact Adaptation stages

associating each task with a specific Fuzzy Task Agent (FTA) which governs the AA for this task. Figure 5.2 shows the aims of the AA which are individually described below in a clockwise fashion starting from "Process Start".

- Device/Environment interaction data capture over period X:
  A main component of AA is continuous data capture from the devices (artefacts) in both times when a user is present and when the AIE is unoccupied. The data is captured in a continuous fashion and processed in discrete intervals (e.g. days).
- Creation of a linguistic-label-centred model of the device information:
  Devices/artefacts are considered less in the sense of physical, individual devices, but more in the sense of their utility/function in the users' point of view. As an example, the current chapter will focus on the linguistic-label-centred model "Ambient Light Level" which reflects the users' perception of the indoor light levels within an AIE and which itself is based on the information of a series of individual devices—in this case light sensors—present in the space.
- Uncertainty information extraction for the created linguistic-label-centred model:
  The third step focuses on the extraction of the uncertainty information related to the established linguistic-label-centred model. In other words, AA aims to identify the uncertainty resulting from the incorporation of data from multiple devices (e.g. light sensors) into one global model (e.g. Ambient Light Level) over a certain period of time (e.g. X = 1 day).
- Creation of a fuzzy (interval type-2) model of the linguistic labels for period X:
  The acquired information on the uncertainty distribution within the captured data is employed to generate interval type-2 fuzzy sets that represent the linguistic-

label-centred model (e.g. Ambient Light Level) over the capturing time X (e.g. 1 day). In other words, information about the linguistic interpretation of the model (provided ahead of time as type-1 fuzzy set prototypes) is fused with the device-based uncertainty captured in the previous step to generate an overall model over the period X for the linguistic-label-centred model.

5.5 Artefact Adaptation Design

In the design of AA, we have chosen fuzzy sets as the basis for the adaptive models of the artefacts/devices within the environment. Fuzzy sets have a series of capabilities which make them ideal for the application within AA, specifically:

- Fuzzy sets are easily adaptable.
  As fuzzy sets can take a wide variety of shapes and forms (triangular, gaussian, etc.), the parameters of which are easily modified, fuzzy sets are easily adaptable to incorporate new information and provide a very flexible representation for AA.
- Fuzzy sets are easily interpretable.
  Fuzzy sets have been designed to model and reflect human reasoning and are generally connected to linguistic labels. For example, a certain device (e.g. a lamp) can be represented using a linguistic variable understandable to human users (e.g. light level) which in turn can be mapped to a series of linguistic labels which are familiar terms for the user (e.g. dim, bright, etc.). Each of these labels can be modelled using a fuzzy set.
- Fuzzy logic sets allow for the modelling and handling of uncertainty.
  Fuzzy logic sets are accepted to have great potential in dealing with real-world uncertainty and have been employed in that regard in a wide variety of applications. Further, recent advancements in fuzzy logic theory, (specifically type-2 fuzzy logic theory) have shown great potential in further advances in dealing with uncertainty [2–6, 15, 18, 21–24].
- Fuzzy logic sets allow for a seamless integration of the AA into one transparent agent which is referred to as Fuzzy Task Agent (FTA) within the ATRACO project.

The different qualities of fuzzy sets are essential to the successful deployment of AA and as a part of ATRACO. Therefore, we will investigate both the deployment of existing fuzzy set models (i.e. type-1 fuzzy logic sets) as well as to research the application of higher order fuzzy sets such as interval type-2 fuzzy systems [18], zSlices-based general type-2 fuzzy sets [25] and the recently introduced Linear General Type-2 Fuzzy Sets [2, 3, 6].

In the following subsections, we will give detailed information regarding the stages involved in the design of the AA as well as the implementation of the AA.

5.5.1 *The Medium/Long-Term Capture of Device/Environment Interaction Data*

The initial approach within the first 2 years of ATRACO allowed the capturing of user/device interactions at runtime. However, it was limited to the capturing of data to the specific times when a user was present and active within the AIE. This "contact time" between the system and a user is often limited (e.g. the user is at work for most of the day) and as part of ATRACO it is directly limited through the relatively short periods of real-world experiments. Further, it is not only the immediate interaction between the devices and the user that yields crucial information about the artefacts/devices (and how the system can adapt to changes in their characteristics). The approach of capturing more general medium and long-term-based information addresses previously ignored factors for artefact adaptation such as outside influences (changes in weather, seasons, etc.) as well as the complexities associated with a low number of user/device interactions which may not be sufficient to accumulate enough data to generate an adequate model for AA.

In order to enable the medium to long-term capturing of user/environment interaction data in the final year prototype of ATRACO, a time-based artefact-information capturing approach is employed, i.e. the information gathered from/available through the physical devices within the AIE (i.e. the iSpace) are captured 24 h, 7 days a week in predefined regular intervals (chosen to be an interval of 6 min). This approach enables the capturing of direct user/device interaction information (when the user is present) as well as indirect interaction information (e.g. the user has covered a sensor by obstructing it) and user-independent device/environment interaction information (e.g. changes in weather, etc.).

5.5.2 *Design of a Human-Centred Model of the Gathered Device Data*

AA in the final year prototype aims to further narrow the gap between the user of an AIE and the actual system. In order to achieve this, individual devices/artefacts are grouped according to their role or the nature of the information that they produce or consume. The grouping itself is motivated and guided by common human concepts (such as humidity, light level, temperature, volume, etc.) which are in turn employed as linguistic labels for the models created for each individual group.

While the conceptual view of this approach is depicted in Fig. 5.3a, an example of this approach is the grouping of the light sensors in an AIE according to the role of the information they produce: information about the light level at the location of the light sensor. By grouping the individual light sensor devices which provide localized light level information, we are able to construct a model about the more general light

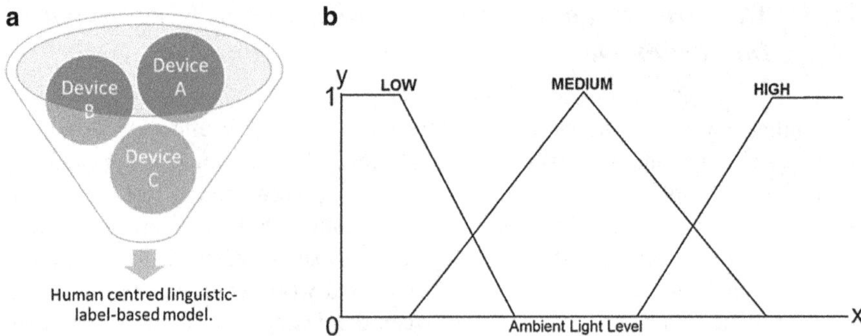

Fig. 5.3 (**a**) From individual device information to a human-centred linguistic-label-based model (**b**) A sample type-1 fuzzy set prototype model

level (for example, within the living room), a common human concept which we refer to and label as "Ambient Light Level". Incidentally, the concept is potentially applicable and useful in abstraction of information in a variety of areas in AIEs, for example, "Energy Consumption", rather than meter readings for individual gas, electricity and oil, etc.

In order to employ such human-centred concepts as models, we rely on the human interpretable feature of fuzzy sets and systems which allow the modelling of variables using a series of linguistic labels (e.g. low, medium, high for brightness). Thus, as part of AA, we employ type-1 fuzzy sets to establish human-centred linguistic-label-based prototype models of human concepts such as Ambient Light Level. An example of such a model—which we refer to as *type-1 prototype model* is depicted in Fig. 5.3b. A type-1 prototype model is designed based on expert knowledge. Subsequently, it is augmented with information gathered on the uncertainty resulting from associating the created model with a series of real-world data sources. This augmentation step of the initial type-1 prototype model is based on employing interval type-2 fuzzy sets which allow the modelling of the uncertainty related to the "distilling" of information from multiple devices into one human-centred model. Further details of this approach are presented in the following subsection.

5.5.3 *Modelling of the Uncertainty Encompassed in the Human-Centred Model for a Specific Timeframe*

Amalgamating the information from multiple (potentially heterogeneous) devices into one human-centred linguistic-label-based model makes it essential to employ a model that can represent the uncertainty resulting from the amalgamation of the different data sources (inter-device uncertainty) as well as the uncertainty associated with each device (intra-device uncertainty). Figure 5.4a depicts an example interval

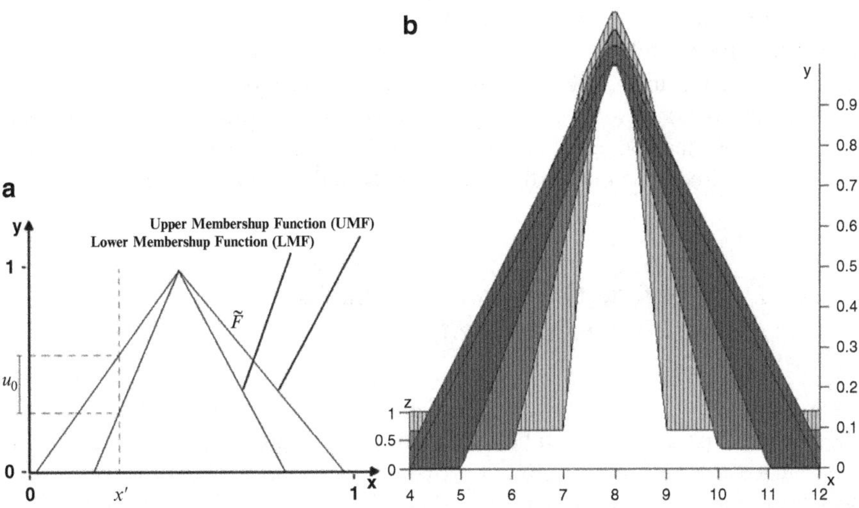

Fig. 5.4 (**a**) An example of an interval type-2 fuzzy set (**b**) An example of a zSlices-based general type-2 fuzzy set used to model agreement (here based on three interval type-2 fuzzy sets)

type-2 fuzzy set that can help to illustrate the use of interval type-2 fuzzy sets in order to create human-centred models from information gathered from a series of devices. As the information is gathered from multiple devices (e.g. the light level sensors), a merging or amalgamation of the information is required. One approach could, for example, be taking the average of the different values retrieved from the individual sensors, resulting in an "average light level" which can consequently be employed to inform a human-centred model for "Ambient Light Level". However, as the information gathered from multiple devices is generally not identical (i.e. some sensors might have different properties, some may be in the shadow, etc.), it is necessary to model the uncertainty created by amalgamating the individual sources of information in a different way.

Considering Fig. 5.4a, the x-axis of an interval type-2 fuzzy set can represent the amalgamated information related to the human-centred model (e.g. the ambient light level). On the other hand, the degree of membership in the fuzzy set is represented by an interval formed as a result of the Upper and Lower Membership Functions which form the bounds of the interval type-2 set. It is the difference between these two membership functions which directly relates to the intervals produced as degrees of membership for a given input that encodes the amount of uncertainty associated with that given input. In other words, if a group of devices (e.g. light sensors) is employed and modelled using a human-centred linguistic-label-based interval type-2 fuzzy set, areas of this model where the information gathered from all devices is equal will result in a very small interval—in fact a single point of membership, whereas areas where the information from the individual devices vary significantly will result in degrees of membership associated with a high level of uncertainty, that is, a large interval.

While employing interval type-2 fuzzy sets can indeed enable the modelling of the uncertainty related to a human-centred model which is based on a series of information sources (devices), the resulting model is directly based on the information (data) captured over a specific timeframe. In order to adapt the model in a continuous fashion over time (as new data becomes available), a more complex model is required which we will introduce in the following subsection.

5.5.4 Design of the Continuous Adaptation of the Human-Centred Model

In order for the continuous adaptation of the human-centred model to be realized, the interval type-2 fuzzy set model referred to in the previous section requires to be changed or extended to allow for a continuous model adaptation mechanism which reinforces persistent states while not disregarding new or irregular states. For example, if an interval type-2 fuzzy set-based model is created for the human-centred model of "Ambient Light Level" for the living room of an AIE based on information gathered for a single day of the week, then this model should be refined with the new information that becomes available on the following day. If, for example, a single sensor is obstructed (e.g. by a piece of clothing) for 1 day, this change in information should not fundamentally change the model; however, it should start influencing it in a limited fashion. As the change persists over days, the influence grows gradually until the change is fully integrated into the model.

As part of ATRACO we have developed what is referred to as *agreement* for zSlices-based general type-2 fuzzy sets which allows for the agreement to be computed over a series of interval type-2 fuzzy sets. As part of AA, each of these interval type-2 fuzzy sets reflect the human-centred model of a specific concept such as Ambient Light Level as created from the data collected from individual devices over a timeframe X (for example, a day). As time progresses and new data become available, several of these interval type-2 fuzzy sets are created (e.g. one per day). Finally, the overall human-centred linguistic-label-based model is created by computing the zSlices-based general type-2 fuzzy set that represents the agreement of the individual interval type-2 fuzzy set models. As new interval type-2 fuzzy set models become available, old ones can be disregarded, thus ensuring a both continuous and stable adaptation to changes in the characteristics of the artefacts/devices. Figure 5.4b depicts an example of a zSlices-based general type-2 fuzzy set employed to model agreement on the basis of three individual interval type-2 fuzzy sets.

Furthermore, we have developed a sliding window approach which takes a whole week (7 days) into consideration in order to construct a final model. Specifically, daily models to be included in the overall agreement-based model are selected based on a sliding window approach as shown in Fig. 5.5, assuring that the model continuously adapts to changes in the artefact/environment interactions. The daily

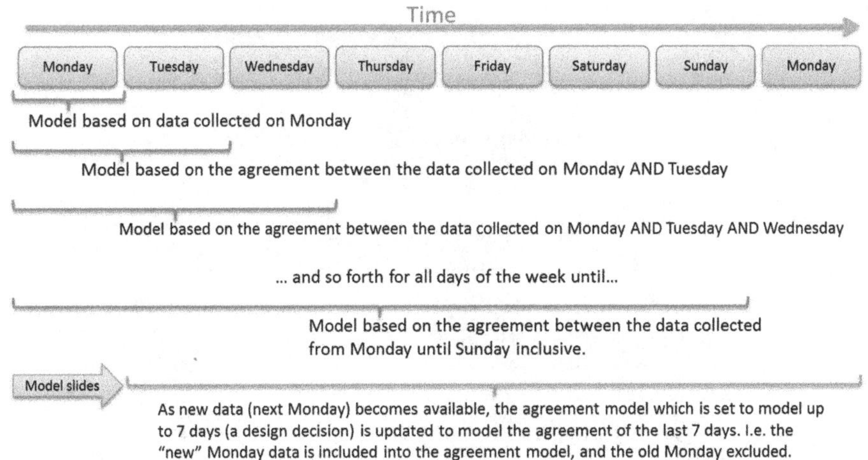

Fig. 5.5 Sliding window-based continuous adaptation in AA

models were found to be the most appropriate in our experiments, however, for other AIEs, this sliding window interval can vary. It should be noted that the increase in the window interval requires an increase in the number of zSlices used. Hence, the designers of the systems need to make an appropriate decision accounting for the tradeoff between the complexity of the model (higher numbers of zSlices may require higher computational power) and the performance of the model.

In the following subsection, we will provide details on the actual implementation of AA, as they were deployed in the final year prototype in ATRACO.

5.5.5 *Artefact Adaptation Implementation*

In this section, we will present the overall information flow in the final year AA component implementation while providing brief descriptions of the individual stages implemented within the AA and importantly, the position of AA within ATRACO. The internal stages of AA are depicted in Fig. 5.6, all of which will be briefly reviewed in the subsections below.

All stages of the AA are implemented using Java as the programming language which is widely employed and provides the important advantage of resulting in software which is executable on a wide variety of platforms (Linux, Windows, Mac…). Communication with the devices is via the well-established UPnP framework, which has been implemented using the Youpi platform, developed as part of ATRACO. Further advantages include the support for multi-core and distributed computing as well as a large amount of freely available open-source software packages allowing simple functionality expansion.

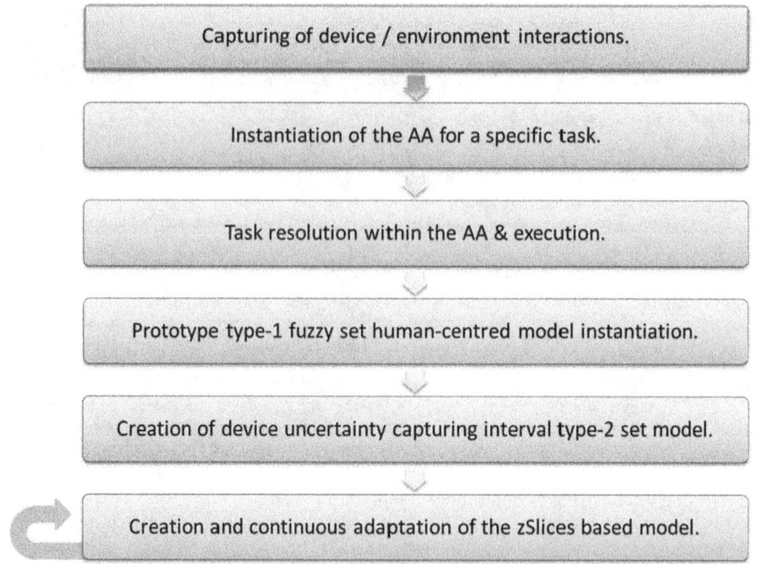

Fig. 5.6 Individual stages within the AA implementation of the final year prototype

```
#uuid = service action argument

cc2fd9ad-94f1-4c1e-a852-ed482116b09f = urn:iwg-org:serviceId:analogue:1 GetValue ResultStatus
41a9d016-66f4-4f16-9b35-9b6eca6eeada = urn:iwg-org:serviceId:analogue:1 GetValue ResultStatus
8d2551fd-7532-4736-8ad4-59e8150468d8 = urn:iwg-org:serviceId:analogue:1 GetValue ResultStatus
659862e5-62b3-48b1-888d-30e8f2760966 = urn:iwg-org:serviceId:analogue:1 GetValue ResultStatus
9c246b10-cf96-11de-8a39-0800200c9a66 = urn:iwg-org:serviceId:analogue:1 GetValue ResultStatus
ce3fd139-427d-490d-acfe-615968eba1c9 = urn:iwg-org:serviceId:analogue:1 GetValue ResultStatus
40f727d2-9d7f-4067-9db6-2c6b7e323860 = urn:iwg-org:serviceId:analogue:1 GetValue ResultStatus
4dd0c713-d560-4bb6-8eeb-a985fcb4ec7b = urn:iwg-org:serviceId:analogue:1 GetValue ResultStatus
```

Fig. 5.7 Logging Component—example of the specification of devices to be logged (i.e. light sensors in the living area in the iSpace)

5.5.5.1 Capturing of Device/Environment Interactions

The first stage within AA has been implemented as an independent subcomponent (which we will refer to as "Logging Component") of the FTA which enables it to be running in the background even when no FTA is currently active. The Logging Component is executed with the AIE (the iSpace) and dynamically starts monitoring and logging information of a set of specified devices. Figure 5.7 shows an example device selection which instructs the Logging Component to start the logging process of all light sensors within the living room area which are identified using their UPnP unique identifiers. The actual information to be gathered from the specified devices is communicated to the Logging Component by defining the respective UPnP parameters (service name, action and argument) for the devices as can be seen in Fig. 5.7.

The actual capturing interval is set at the execution time of the Logging Component and is identical for all devices within the AIE. In other words, all devices are queried at the same time (respecting network and synchronization constraints), thus providing a coherent snapshot of the environment at the device level at that particular time. The information captured is date and time-stamped and stored for further evaluation by the AA component as will be described in the following subsections.

5.5.5.2 Instantiation of the AA for a Specific Task

The AA is instantiated in the context of a given sphere and a given task as part of the FTA by the Sphere Manager. Specifically, the FTA (and thus AA) is provided with a series of handles which allow it to communicate with the specific devices in question. The Sphere Manager groups the devices passed to the FTA into three distinct categories as follows:

- Sensor inputs: sensors available within the environment, for example, light sensors, temperature sensors, humidity sensors, etc.
- User input devices: the user input devices are abstracted to a virtual "user will" device which is generated by the Interaction Agent. It is the interaction agent that subsequently communicates the users' will (which the user expresses through any interface such as voice, touch…) through the Sphere Manager to the FTA.
- Output devices: all devices which actively affect the environment, for example heaters, lights, automated window blinds, etc.

5.5.5.3 Task Resolution Within the AA and Execution

A task, which can involve a large number of devices, is resolved within AA in the following fashion:

1. All sensor devices are grouped in preparation for the amalgamation of their data as part of a human-centred model.
2. Similarly, output devices are grouped to create a single output device. (If they have not been previously grouped by the AIE/Sphere Manager). It should be noted that, as part of a specific task assigned to an FTA, the planning component ensures that outputs are compatible. For example, one task would generally not include a heating device as well as a ceiling lamp; both devices would be part of separate FTA (and thus AA) tasks.
3. The input and output groups are associated with the virtual "User Will" device as part of the FTA.

After all the artefacts/devices have been resolved, the respective device groups are associated with type-1 fuzzy set prototype models as described in the following subsection.

5.5.5.4 Prototype Type-1 Fuzzy Set Human-Centred Model Instantiation

Overview

As the generation of a prototype type-1 fuzzy set human-centred model involves a series of steps, we start by providing an overview followed by detailed explanations of the individual steps in corresponding subsections.

Each group of devices is associated with a predefined type-1 fuzzy set model which is based on basic knowledge about the specific human-centred concept. For example, for a series of heaters, the concept of *Indoor Temperature* is associated with a series of type-1 membership functions encapsulating the meaning of "low", "medium" and "high" in terms of *Indoor Temperature*. The prototype sets are either predefined (by an expert) or they are dynamically created to cover the input/output range of the human-centred model. As part of our current research, the human-centred model input/output is generated as the average of the respective input/output devices. In order to illustrate this step of the implementation, we will refer to the example of ambient light level which has been employed throughout to establish and experimentally verify the proposed concepts and architecture as the "immediate effect" of light level changes facilitates experimentation (compared to, for example, heating, which varies at a comparatively slow speed in an indoor environment).

In the following subsections, we will provide overview of the specific steps involved in the generation of a type-1 fuzzy set human-centred model based on the example of the light sensors deployed in the living room area of the iSpace at the University of Essex.

Setup of the Input Variable to the Human-Centred Model from the Physical Devices

As AA associates groups of physical devices with a human-centred concept, it is essential to map the input to/from all the devices to a single input for the resulting model. While a variety of "amalgamation methods" could be explored (for example, in terms of sensors, a vast amount of research exists on the approach to sensor fusion), we have focussed on employing the arithmetic mean (i.e. average) of all the physical device inputs to serve as the input to the human-centred model. This approach allows for a straightforward amalgamation of the information of the individual devices and provides the additional advantage of exploring the variation around the mean, which is directly employed to determine the uncertainty at given measurement points. An example, for the light sensors in the living room area of the iSpace is provided in Fig. 5.8.

Date Time	living-fridge-lux	living-plasma-lux	kitchen-cupboard-right-lux	kitchen-cupboard-centre-lux	kitchen-cupboard-left-lux	living-above-samsung-lux	living-control-wall-lux	living-photoframe-lux	**ARITHMETIC MEAN**
2010-10-07_08-26-33	48	40	24	54	8	50	80	113	**52.125**
2010-10-07_08-32-33	68	60	34	124	8	60	100	133	**73.375**
2010-10-07_08-38-33	158	160	124	484	28	140	220	273	**198.375**
2010-10-07_08-44-33	168	170	154	431	38	150	230	293	**204.25**
2010-10-07_08-50-33	278	280	364	621	48	240	350	443	**328**

Fig. 5.8 A real-world data sample collected from the light sensors in the living room area in the iSpace at the University of Essex

Instantiation of the Type-1 Fuzzy Set Prototype Models

In order to instantiate the type-2 fuzzy set prototype models, first and foremost, the domain $D=[l, r]$ of the human-centred model is established by setting l equal to the minimum of the potential input values of the physical devices and r to the maximum of the potential input values of the physical devices. The ranges of the physical device inputs can be retrieved though the Sphere Manager device handles. After the domain is established, a predefined number of type-1 fuzzy sets is created to model the entire domain. If no type-1 fuzzy sets have been predefined for a specific group of physical devices, the type-1 fuzzy sets are spread evenly over the domain. This latter case applies to the human-centred model of Ambient Light Level for the living room. The type-1 sets generated are depicted in Fig. 5.9.

It should be noted that the actual role of the prototype sets is limited: they directly and only aim at reflecting a logical basis of the concepts represented, i.e. in terms of Ambient Light Level, the sets with labels "Very Low", "Low", "Medium", "High", "Very High" serve to imitate the perception and order that humans associate with those labels in the real world (e.g. a low brightness is less bright than a high brightness). After this logical basis has been established through the type-1 sets, these sets are then augmented with the uncertainty extracted from the heterogeneous sources of information (e.g. different sensors) collected. We will describe this step in the following subsection.

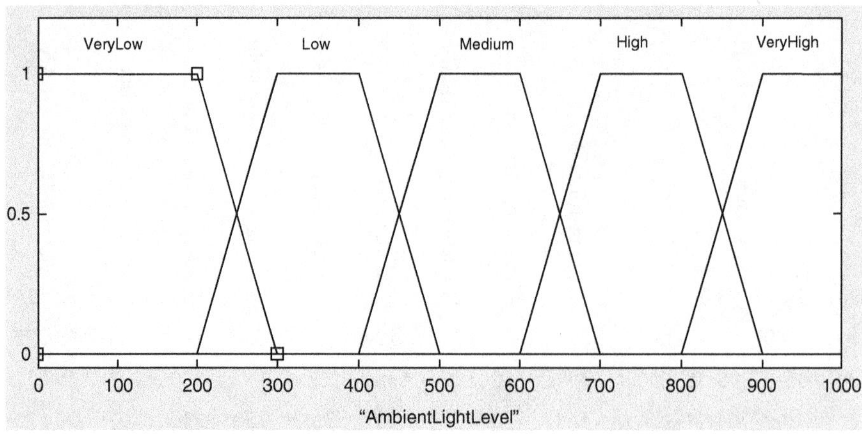

Fig. 5.9 Type-1 fuzzy sets—prototype for the Ambient Light Level model

Creation of Device Uncertainty Capturing Interval Type-2 Set Model

In order to capture the uncertainty resulting from the amalgamation of information from a series of physical devices (where each device itself may be subject to uncertainties such as sensor noise, environmental impact, etc.), the type-1 prototype sets discussed in the previous section are used as blueprints for interval type-2 sets which—in addition to the information provided by the type-1 sets—encode uncertainty information gathered from the device information. The uncertainty information is extracted and encoded as follows:

- For each data sample S^i, where $i \in [1, I]$ and I is the number of samples (i.e. rows in Fig. 5.8):
  - Compute the maximum $S^i_{\max}$ and minimum $S^i_{\min}$ values read from all devices.
  - Compute the difference $S^i_d = S^i_{\max} - S^i_{\min}$ as a metric of the variation in the data at this sample.
- Sort all samples S^i according to their pre-computed mean values S^i_{mean}.
- Uncertainty information in relation to the mean (which is the input variable to the human-centred model e.g. Ambient Light Level) is encoded in the computed variation around the mean, i.e. the higher the variation around a specific mean and as such the higher the variation which exists in the values retrieved from all physical devices at a specific sample time, the higher the associated uncertainty. An example of the extracted uncertainty information for the light sensors in the living room in the iSpace on a single day is provided in Fig. 5.10a.
- Finally, a linear equation is fitted to the uncertainty information (see Fig. 5.10b). The reason for the fitting of a linear model is twofold:

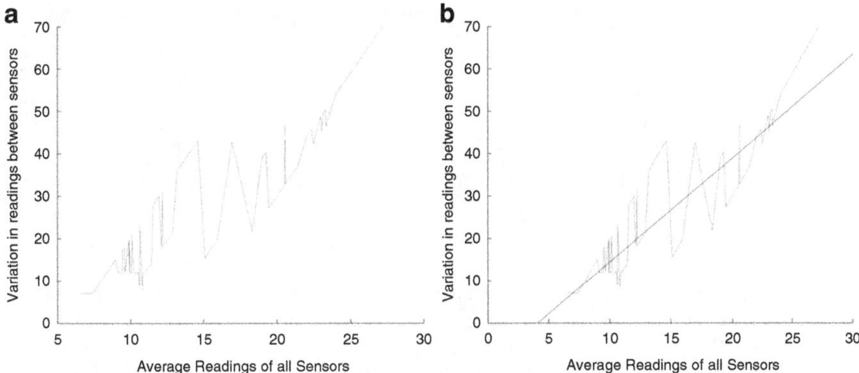

Fig. 5.10 (**a**) Example of the uncertainty information as extracted from the data from multiple light sensors in the living area of the iSpace at the University of Essex (**b**) Example of the uncertainty information as extracted from the data from multiple light sensors in the living area of the iSpace at the University of Essex (*blue*) as well as the linear approximation of the information (*red*)

- The fitting of a model to the uncertainty information is essential to allow for a prediction of the uncertainty information for mean device values, which have not been witnessed yet as part of the data collection. As an example, consider mean values in Fig. 5.10a which are bigger than 30. In order to associate such values with an uncertainty distribution, the creation of a model of the existing data is essential.
- The choice of a linear model of the uncertainty information is based on our experience with the devices currently in the iSpace which seem to indicate a linear increase of uncertainty as the mean value of the individual devices increases. Thus, a linear model provides a good approximation with minimal complexity. Naturally, more complex models (such as logistic models) can be employed as part of future research.

As a final step, the uncertainty information, which is encoded as a linear equation in our case, is employed to augment the prototype type-1 fuzzy sets. In other words, the uncertainty information is employed to inform the creation of interval type-2 fuzzy sets from the initial type-1 fuzzy sets. An example of the resulting interval type-2 fuzzy sets based on the prototype sets shown in Fig. 5.9 and the uncertainty information (respectively its linear approximation) shown in Fig. 5.10b is shown in Fig. 5.11. Considering Fig. 5.11, it is clear how the continuing increase in uncertainty as the mean value of the individual sensors increases results in a continuous growing of the Footprint of Uncertainty (FOU) of the interval type-2 fuzzy sets, i.e. the difference between the upper and lower membership functions. It is this increasing FOU that subsequently allows the human-centred model of Ambient Light Level to represent the uncertainty present in its inputs gathered from the real-world sensors.

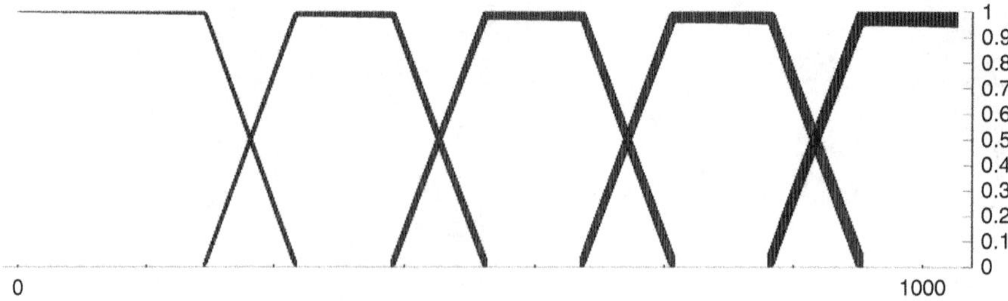

Fig. 5.11 Interval type-2 fuzzy sets modelling the human-centred concept of Ambient Light Level for a specific timeframe (here: 1 day). The model results from the combination of the type-1 prototype sets (Fig. 5.9) and the uncertainty information extracted from Fig. 5.10b. (Sets depicted *from left to right* represent the linguistic labels "Very Low", "Low", "Medium", "High" and "Very High", respectively)

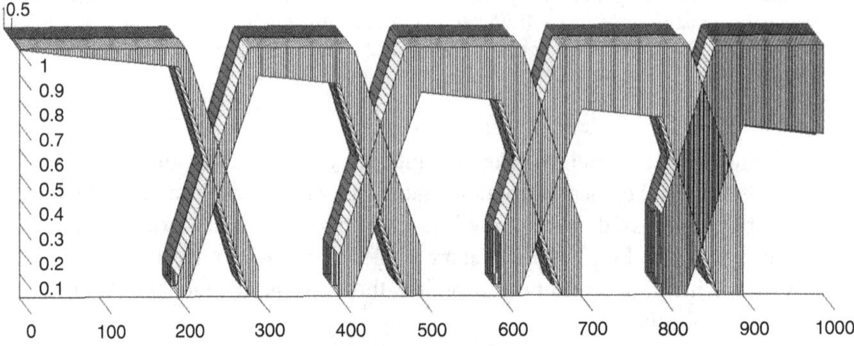

Fig. 5.12 zSlices-based model for Ambient Light Level based on data captured from two consecutive days. (Sets depicted *from left to right* represent the linguistic labels "Very Low", "Low", "Medium", "High" and "Very High", respectively)

Continuous Adaptation and Creation of the zSlices-Based Model

In Sect. 5.5.4, we described how agreement-based modelling using zSlices-based general type-2 fuzzy sets (see [24] and [26]) can be employed to continuously adapt to changes in the device characteristics and inter-relationship of the individual physical devices. Importantly, the approach also allows to progressively integrate such changes in the real world into the model while quickly eliminating short-term variations (such as the temporary obstruction of a light sensor by a piece of clothing). The theory describing the operations required to generate zSlices-based models of multiple interval type-2 fuzzy set-based models is described in detail in [26] and herein, we will provide real-world example of the technique. Further, Fig. 5.12 provides an example of a zSlices-based model generated using two datasets captured on consecutive dates (incorporating the model shown in Fig. 5.11).

As can be seen from Fig. 5.12, the agreement (visualized as overlap) between the two individual slices (displayed as yellow and pink) is very large. This is expected and indeed desired as the data was collected in two consecutive days. Also, the changes in the characteristics of the physical devices are minimal, appearing gradually (such as changes in the weather). Minimal changes are already visible in the discrepancies of the both zSlice levels (pink and yellow).

As part of the final year prototype of AA in ATRACO, we have investigated the use of up to seven zSlice levels, allowing accurate continuous modelling while maintaining a reasonable processing cost both in terms of creating as well as evaluating the resulting zSlices-based models. Several studies into the generation of human-centred zSlices-based models were conducted around the concept modelling the human-centred model for "Ambient Light Level". We will give further details to experiments in the following section.

5.6 Experiments and Results

The real-world experiments were conducted in the iSpace at the University of Essex and focused on adapting the lighting conditions in the main living area. Herein, we will provide three individual experiments, which were chosen for their significantly different contexts in terms of time of year and iSpace occupation. We commence by providing a brief of the experimental setup, focusing on the actual hardware devices employed. Subsequently, we will review the context of all experiments referred to as Experiment A, Experiment B and Experiment C. Also, details on the dynamic creation of the adaptive human-centred models based on the capturing of daily device/environment interaction datasets as well as the continuous creation (and adaptation) of an adaptive model over the whole experiment period will be provided.

5.6.1 Overview of the Experimental Setup

Experiment A and Experiment B are located in the open-plan living area of the iSpace whereas Experiment C takes place in the other locations of the iSpace, i.e. the study and the bedroom allowing a direct comparison between the individual locations. Figures 5.13 and 5.14 show the open-plan living area from a variety of angles. Likewise, the left and right views in reference to the entrance door of the room for the study are displayed in Fig. 5.15a. Additionally, the bedroom is shown in Fig. 5.15b. The sensors are numbered as follows:

1. Living-photoframe-lux
2. Living-above-samsung-lux
3. Living-control-wall-lux
4. Living-fridge-lux
5. Living-plasma-lux

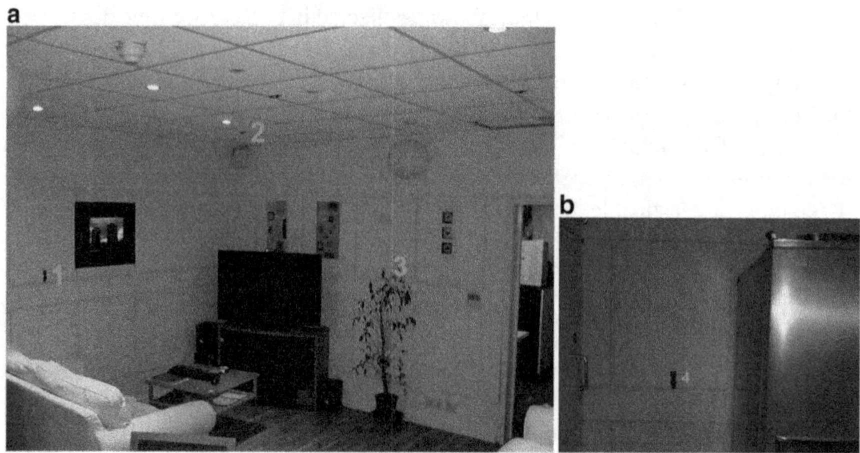

Fig. 5.13 (**a**) The couch and TV area in the iSpace (*front view*) (**b**) The light sensor located next to the fridge in the iSpace

Fig. 5.14 (**a**) The couch and TV area in the iSpace (*side view*) (**b**) The kitchen area—part of the open-plan living room in the iSpace (**c**) The kitchen area—part of the open-plan living room in the iSpace (*right corner*)

6. Kitchen-cupboard-left-lux
7. Kitchen-cupboard-centre-lux
8. Kitchen-cupboard-right-lux
9. Study-desk-light
10. Study-entrance-light
11. Bedroom-entrance-light
12. Bedroom-display-light
13. Bedroom-bed-light

The data capture for all experiments was conducted through UPnP, relying on the Youpi libraries developed as part of ATRACO. While an Apple machine running MAC OS was employed for the data capture, all further processing was done on a PC running Windows 7 and a MacBook running Mac OS 10.5.x.

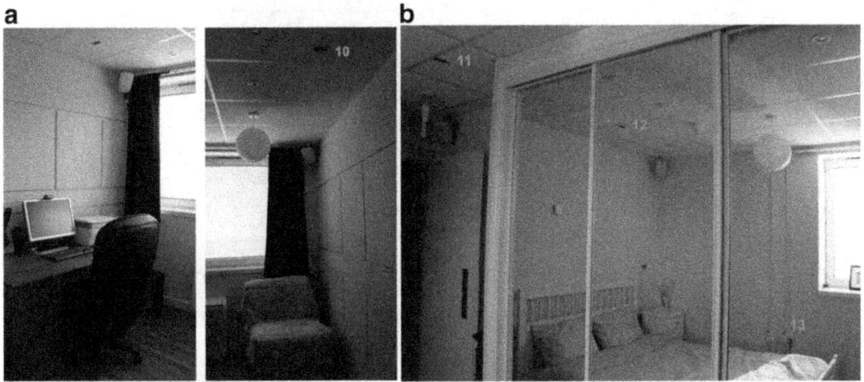

Fig. 5.15 (**a**) The study room in the iSpace (*left and right views*) (**b**) The bedroom in the iSpace (including *right view* from the mirror)

5.6.2 Experiment A

5.6.2.1 Context

Experiment A was conducted from October 7th 2010 until October 13th 2010 inclusive. During the experiment, the iSpace was solely used for development purposes during (more or less) office hours. In other words, a variety of people temporarily occupied the AIE and altered the environment—in terms of light levels mainly in the following ways:

- Use of the lamps and lights available in the iSpace.
- Obstruction of light sensors through the positioning of equipment, furniture and other items (e.g. consider the partial obstruction of the control-wall-lux light sensor by an indoor plant visible in Fig. 5.13a).
- By adjusting the window blinds (see Fig. 5.14a).

It should be noted that Experiment A focuses only on device/environment interactions in order to extract the uncertainty levels, which may change within time. Hence, for Experiment A, there is no need to identify the individuals using the iSpace.

Figure 5.16 provides an overview of the captured device information as well as the sun hours measured for Colchester, UK (the location of the iSpace). The sun hours have a strong and direct impact on the adaptation requirements through AA and Fig. 5.16 indicates its correlation to the captured indoor light levels. The correlation is of particular importance as it highlights the impact of natural light on AIEs and AA in particular.

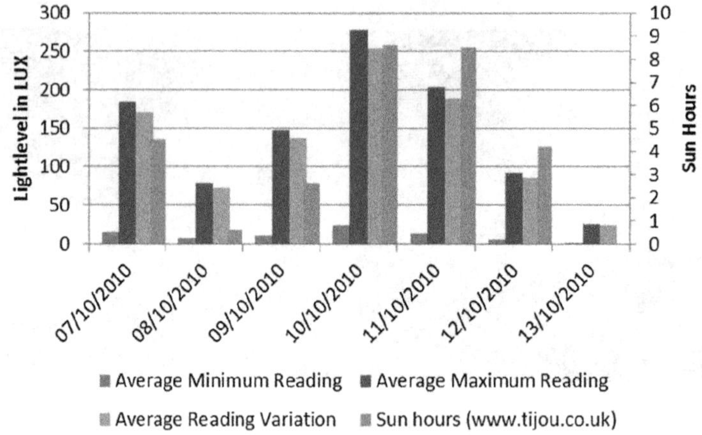

Fig. 5.16 Experiment A—average indoor light level readings and outdoor sun hours

5.6.2.2 Daily Device/Environment Data Capture and Human-Centred Model Creation

The devices were monitored continuously and their information was captured every 6 min over 7 days. The information captured each day was reviewed and we have observed that there is strong variation in the data collected, which is related to changes in the characteristics of the devices and the AIE in general (such as a significant impact of changing outside weather conditions (e.g. number of daily sun hours) as shown in Fig. 5.16). The information captured each day was processed in order to give rise to a human-centred model of "Ambient Light Level", which is based on amalgamation of the information of all individual light sensors. In order to achieve this, each day, the uncertainty information resulting from the amalgamation of all the living room sensors was extracted and the parameters for the linear model of the uncertainty variation across the human-centred input were computed. It should be noted that while the information of all the sensors which were physically located within the open-plan living room was collected, the model created was based on the subset of the actual living room sensors (i.e. excluding the kitchen sensors) in order to provide a representative model for the living room area specifically. Figure 5.17 provides an overview of the information extracted from the daily data (only the living room sensor data has been employed). Interestingly, it can be seen that while there are strong variations in the data, the uncertainty information in Fig. 5.17 indicates a stable and continuous increase of the uncertainty over the input to the human-centred model. This notion is further strengthened by the visualization of the uncertainty information—both in terms of daily real data and the fitted linear models in Fig. 5.18.

During the experiment, an interval type-2 fuzzy set was created each day based on the uncertainty information extracted (shown in Fig. 5.17). An example of the model created after the first day (7th October 2010) is given in Fig. 5.19. Finally, it

	Average Minimum Reading	Average Maximum Reading	Average Reading Variation	Uncertainty variation model: slope	Uncertainty variation model: intercept
07/10/2010	14.61666667	184.3125	169.6958333	0.0008	0.0051
08/10/2010	6.904166667	78.525	71.62083333	0.0007	0.0111
09/10/2010	11.17916667	147.8	136.6208333	0.0008	0.0157
10/10/2010	24.58333333	277.85	253.2666667	0.0007	0.0272
11/10/2010	14.39166667	202.8291667	188.4375	0.0008	0.0046
12/10/2010	5.854166667	91.54166667	85.6875	0.0008	0.0043
13/10/2010	0.8875	25.18333333	24.29583333	0.0007	0.0026

Fig. 5.17 Information extracted/computed during Experiment A

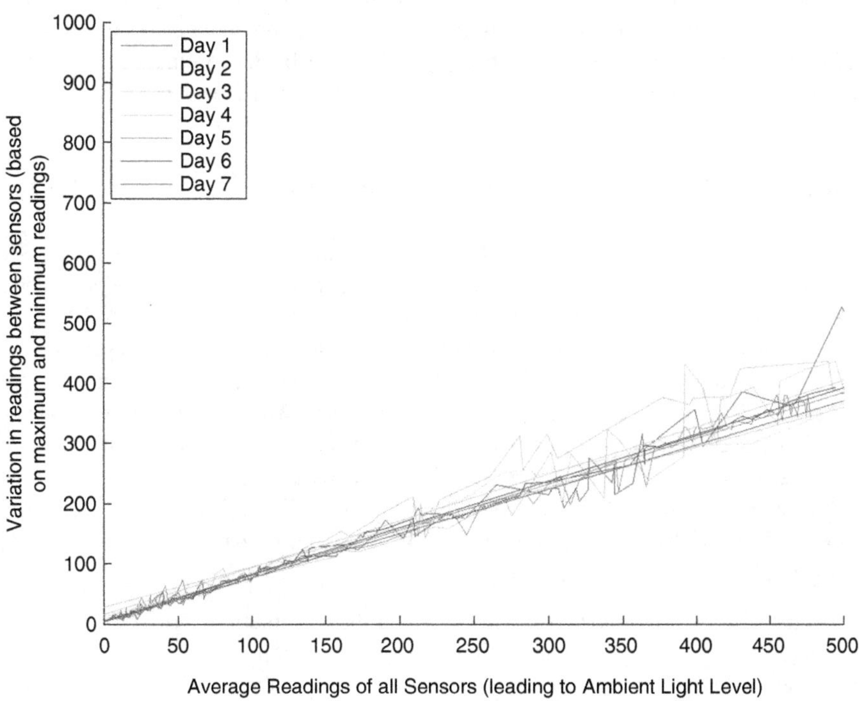

Fig. 5.18 Uncertainty information based on variation around the measured mean (daily linear model plotted in *straight line*) for Experiment A

Fig. 5.19 Ambient Light Level model for the 7th October 2010. Linguistic labels *from left to right*: "Very Low", "Low", "Medium", "High" and "Very High"

should be noted that while the human-centred model actually employed and created by AA is based on the continuous incorporation of the models generated over consecutive days. The following subsection will describe the continuous integration of the daily models and the resulting zSlices-based general type-2 fuzzy sets.

5.6.2.3 Continuous and Gradual Adaptation of Model over Multiple Days

The Ambient Light Level model created based on the data captured on a daily basis is further processed each consecutive day in order to create an adaptive model, which is able to reflect changes in the artefacts' characteristics. In order to illustrate the "growth" and creation of the adaptive model as part of this experiment, we will be showing visualizations of the individual linguistic label "Medium Ambient Light Level" as its model is adapted from day to day. We are focusing on showing one individual label ("Medium") for ease of visualization. However, we include the final output, i.e. the adaptive model created based on the seven consecutive experiment days for informative purposes as part of Fig. 5.20. The model of the label "Medium" for the human-centred overall model of Ambient Light Level is shown in Figs. 5.21, 5.22, 5.23, and 5.24:

- Figure 5.21a: The model is based solely on the information captured on the 7th October 2010 (1 day). No agreement modelling is applied.
- Figure 5.21b: Two individual datasets from two consecutive days are available. Individual models are created for both days. Finally, the agreement between both models is computed and a final zSlices agreement-based general type-2 fuzzy set is created.
- Figures 5.22, 5.23, and 5.24: The agreement models are recomputed with the corresponding number of available daily sets as an input.

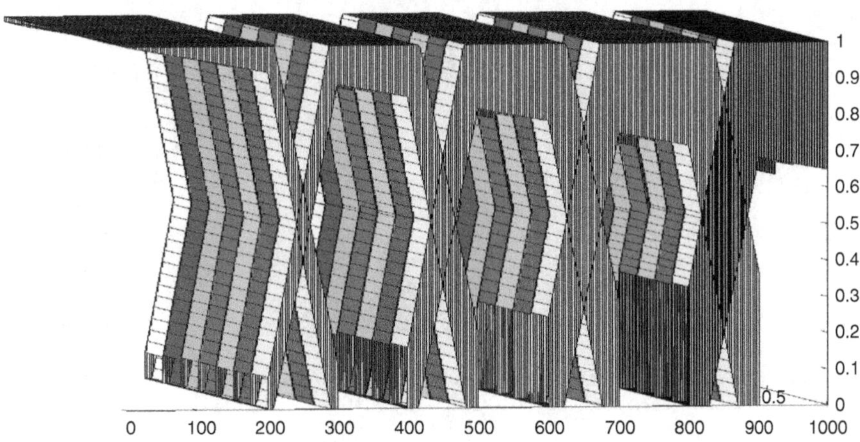

Fig. 5.20 Ambient Light Level model—full model, 07/10/2010–13/10/2010

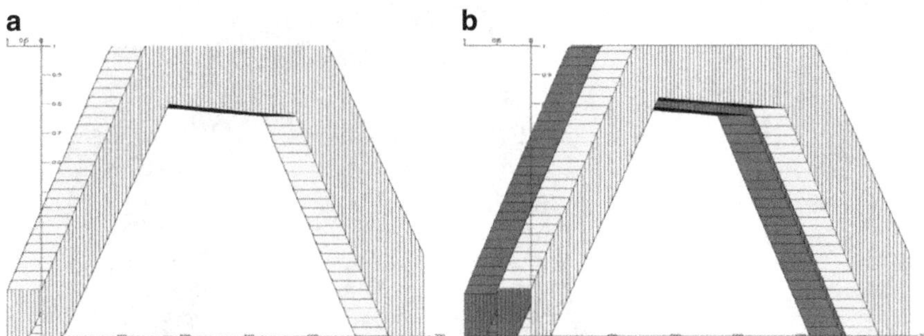

Fig. 5.21 (**a**) Ambient Light Level model—label: medium, 07/10/2010 (**b**) Ambient Light Level model—label: medium, 07/10/2010–08/10/2010

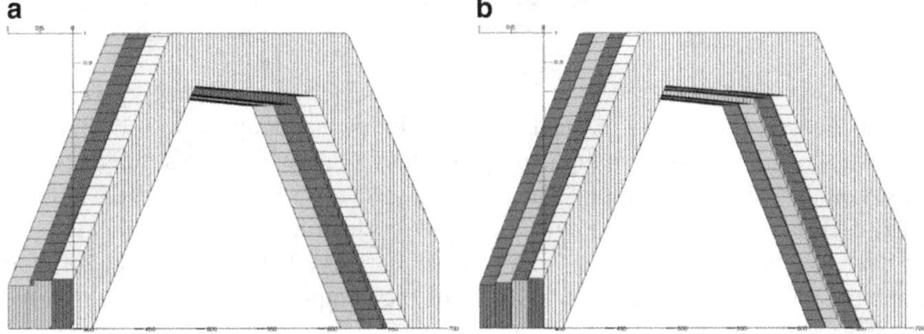

Fig. 5.22 (**a**) Ambient Light Level model—label: medium, 07/10/2010–09/10/2010 (**b**) Ambient Light Level model—label: medium, 07/10/2010–10/10/2010

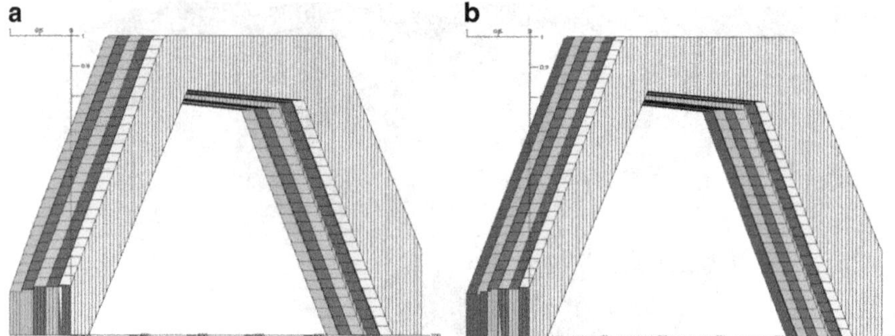

Fig. 5.23 (**a**) Ambient Light Level model—label: medium, 07/10/2010–11/10/2010 (**b**) Ambient Light Level model—label: medium, 07/10/2010–12/10/2010

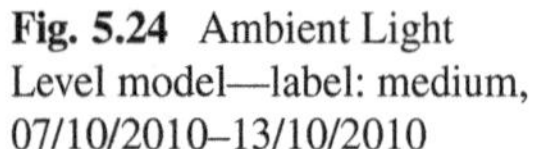

Fig. 5.24 Ambient Light Level model—label: medium, 07/10/2010–13/10/2010

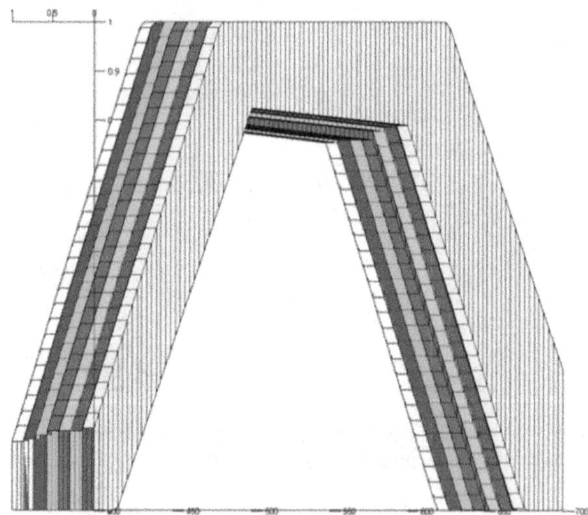

The agreement-based models (shown in Figs. 5.21, 5.22, 5.23, and 5.24) are not based on a *concatenation* (i.e. stacking up) of individual daily models. Each of the models shown is a fully recomputed agreement model, which takes the given sets based on the daily data as input. The resulting agreement-based sets have a number of zLevels that is equal to the number of input sets in order for the agreement model to have the ability to model all "opinions" (in our case daily sets) and any of their potential level of agreement. Full details of the computation of agreement-based models are available in [26].

5.6.2.4 Discussion

As part of Experiment A, we have shown the continuous adaptation provided by AA. The proposed techniques based on the capturing of information on device/environment interactions (over the short, medium and longer term) and the creation of a human-centred model based on the amalgamation of the data from multiple physical devices were demonstrated, both in concept and in practice. Further, the uncertainty extraction from daily models was demonstrated as well as the agreement-based continuous adaptation of the resulting model. Importantly, the proposed approach is applicable to long term deployment in real-world environments through its "sliding-window" design.

In the following section, we will present the results of a similar experiment (Experiment B), which was conducted in a significantly different context in order to show the capability of adapting to a range of conditions and changes.

5.6.3 *Experiment B*

This section provides the outstanding details of Experiment B, including an overview of the context of the experiment which is of primary importance in order to demonstrate the adaptive behaviour of AA.

5.6.3.1 Context

Experiment B was conducted from January 25th 2011 until January 31st 2011 inclusive. In contrast to Experiment A, where iSpace was solely used for development purposes during office hours, Experiment B was conducted over a period of continuous inhabitancy of the iSpace, combined with development work during office hours on some days. In other words, while a variety of people temporarily occupied the AIE and altered the environment, the majority of the human intervention is caused by the inhabitant. In terms of light level, the changes can be those similar to Experiment A described in Sect. 5.6.2.1.

Significantly, in contrast to Experiment A, the impact of a near-permanent human presence in the iSpace allowed us to test the AA in conditions even closer to real-world deployment conditions and maximized the human–system interaction. Figure 5.25 provides an overview (based on averages of the metrics employed during model creation) of the captured device information as well as the sun hours as measured for Colchester, UK (the location of the iSpace). When comparing the same metrics from Experiment A in Fig. 5.16, it can be noted that the amount of sun (in terms of sun hours) over the duration of Experiment B was significantly lower than in Experiment A, allowing us to evaluate the adaptation of the AA component to significantly different conditions.

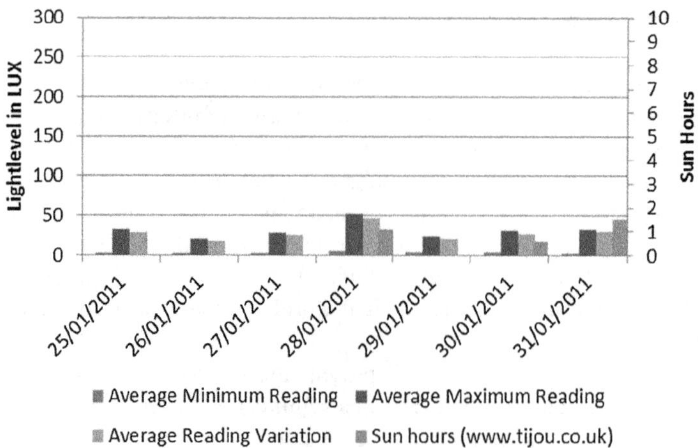

Fig. 5.25 Experiment B—average indoor light level readings and outdoor sun hours

5.6.3.2 Daily Device/Environment Data Capture and Human-Centred Model Creation

The devices were monitored continuously and their information was captured every 6 min over 7 days. The captured data was in the same format as in Experiment A. The data collected each day of the Experiment B has been reviewed and they clearly suggest that there is strong variation that is related to changes in the characteristics of the devices and the AIE in general (such as a significant impact of changing outside weather conditions (e.g. number of daily sun hours) as shown in Fig. 5.25). Also, the captured indoor light levels were overall significantly lower than in Experiment A. This effect is based on two major differences between both experiments:

- The lower number of sun hours overall (see Fig. 5.25).
- The fact that the iSpace was inhabited permanently. This is significant as occupants generally close the blinds, thus preventing a large amount of the natural light from entering the AIE.

In particular, the latter point is an ideal example of the device/environment interactions which occur and are not directly "executed" by the user. Instead, the users' actions have significant effects (both in their presence and absence) to which the device models need to be adapted.

Figure 5.26 provides an overview of the information extracted from the daily data captured for Experiment B. Comparing Figs. 5.18 and 5.26, in terms of the uncertainty variation model parameters, it is clear that as part of Experiment B, the data captured from the individual days are significantly more varied than in Experiment A. This notion is graphically visible in Fig. 5.27, where comparing, for example, the uncertainty models of Day 5 and Day 6 clearly indicates the

	Average Minimum Reading	Average Maximum Reading	Average Reading Variation	Uncertainty variation model: slope	Uncertainty variation model: intercept
25/01/2011	3.716667	31.90417	28.1875	0.0009	0.0047
26/01/2011	2.595833	20.62917	18.03333	0.0023	-0.0010
27/01/2011	2.745833	28.23333	25.4875	0.0009	0.0099
28/01/2011	5.808333	52.2625	46.45417	0.0015	0.0001
29/01/2011	3.908333	23.875	19.96667	0.0025	-0.0101
30/01/2011	4.533333	31.65	27.11667	0.0009	0.0199
31/01/2011	3.295833	33.01667	29.72083	0.0018	0.0013

Fig. 5.26 Information extracted/computed during Experiment B

significant difference in the data captured for both days. Additionally, Fig. 5.27 further visualizes the overall much lower light levels (compared to Experiment A) witnessed in Experiment B. Considering the visualizations of the actual data (variation around the mean) in Fig. 5.27 also further clarifies the need for the construction of a (in our case linear) model in order to be able to construct a final human-centred model that is able to deal with input values above the witnessed data (in our case about 350 LUX and more).

As in Experiment A, over the duration of Experiment B, an interval type-2 fuzzy set was created each day based on the uncertainty information extracted from the living room light sensors (shown in Fig. 5.26).

5.6.3.3 Continuous and Gradual Adaptation of Model over Multiple Days

The Ambient Light Level model created based on the data captured on a daily basis is further processed each consecutive day in order to create an adaptive model, which is able to reflect changes in the artefacts' characteristics. We will demonstrate the adaptive capabilities of AA by showing the same set as in Experiment A in order to provide comparable models. The model of the label "Medium" for the human-centred overall model of Ambient Light Level is shown in Figs. 5.28, 5.29, 5.30, and 5.31 together with the full model (Fig. 5.32) in a similar fashion as employed in Experiment A, therefore the remarks will not be repeated here.

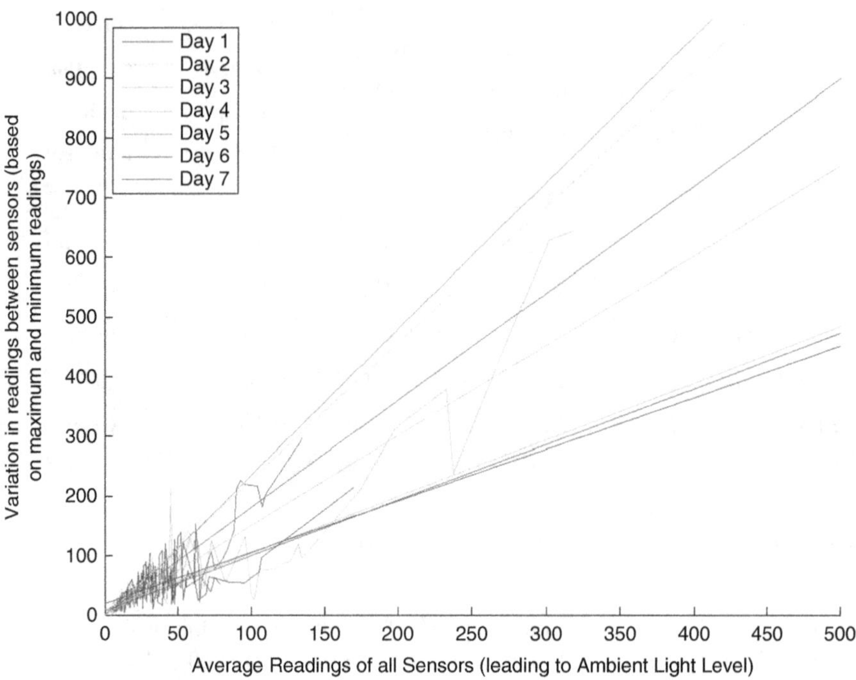

Fig. 5.27 Uncertainty information based on variation around the measured mean (daily linear model plotted in *straight line*)

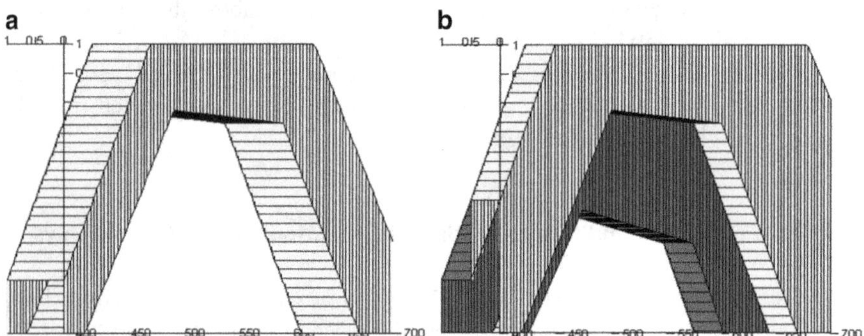

Fig. 5.28 (**a**) Ambient Light Level model—label: medium, 25/01/2011 (**b**) Ambient Light Level model—label: medium, 25/01/2011–26/01/2011

5.6.3.4 Discussion

As part of Experiment B, an adaptive model for the human-centred model of Ambient Light Level was created based on the information captured from the (living room) light sensors situated in the iSpace living room area. We have shown, by

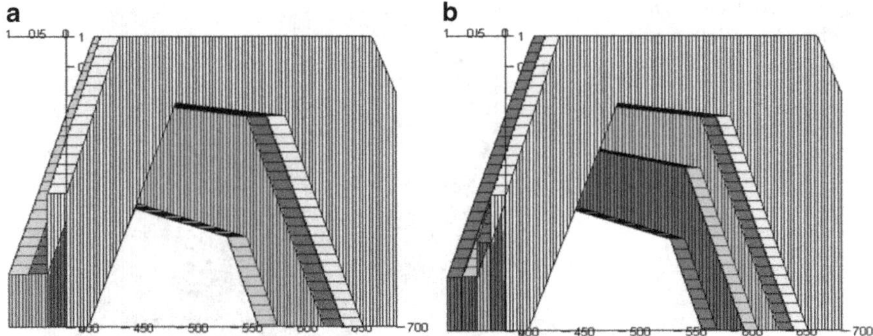

Fig. 5.29 (**a**) Ambient Light Level model—label: medium, 25/01/2011–27/01/2011 (**b**) Ambient Light Level model—label: medium, 25/01/2011–28/01/2011

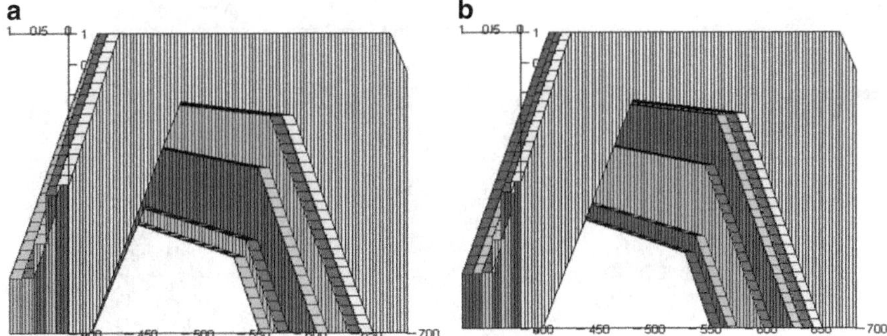

Fig. 5.30 (**a**) Ambient Light Level model—label: medium, 25/01/2011–29/01/2011 (**b**) Ambient Light Level model—label: medium, 25/01/2011–30/01/2011

relying on the example of "Medium" Ambient Light Level, how agreement-based zSlices general type-2 fuzzy sets can be employed to generate a model of the information which is able to adapt to short, medium and long term changes in the device and environment characteristics.

In contrast to Experiment A, the data in Experiment B was much more limited (specifically in terms of the range of the information). The resulting increase in uncertainty is shown directly in the extracted uncertainty information models (linear models in Fig. 5.27) as well as in the resulting models for Ambient Light Level shown in Figs. 5.28, 5.29, 5.30, and 5.31. Overall, as can be seen from Fig. 5.32, the uncertainty has directly been incorporated into the overall model of Ambient Light Level and is particularly large at high levels of agreement—which is intuitive as large number of uncertain sets (opinions) naturally result in only a very uncertain agreement (agreed opinion). In other words, the agreement model employed seeks to "home in" on the actual meaning of the human-centred label "medium" which is reflected in the "narrowing" of the individual zSlices as the zLevel increases. However, as the agreement model is based on the set-theoretical operation of

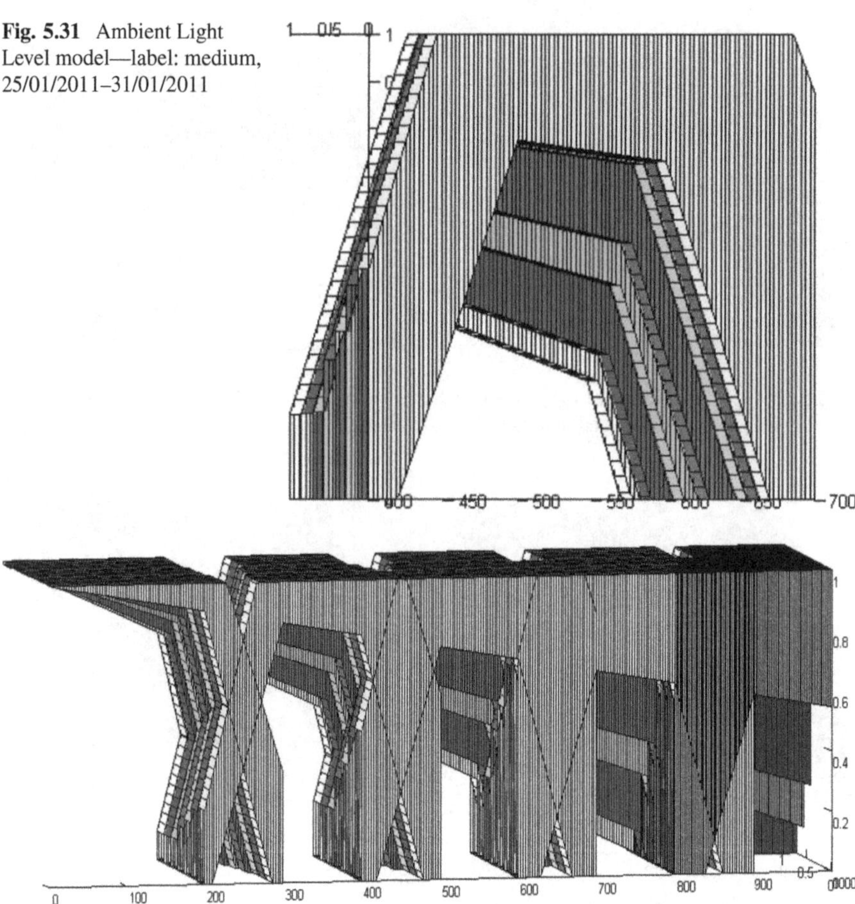

Fig. 5.31 Ambient Light Level model—label: medium, 25/01/2011–31/01/2011

Fig. 5.32 Ambient Light Level model—full model, 25/01/2011–31/01/2011

intersection (i.e. logical AND), the minimum level of certainty is preserved as part of the higher levels of agreement. This is a necessary consequence of the logical AND operation: IF sets (opinions) are aggregated that include very low degrees of membership—THEN those low degrees of membership are a possibility; thus, they need to be part of the agreement model. More discussion can be found in [26].

5.6.4 *Experiment C*

This section provides the outstanding details of Experiment C, including an overview of the context of the experiment which is of primary importance in order to explore if the proposed AA mechanisms actually expose adaptive behaviour based on different physical locations in addition to short, medium and long term changes. In Experiment C, the daily capturing of device/environment interactions differ from Experiment A and Experiment B in terms of the sensors located in different rooms within the iSpace. As the creation of human-centred models in each particular room of the iSpace incorporates the same approach as in the previous Experiments A and B; the focus of Experiment C will be on the comparison of the created final models (full) as well as the final individual labels for the Ambient Light Level for consistency and realization purposes. The continuous and gradual adaptation of the final model, which has been demonstrated in the previous Experiments A and B, is investigated further with regards to the sliding approach where we provide details and comparisons on the two final models which adapt over additional days beyond the weekly model.

5.6.4.1 Context

Experiment C was conducted from March 28th 2011 until April 4th 2011 inclusive. It should be noted that the 8 day period will be referred to as "Period 1" for the first 7 days (28/03/2011–03/04/2011) and "Period 2" for the last 7 days (29/03/2011–04/04/2011) in order to facilitate the discussion/illustration of the continuous adaptation beyond the 7-day model.

Experiment C was conducted over a period of mixed inhabitancy in the iSpace, including the duration of evaluation sessions where multiple participants (i.e. lay users) temporarily occupied the AIE and altered the environment. In terms of the changing light levels, the effects can be summarized (similarly to previous Experiments A and B) as follows:

- Use of the lamps and lights available in the iSpace with the distinction of occupancy in multiple rooms compared to the previous Experiments
- Obstruction of light sensors through the positioning of equipment, furniture and other items (e.g. consider the partial obstruction of the control-wall-lux light sensor by an indoor plant visible in Fig. 5.13a.)
- By adjusting the window blinds and the curtains (see Figs. 5.14a and 5.15).

Significantly, the enhancement of the occupied space in terms of all available rooms such as the bedroom and the study in addition to the living room in the iSpace allowed us to test the AA in conditions where human–system interaction is scaled up to the whole flat thereby incorporating a more realistic scenario with regards to the maximization of functionality and practicality.

Figures 5.33, 5.34, and 5.35 provide an overview (based on the averages of the metrics employed during model creation) of the captured device information in each room of the iSpace as well as the sun hours as measured for Colchester, UK (the location of the iSpace). When comparing the same metrics both from Experiment A and Experiment B for the open-plan living area of the iSpace with the readings illustrated in Fig. 5.33, it can be noted that the amount of sun (in terms of sun hours) has increased as well as the light levels in LUX, allowing us to evaluate the

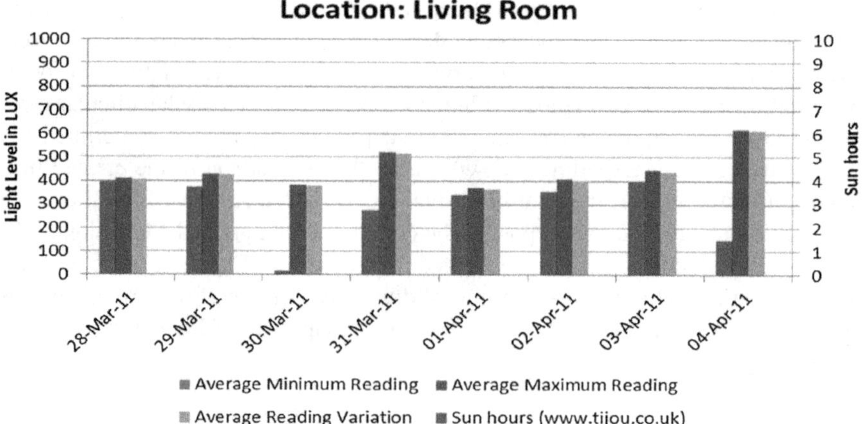

Fig. 5.33 Experiment C—Average indoor light level readings and outdoor sun hours for living room

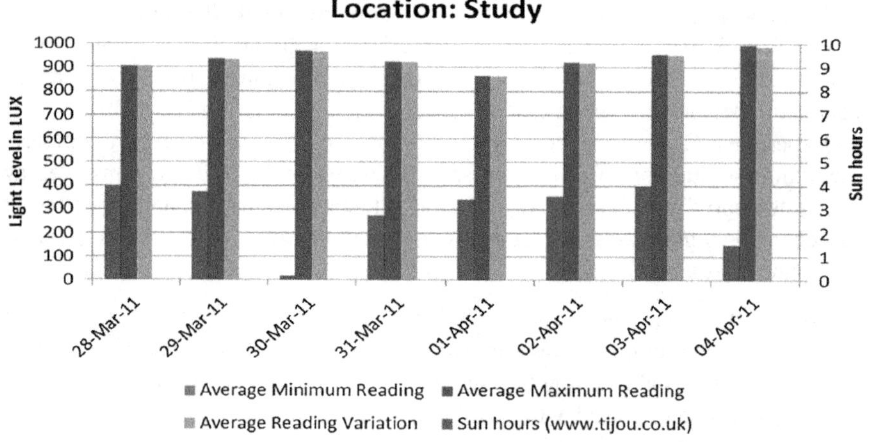

Fig. 5.34 Experiment C—Average indoor light level readings and outdoor sun hours for study

behaviour of the AA component in significantly different conditions. Additionally, the change in season from winter to spring between the different sets of experiments (resulting in the increase in sun hours and sun impact) provides additional supportive information in terms of the long term adaptation capabilities of AA.

Furthermore, it is noteworthy that the light levels in LUX recorded within the various locations in the iSpace can be quite different as illustrated in Fig. 5.34 in comparison to Fig. 5.35, bringing about another significantly distinct condition for the evaluation of AA, highlighting the impact of the geographical location of the specific rooms and the need for individual models.

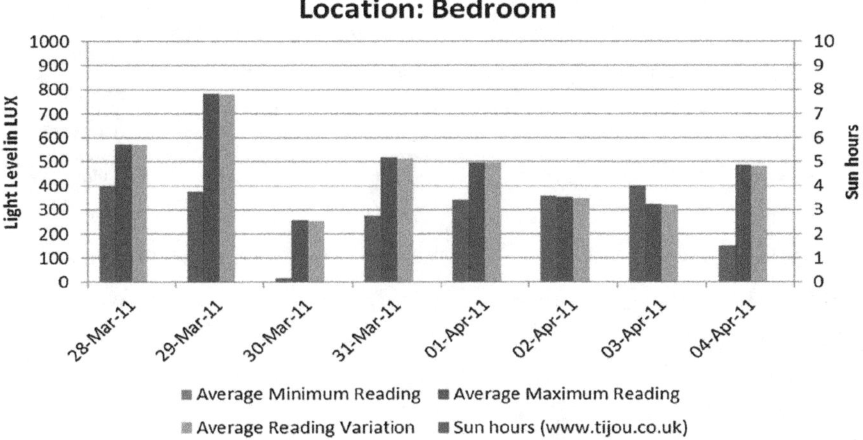

Fig. 5.35 Experiment C—average indoor light level readings and outdoor sun hours for bedroom

5.6.4.2 Daily Device/Environment Data Capture and Human-Centred Model Creation

The devices in the three separate locations (the living room, the study and the bedroom) were monitored continuously and their information was captured every 6 min over 7 days. The captured data was in the same format as for Experiment A and Experiment B with the additional sensor data captured by the light sensors in the study and the bedroom.

The current section will include the comparison of the information captured in each location on a randomly selected day which is 29th March 2011. A specific date was chosen to conserve space and provide more clarity. Additionally, the uncertainty information extracted from the individual daily datasets for each location, essential in the creation process of the human-centred model, is provided.

The information captured each day was processed for each three rooms in order to create three location-specific human-centred models of "Ambient Light Level" each of which is based on an amalgamation of the information from individual light sensors based in the respective room within the period of the 7 days (29/03/2011–04/03/2011) which we will refer to as *Period 2*. The reason why we present results for Period 2 is to investigate the behaviour of the AA with the most up-to-date data (we will later refer to *Period 1* which chronologically predates Period 2 as it starts on the 28th March).

Figures 5.36, 5.37, and 5.38 provide an overview of the information extracted from the daily data according to the different locations in the iSpace. Comparing Fig. 5.37 to Fig. 5.36 and Fig. 5.38, according to two particular parameters of the uncertainty variation model, which are the average maximum reading and the average maximum variation, it is clear that the results for the data captured from the study room are significantly more varied (having recorded values between 862

	Average Minimum Reading	Average Maximum Reading	Average Reading Variation	Uncertainty variation model: slope	Uncertainty variation model: intercept
29/03/2011	5	430.8	425.8	0.0017	0.0029
30/03/2011	5	383.4	378.4	0.0017	-0.0011
31/03/2011	7.4	522.8	515.4	0.0016	0.0102
01/04/2011	6.8	370.6	363.8	0.0018	-0.0032
02/04/2011	6.8	406.8	400	0.0017	0.0002
03/04/2011	7.2	443.2	436	0.0017	0.0038
04/04/2011	7	619.2	612.2	0.0015	0.0185

Fig. 5.36 Information extracted/computed during Experiment C for living room (Period 2)

	Average Minimum Reading	Average Maximum Reading	Average Reading Variation	Uncertainty variation model: slope	Uncertainty variation model: intercept
29/03/2011	1.5	935	933.5	0.0011	0.0182
30/03/2011	3.5	970	966.5	0.0011	0.0094
31/03/2011	3.5	925.5	922	0.0011	0.0154
01/04/2011	5	867	862	0.0012	0.0056
02/04/2011	3.5	923	919.5	0.0011	0.0161
03/04/2011	3.5	956	952.5	0.0011	0.0210
04/04/2011	5	993.5	988.5	0.0011	0.0205

Fig. 5.37 Information extracted/computed during Experiment C for study (Period 2)

and 993.5) than the results for the data captured from the living room and the bedroom. Figures 5.39, 5.40, and 5.41 give a graphical overview on the uncertainty information for the three locations.

As in Experiment A and Experiment B, over the duration of Experiment C, an interval type-2 fuzzy set human-centred model for each location was created each day based on the uncertainty information extracted. Next section puts emphasis on the gradually and continuously adapted overall models, which are the resulting zSlices-based general type-2 fuzzy sets, for different locations.

	Average Minimum Reading	Average Maximum Reading	Average Reading Variation	Uncertainty variation model: slope	Uncertainty variation model: intercept
29/03/2011	5	781.6666667	776.6666667	0.0013	-0.0148
30/03/2011	4.666666667	258	253.3333333	0.0011	0.0022
31/03/2011	4.666666667	516	511.3333333	0.0011	0.0044
01/04/2011	3.333333333	497.3333333	494	0.0011	0.0070
02/04/2011	3.333333333	351	347.6666667	0.0011	0.0031
03/04/2011	3.666666667	324.6666667	321	0.0011	0.0069
04/04/2011	4.333333333	486	481.6666667	0.0011	0.0024

Fig. 5.38 Information extracted/computed during Experiment C for bedroom (Period 2)

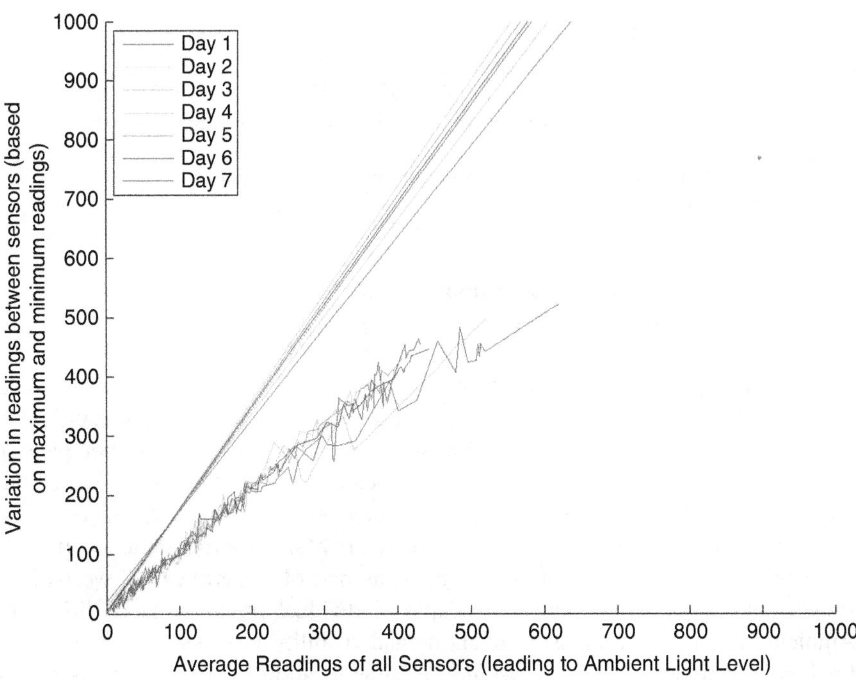

Fig. 5.39 Uncertainty information based on variation around the measured mean for living room (daily linear model plotted in *straight line*)

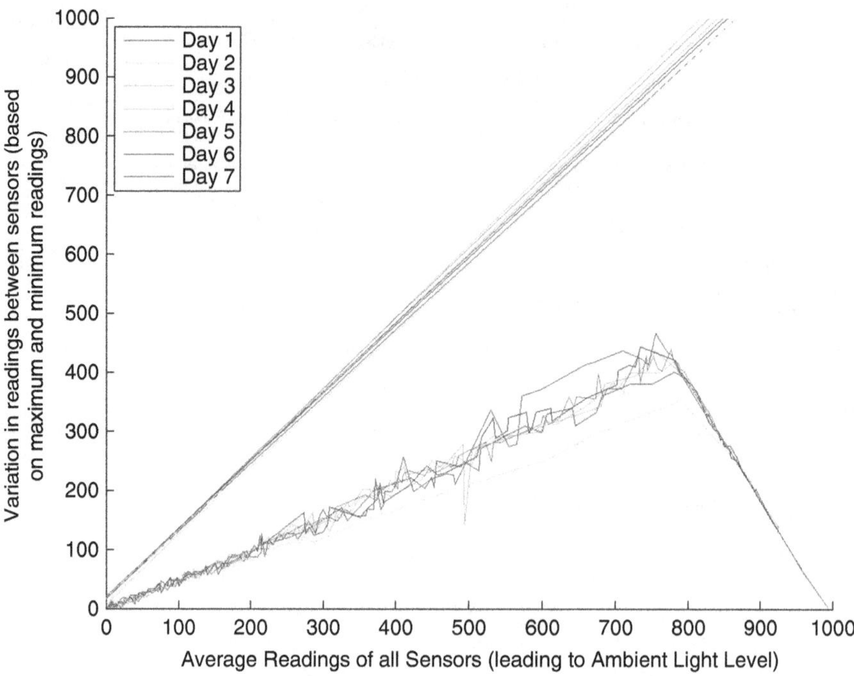

Fig. 5.40 Uncertainty information based on variation around the measured mean for study (daily linear model plotted in *straight line*)

5.6.4.3 Comparison of the Continuously Adapted Model over Multiple Locations

We have detailed the creation process of the final adaptive model in the previous Experiment A and Experiment B. In the current section, we will not revisit this creation process but focus on the comparison between the final (i.e. complete 7 day) adaptive models belonging to different locations. Further, we will contrast the models for two different time frames, i.e. the models created in January in a specific location will be compared to the models created in March for that same location.

In order to compare the adaptive models as part of Experiment C, we will be showing visualizations of the final adapted individual linguistic label "Medium Ambient Light Level" for reasons of clarity and visibility. First, we will examine the final individual label "Medium" within the same location as in Experiments A and B, i.e. the living room. Then, we will present the same final adapted linguistic label ("Medium") with respect to different locations (living room, bedroom and study) and illustrate the variation based on the context (different location with different uses, properties and device-environment interactions).

We will include the final outputs for each location, i.e. the adaptive models created based on the seven consecutive days period for comparison purposes.

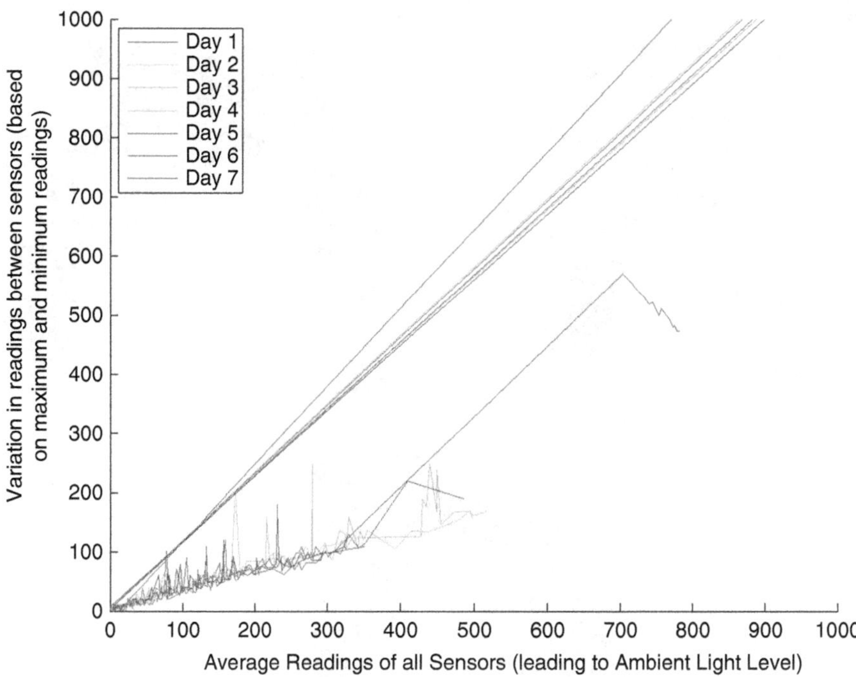

Fig. 5.41 Uncertainty information based on variation around the measured mean for bedroom (daily linear model plotted in *straight line*)

Specifically, two *full* instances of the adaptive models for each location will be provided in order to demonstrate the sliding approach. As such, the 8 day period of Experiment C will be investigated in two parts which we will refer to as Period 1 (28/03/2011–03/04/2011) and Period 2 (29/03/2011–04/04/2011).

Focus on the Final Individual Label "Medium" for all Three Locations

The individual label "Medium" for the human-centred overall model of Ambient Light Level belonging to the living room of the iSpace for Period 2 is shown in Fig. 5.42a. Comparing it to the model shown in Fig. 5.31 (created as part of Experiment B), it is visible that the agreement zSlices have relatively high variation between each other whereas those of the current model shown in Fig. 5.42a are quite regularly distributed. This shows that the measured light level information during Experiment B (within January 2011) was subject to more significant variations in uncertainty whereas the light levels witnessed during Experiment C (within March–April 2011) resulted in a relatively even uncertainty distribution.

As mentioned before, another purpose of this experiment is to perform a comparison of the adapted models based on their context/location. Analogous to

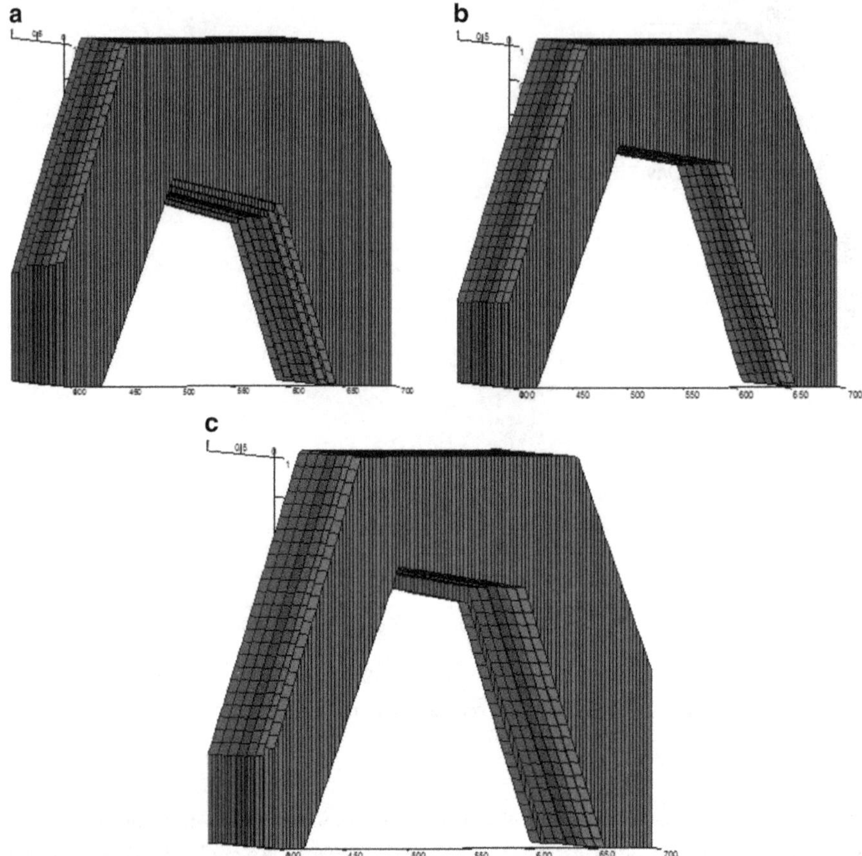

Fig. 5.42 (**a**) Ambient Light Level model for living room—label: medium, Period 2 (29/03/2011–04/04/2011) (**b**) Ambient Light Level model for study—label: medium, Period 2 (29/03/2011–04/04/2011) (**c**) Ambient Light Level model for bedroom—label: medium, Period 2 (29/03/2011–04/04/2011)

Fig. 5.42a, b shows the individual label "Medium" for the human-centred overall model of Ambient Light Level belonging to the study of the iSpace for Period 2 while Fig. 5.42c shows the individual label "Medium" for the human-centred overall model of Ambient Light Level belonging to the bedroom of the iSpace for Period 2.

Comparing all three models which were created for the same timeframe, the variation between the individual models is still apparent. This is intuitive and reflects the uncertainty introduced by the individual sensors deployed in the specific rooms. All sensors are subject to their specific sources of uncertainty resulting from, for example, their position in the respective room, the device-environment interactions of the specific room, the amount of human use of the specific room, etc.

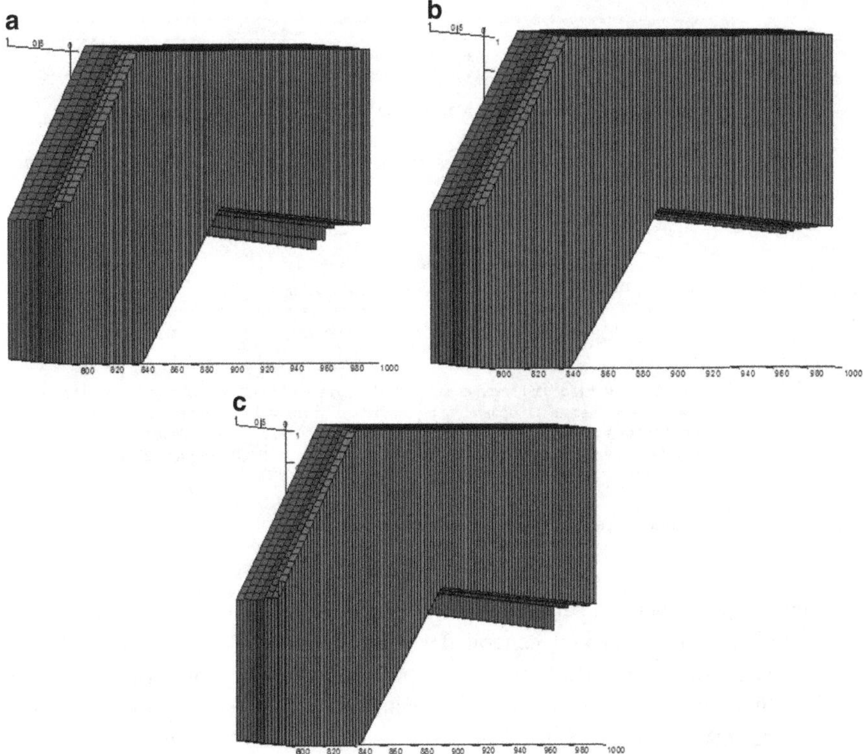

Fig. 5.43 (**a**) Ambient Light Level model for living room—label: very high, Period 2 (29/03/2011–04/04/2011) (**b**) Ambient Light Level model for study—label: very high, Period 2 (29/03/2011–04/04/2011) (**c**) Ambient Light Level model for bedroom—label: very high, Period 2 (29/03/2011–04/04/2011)

Focus on the Final Individual Label "Very High"

In this subsection, we will demonstrate the individual label "Very High" where the variations in the uncertainty levels are comparatively more visible than the rest of the individual labels, thus allowing us to clearly show the variation between the complex, overall, room-specific models.

Markedly, one of the major differences that can be observed between the individual labels "Medium" in Fig. 5.42 and "Very High" in Fig. 5.43 is the increase in the uncertainty *boundaries* (in fuzzy logic referred to as the Footprint of Uncertainty) as the model input increases along the *x*-axis. This is a direct result of the uncertainty model. It should be noted that the increase in the uncertainty from lower to higher levels of agreement is due to the nature of multi-level agreement [26]. Based on the variation in the models shown as part of this section, the adaptation capabilities of AA are clearly exemplified: while all the sensors are of the same make and type and most of the conditions were highly similar for each location (such as natural

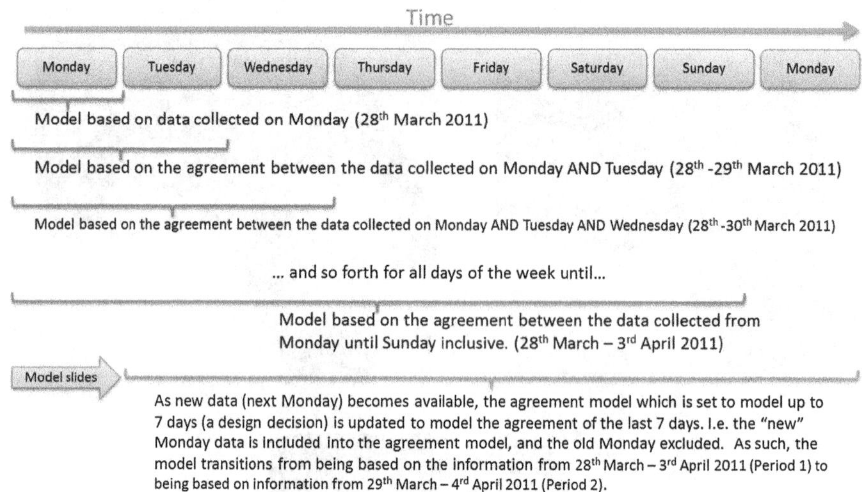

Fig. 5.44 Sliding window approach: the movement and creation

sun light impact over the same days), the local context of the individual devices (including their position and general device-environment interactions) and their usage are heterogeneous, resulting in the expected subtle, but visible variation in the models. In the following section, we will further focus on illustrating the continuous adaptation of the complete 7-day model as new information becomes available (i.e. data from a new week is incorporated).

Focus on the Sliding Window Approach

As noted before, as part of Experiment C, we aim to visualize the sliding window design. Until this point in the current section, each individual model that has been visualized and all related data has been based on Period 2 (29/03/2011–04/04/2011). In order to illustrate the sliding window mechanism, we illustrate the continuous adaptation beyond the 7 day model window here by visualizing the results from the transition of Period 1 to Period 2. Figure 5.44 summarizes the creation of the overall model for Period 1 and shows (in orange) the transition to Period 2.

To illustrate the mechanism of the sliding window approach, we concentrate on the two overall models created for each location for each specific *modeling window* in Fig. 5.44. Comparing Figs. 5.45 and 5.46, where the new Monday (04/04/2011) has replaced the old one (28/03/2011) and where exactly the same process of computations has been re-performed on the new data belonging to the new period (Period 2 in this case), the final model has adapted itself to the changing conditions with regards to the device/environment interactions.

While the difference between the two zSlices-based models is difficult to judge by eye, we direct the reader to the visual differences which can be seen most

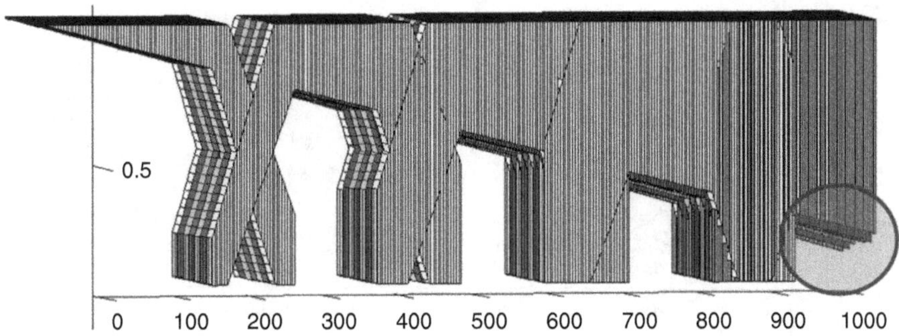

Fig. 5.45 Ambient Light Level model for living room—full model for Period 1 (28/03/2011–03/04/2011) (*circle* depicts one area of interest)

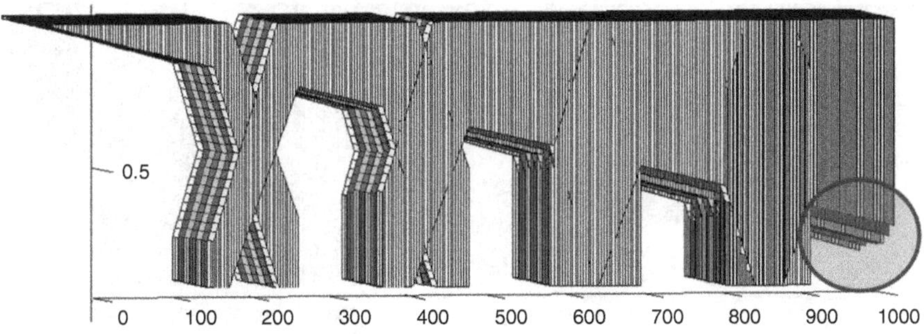

Fig. 5.46 Ambient Light Level model for living room—full model for Period 2 (29/03/2011–04/04/2011) (*circle* depicts one area of interest)

clearly in the individual labels "Medium", "High" and "Very High". In Figs. 5.45 and 5.46, we have highlighted a specific area of interest which upon comparison shows the adaptation within the models between the two Periods. It is predictable and reasonable that these *higher* models exhibit a *higher* visual variation as the uncertainty models for the human-centred concepts indicate a persistently growing level of uncertainty as the input increases.

Figures 5.47 and 5.48 illustrate the 7-day models for the Ambient Light Level for the study room in the iSpace. The "slide-d" model, which is basically the full model for Period 2, reveals the consistency between the device/environment interactions over the 7 days and hence contributes to the continuous and gradual adaptation behaviour of the model with slight changes. This similarity between the two models can also be interpreted as a pattern of inhabitant preferences that are applied in the study and might also be the reflection of a routine (e.g. working between 8 a.m. and 5 p.m. under office conditions with high levels of light) on the final human-centred model.

In contrast to the overall models for the study, Figs. 5.49 and 5.50, which illustrate the full models for the bedroom, have visible variations over the period of

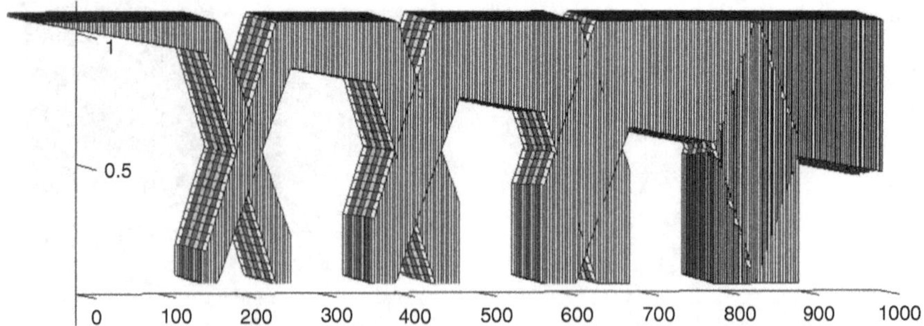

Fig. 5.47 Ambient Light Level model for study—full model for Period 1 (28/03/2011–03/04/2011)

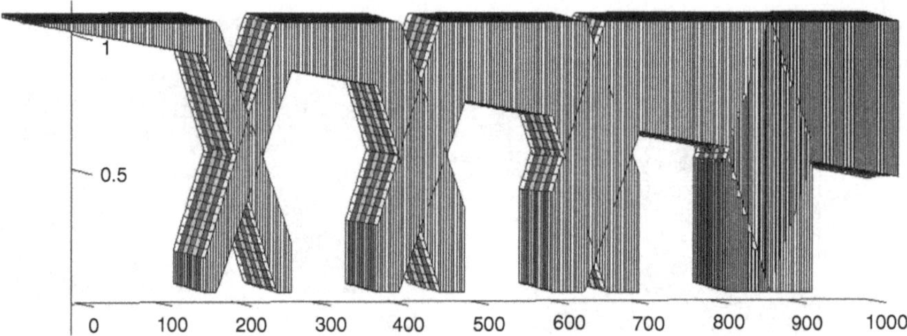

Fig. 5.48 Ambient Light Level model for study—full model for Period 2 (29/03/2011–04/04/2011)

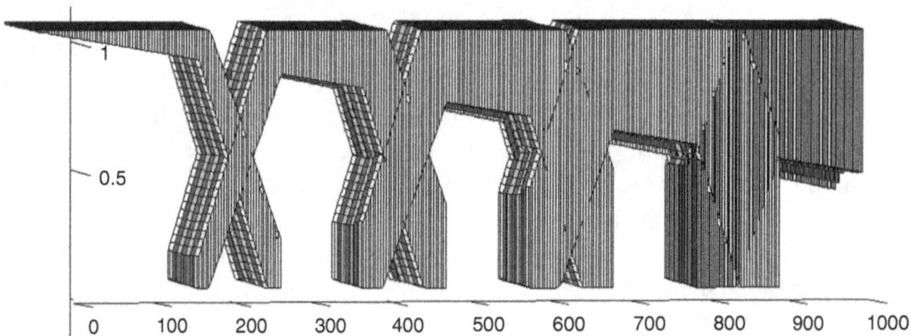

Fig. 5.49 Ambient Light Level model for bedroom—full model for Period 1 (28/03/2011–03/04/2011)

the experiment. Similar to the analysis for the final models belonging to the living room, which also show variation over the days, it can be seen that the model for the bedroom has adapted itself accordingly.

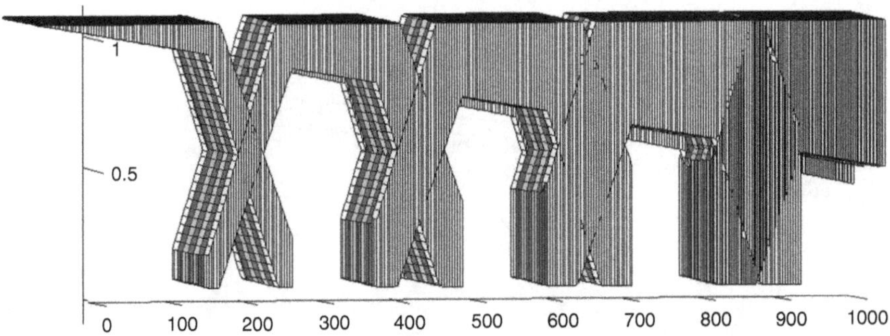

Fig. 5.50 Ambient Light Level model for bedroom—full model for Period 2 (29/03/2011–04/04/2011)

In the current section, we have presented the final individual models as well as final full models with respect to differing context generated as part of Experiment C. We proceed in the following section by providing a deeper discussion of the results.

5.6.4.4 Discussion

As part of Experiment C, adaptive models for the human-centred model of Ambient Light Level for different locations within the test bed were created based on the information captured from the light sensors situated in the corresponding areas. Essentially, we have compared the overall individual linguistic labels "Medium" and "Very High" to demonstrate the variation in the uncertainty levels, detailing how agreement-based zSlices general type-2 fuzzy sets can adapt to short, medium and long term changes in the device and environment characteristics. Also, we have put particular emphasis on the changes in the geographical location besides the seasonal changes.

Furthermore, we have also visually demonstrated the sliding window approach for a better understanding of how the sets perform the continuous and gradual adaptation by extending our experiment period to 8 days, thereby allowing a new day to be introduced and updating the human-centred model by repeating the process of calculations done for the previous period.

Unlike Experiment A and Experiment B, Experiment C has been conducted over a period of mixed inhabitancy. In other words, there were evaluation sessions taking place where the participants were altering the environment mainly in the living room and the other rooms were being utilized as a control room (i.e. bedroom) and as a presentation room (i.e. study). Also, the test bed was remaining idle in intermittent periods as there were no permanent inhabitants occupying the flat. Additionally, in terms of the context of Experiment C, it is important to realize that the time of the year that Experiment C has been conducted in, involves longer days (in terms of sunshine exposure) than the time of the year that the previous experiments were

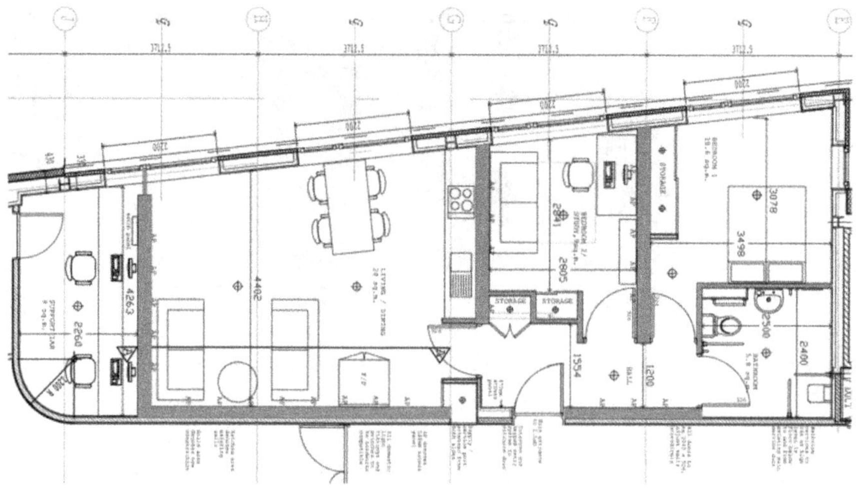

Fig. 5.51 Architectural layout of the iSpace

carried out. As such, Experiment C allowed for an investigation of AA in a strongly varied context resulting both from environmental conditions (e.g. seasonal changes) as well as a different style of user occupancy.

Additionally, Experiment C explored location-specific contrasts within the human-centred models. By location-specific, we refer to sensors in differing locations and hence various device/environment interactions. As the location of the sensors is directly related to the uncertainty of the information that can be gathered from each respective sensor (in turn related to the device/environment interactions), it is clear that a specific focus on the variation between models created for different locations is worthwhile. As expected, the results suggest that locational variation causes visible changes on the adapted final individual models, and hence on the final overall models for the Ambient Light Level.

In order to illustrate the physical setup of the iSpace and the location of the individual locations referred to as part of the experiment, the architectural layout for the test bed has been given in Fig. 5.51. From the blueprints, it can also be seen that the living room, study and the bedroom are exposed to the sunlight/daylight from the same side within the building, facilitating experiment conditions as all rooms are faced by the sun at similar points during the day. Naturally, apart from the light level conditions, differing locations mean that inhabitants perform different activities which also have quite significant effect on device/environment interactions depending on the activities carried out. Throughout, the results we have achieved are coherent with the concept of human-centred models where the uncertainties are dependent on the device/environment interactions. The interpretations of the overall models, their comparison to those created in previous experiments as well as the specific focus on model variation dependent on model-location as well as a *close-up* view of the sliding model mechanism complete the significant amount of information provided as part of Experiment C.

5.7 A Novel User-Centred Approach to Adaptive Modelling in AIEs

One of the most important features of AA is to be able to provide adaptive modelling as has been mentioned previously. During our research, the advancements in fuzzy logic theory have enabled us to explore the modelling of uncertainty present in AIEs in an adaptive way. In addition, we have focussed on achieving human interpretable and user-centred models to be used in AIEs and fuzzy logic sets have provided the grounds to obtain adaptivity and interpretability. Furthering our research, we have developed a novel approach for realizing the paradigm of Computing With Words (CWWs), which is coined by Zadeh [28] to be reasoning by words rather than numbers. From an AIE perspective, CWWs is a potential way forward to establish natural communication between the advanced technologies and human users. Within this approach, we have improved the modelling of the linguistic labels (words) in a user-centred fashion using Linear General Type-2 (LGT2) Fuzzy Sets [3], which offer adaptive modelling based on zSlices [25]. Herein this section, we will briefly demonstrate the word models for Ambient Light Level created based on the data from Period 2 as described in Sect. 5.6.4.

5.7.1 Creation of LGT2 Models for Ambient Light Level

Linear General Type-2 Fuzzy Sets have been introduced in [3] and have also been deployed in real-world experiments in the iSpace [2, 5, 6] where it has been shown that LGT2 fuzzy sets can outperform interval type-2 (IT2) fuzzy sets in terms of processing time, output accuracy/coverage and user satisfaction.

In contrast to the human-centred "Ambient Light Level" models demonstrated in Sect. 5.6, which were created using seven zSlices (due to the weekly model and sliding window approach), we have used five zSlices. The reason for this is the fact that we have developed another approach that exploits the third dimension of a general type-2 fuzzy set in a different way. More specifically, LGT2 fuzzy sets model the two contrasting linguistic labels (in other words, the user perception) (eg. dark vs. bright, low vs. high, etc.) using the primary membership functions whereas the linguistic modifiers (very, extremely, fairly, etc.) are modelled in the third dimension using secondary membership functions. As visualized in Fig. 5.52, the number of zSlices represents the uncertainty modelling of the linguistic modifiers of the linguistic variable. However, the distinct approach in modelling the third dimension of LGT2 fuzzy sets does not change the amount of data used to model the human-centred "Ambient Light Level" for AA. Figures 5.53, 5.54, 5.55, and 5.56 show the day-by-day LGT2 models created based on the data from Period 2 as described in Sect. 5.6.4.

The important thing to note is that LGT2 models enable for a more efficient design of the linguistic variable, hence, we present the full model for Ambient Light

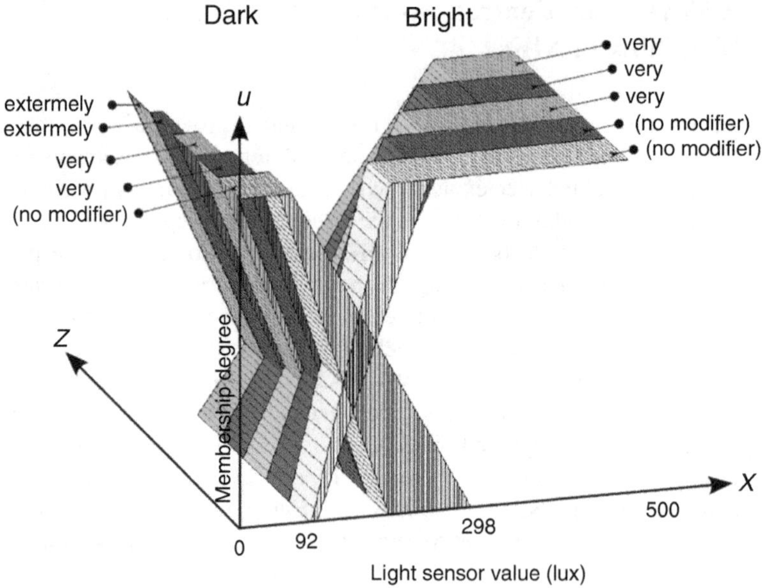

Fig. 5.52 An example of Linear General Type-2 Fuzzy Sets showing the modelling in the third dimension

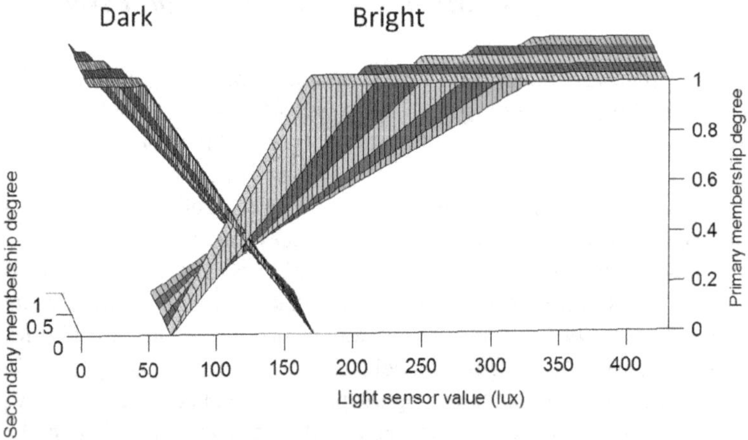

Fig. 5.53 LGT2 model of human-centred Ambient Light Level for living room on 29/03/2011

Level rather than dividing it into individual linguistic labels (eg. Medium, High, etc.). Figure 5.53 is created using 1 day of data captured on 29/03/2011. As can be seen, the domain of the x-axis in Figs. 5.53, 5.54, 5.55, and 5.56 changes, in other words, adapts to the environment as part of fulfilling AA requirements.

Furthermore, Fig. 5.54 presents the adaptation on the domain of the x-axis as the light levels recorded on the third day (31/03/2011) show an increase, which is

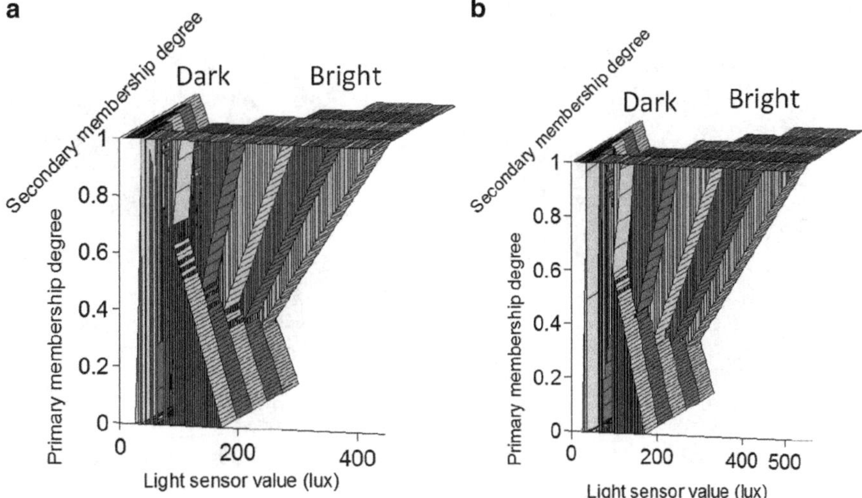

Fig. 5.54 (**a**) LGT2 model of human-centred Ambient Light Level for living room for 29-30/03/2011 (**b**) LGT2 model of human-centred Ambient Light Level for living room for 29-31/03/2011

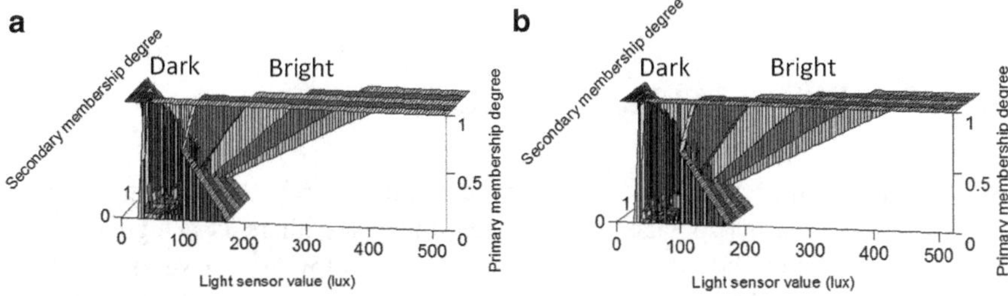

Fig. 5.55 (**a**) LGT2 model of human-centred Ambient Light Level for living room for 29/03/2011–01/04/2011 (**b**) LGT2 model of human-centred Ambient Light Level for living room for 29/03/2011–02/04/2011

immediately reflected in the design and represented by the Ambient Light Level model. It should also be noted that, the nested FOUs (Footprint of Uncertainty) change in thickness, suggesting that the data has become more varied. This fact is further illustrated in Fig. 5.55.

Figure 5.56 includes the final model created using the 7 day data for Period 2. The benefits of using LGT2 fuzzy sets lie in their efficient design and ability to adapt on the fly. Hence, as new data becomes available, the LGT2 fuzzy sets can be created accordingly, without the need to extract uncertainty information. This feature makes the AA to be more independent as the model directly relies on the available data without the need of in between calculations.

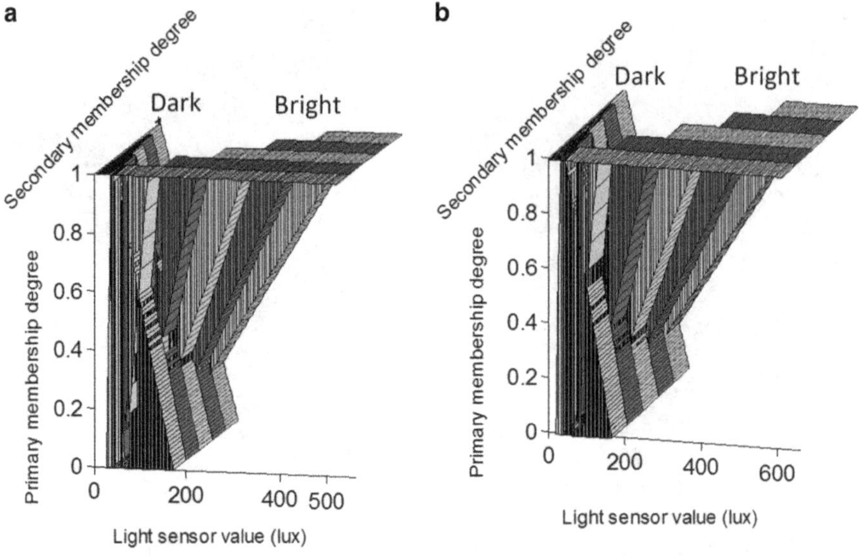

Fig. 5.56 (**a**) LGT2 model of human-centred Ambient Light Level for living room for 29/03/2011–03/04/2011 (**b**) LGT2 model of human-centred Ambient Light Level for living room for Period 2 (29/03/2011–04/04/2011)

5.8 Conclusions and Future Work

Artefact Adaption (AA) is a crucial component at the heart of AIEs. As the number and variety of artefacts/devices within AIEs increases, there is a high need for dynamic, automatic and adaptive mechanisms to employ and make sense of the information provided by this variety of potentially heterogeneous devices. As part of ATRACO, we have investigated AA over multiple years and have adopted two main approaches which we refer to as Stage 1 [13] and Stage 2 approaches.

To briefly recapitulate, the Stage 1 approach focuses on the dynamic creation of adaptive device models based on mainly short-term, direct user-device interaction information. We have shown the successful application of the approach in [13]. After the initial phase of the social evaluation in ATRACO was concluded, the information gained from participants as well as our own experience was combined in order to address some of the perceived shortcomings, in particular the ability to provide stable AA over medium and long periods as well as the need to further reconcile the system view of the resulting AA models and the user view. The latter was of importance, not just to AA, but particularly to UBA (User Behaviour Adaptation) which directly employs the models generated as part of AA to provide user behaviour adaptation.

As a result, the Stage 2 approach to AA was established. It directly focuses on medium to long term adaptation both in the presence and in the absence of the user. The need for the ability to provide AA in the absence of the user was realized to be far more essential in adapting to indirect effects of the user and the environment such as the obstruction of sensors by pieces of clothing, the change of weather or seasons, etc. Furthermore, the Stage 2 approach incorporates the creation of a human-centred model from the information gathered from a series of individual physical artefacts/devices. We have provided in-depth information on how the uncertainty resulting from this information amalgamation is captured and represented in individual timeframe-based models (in our case: days) and finally, how the individual models are combined based on agreement modelling and a sliding window approach in order to provide continuous and stable AA that is able to deal with both short and long term changes in device characteristics. Recently, we have furthered the continuous adaptation approach by employing Linear General Type-2 Fuzzy Sets [3].

Concluding, it is clear that the Stage 1 and Stage 2 are complementary solutions to the complex problem of AA to allow for continuous artefact adaptation over short, medium and long periods, with the users presence and absence while maintaining a user-centred modelling approach creating transparency and thus furthering user-understanding and acceptance. As part of our research in AA, we have achieved significant advancement of the state of art in terms of developing an approach to artefact adaptation in specific, and uncertainty handling/modelling and general type-2 fuzzy sets and systems in general where these advancements could be summarized as follows:

- Novel Artefact Adaptation systems that allowed for the first time the artefacts to adapt to the changes in the artefacts characteristics over short- and long-term intervals which have never been addressed in the literature.
- Novel approaches for general type-2 fuzzy systems based on zSlices as reported in [11, 22–25], and [3] have been identified to allow the adaptation of artefacts to changes of their characteristics and user preferences for the use of these devices.
- Further, multiple devices were conceptualized in terms of their role in contributing to a *wider* human-centred concept (ambient light level) and based on this concept, uncertainty information extracted from the data was combined with prototype type-1 sets to create zSlices-based general type-2 fuzzy sets for the human-centred model.
- A completely novel notion of agreement has been developed, introduced and demonstrated allowing the continuous adaptation of device models over short, medium and long periods [26].
- We have shown the crucial role that the third dimension offered by general type-2 fuzzy sets can play and have demonstrated it both in concept as well as in practice (real world, data-based zSlices general type-2 sets).
- We have further developed advanced uncertainty modelling approaches by using Linear General Type-2 Fuzzy Sets [3] which have been successfully employed in various AIE applications [4–6] outperforming its IT2 counterparts in terms of processing time, output coverage and user satisfaction.

- By developing the mechanisms for employing zSlices-based human-centred concepts (based on information from many devices), we have enabled the creation of significantly more transparent (easier to interpret) rule bases, which is a requirement for user understanding and thus acceptance of the system.

Lastly, we have established AA as a dynamic component, which is encapsulated with UBA (User Behaviour Adaptation) and which dynamically communicates through the SM with the Ontology Manager, thus enabling the dynamic enrichment of the individual users' ontology with knowledge accumulated by the AA. This dynamic handling of user related information has the additional advantage of enabling the maintenance of privacy and access control to the data employed as part of AA.

References

1. Abowd, G.D., Mynatt, E.D., Rodden, T.: The human experience. IEEE Pervasive Comput. **1**(1), 48–57 (2002)
2. Bilgin, A., Hagras, H., Malibari, A., Alhaddad, M.J., Alghazzawi, D.: A general type-2 fuzzy logic approach for adaptive modeling of perceptions for computing with words. In: 2012 12th UK Workshop on Computational Intelligence (UKCI), pp. 1–8. IEEE, Edinburgh (2012)
3. Bilgin, A., Hagras, H., Malibari, A., Alhaddad, M.J., Alghazzawi, D.: Towards a general type-2 fuzzy logic approach for computing with words using linear adjectives. In: 2012 IEEE International Conference on Fuzzy Systems (FUZZ-IEEE), pp. 1–8. IEEE, Brisbane (2012)
4. Bilgin, A., Hagras, H., Malibari, A., Alghazzawi, D., Mohammed, J.: A computing with words framework for ambient intelligence. In: 2013 IEEE International Conference on Systems, Man, and Cybernetics (SMC), pp. 2887–2892. IEEE, Manchester (2013)
5. Bilgin, A., Hagras, H., Malibari, A., Alhaddad, M.J., Alghazzawi, D.: An experience based linear general type-2 fuzzy logic approach for computing with words. In: 2013 IEEE International Conference on Fuzzy Systems (FUZZ), pp. 1–8. IEEE, Hyderabad (2013)
6. Bilgin, A., Hagras, H., Malibari, A., Alhaddad, M.J., Alghazzawi, D.: Towards a linear general type-2 fuzzy logic based approach for computing with words. Soft Comput. **17**(12), 2203–2222 (2013)
7. Cook, D., Das, S.: Smart Environments: Technology, Protocols and Applications, vol. 43. Wiley, New York (2004)
8. Coupland, S., Mendel, J.M., Wu, D.: Enhanced interval approach for encoding words into interval type-2 fuzzy sets and convergence of the word fous. In: 2010 IEEE International Conference on Fuzzy Systems (FUZZ), pp. 1–8. IEEE, Barcelona (2010)
9. Davidsson, P., Boman, M.: Distributed monitoring and control of office buildings by embedded agents. Inf. Sci. **171**(4), 293–307 (2005)
10. Doctor, F., Hagras, H., Callaghan, V.: A fuzzy embedded agent-based approach for realizing ambient intelligence in intelligent inhabited environments. IEEE Trans. Syst. Man Cybern. Syst. Hum **35**(1), 55–65 (2005)
11. Hagras, C.W.H.: Novel methods for the design of general type-2 fuzzy sets based on device characteristics and linguistic labels surveys. In: Proc. Int. Fuzzy Syst. Assoc. World Congr. 2009 Eur. Soc. Fuzzy Logic Technol. Conf., pp. 537–543 (2009)
12. Hagras, H.: Type-2 flcs: a new generation of fuzzy controllers. IEEE Comput. Intell. Mag. **2**(1), 30–43 (2007)

13. Hagras, H., Wagner, C.: Artefact adaptation in ambient intelligent environments. In: Minker, W., Heinroth, T. (eds.) Next Generation Intelligent Environments, pp. 127–151. Springer, New York (2011). doi:10.1007/978-1-4614-1299-1_4. http://dx.doi.org/10.1007/978-1-4614-1299-1_4
14. Hagras, H., Callaghan, V., Colley, M., Clarke, G., Pounds-Cornish, A., Duman, H.: Creating an ambient-intelligence environment using embedded agents. IEEE Intell. Syst. **19**(6), 12–20 (2004)
15. Hagras, H., Doctor, F., Callaghan, V., Lopez, A.: An incremental adaptive life long learning approach for type-2 fuzzy embedded agents in ambient intelligent environments. IEEE Trans. Fuzzy Syst. **15**(1), 41–55 (2007)
16. Kidd, C.D., Orr, R., Abowd, G.D., Atkeson, C.G., Essa, I.A., MacIntyre, B., Mynatt, E., Starner, T.E., Newstetter, W.: The aware home: a living laboratory for ubiquitous computing research. In: Cooperative Buildings. Integrating Information, Organizations, and Architecture, pp. 191–198. Springer, Heidelberg (1999)
17. Kushwaha, N., Kim, M., Kim, D.Y., Cho, W.D.: An intelligent agent for ubiquitous computing environments: smart home ut-agent. In: Proceedings. Second IEEE Workshop on Software Technologies for Future Embedded and Ubiquitous Systems, 2004, pp. 157–159. IEEE, Piscataway (2004)
18. Mendel, J.M.: Uncertain Rule-Based Fuzzy Logic System: Introduction and New Directions. Prentice–Hall PTR, Upper Saddle River (2001)
19. Minar, N., Gray, M., Roup, O., Krikorian, R., Maes, P.: Hive: distributed agents for networking things. In: Proceedings. First International Symposium on Agent Systems and Applications, 1999 and Third International Symposium on Mobile Agents, pp. 118–129. IEEE, Palm Springs (1999)
20. Rutishauser, U., Joller, J., Douglas, R.: Control and learning of ambience by an intelligent building. IEEE Trans. Syst. Man Cybern. Syst. Hum. **35**(1), 121–132 (2005)
21. Wagner, C., Hagras, H.: A genetic algorithm based architecture for evolving type-2 fuzzy logic controllers for real world autonomous mobile robots. In: IEEE International Fuzzy Systems Conference, 2007 (FUZZ-IEEE 2007), pp. 1–6. IEEE, London (2007)
22. Wagner, C., Hagras, H.: zslices—towards bridging the gap between interval and general type-2 fuzzy logic. In: IEEE International Conference on Fuzzy Systems, 2008 (FUZZ-IEEE 2008) (IEEE World Congress on Computational Intelligence), pp. 489–497. IEEE, Hong Kong (2008)
23. Wagner, C., Hagras, H.: zslices based general type-2 flc for the control of autonomous mobile robots in real world environments. In: IEEE International Conference on Fuzzy Systems, 2009 (FUZZ-IEEE 2009), pp. 718–725. IEEE, Jeju Island (2009)
24. Wagner, C., Hagras, H.: An approach for the generation and adaptation of zslices based general type-2 fuzzy sets from interval type-2 fuzzy sets to model agreement with application to intelligent environments. In: 2010 IEEE International Conference on Fuzzy Systems (FUZZ), pp. 1–8. IEEE, Barcelona (2010)
25. Wagner, C., Hagras, H.: Toward general type-2 fuzzy logic systems based on zslices. IEEE Trans. Fuzzy Syst. **18**(4), 637–660 (2010)
26. Wagner, C., Hagras, H.: Employing zslices based general type-2 fuzzy sets to model multi level agreement. In: 2011 IEEE Symposium on Advances in Type-2 Fuzzy Logic Systems (T2FUZZ), pp. 50–57. IEEE, Paris (2011)
27. Youngblood, G.M., Cook, D.J., Holder, L.B.: Managing adaptive versatile environments. Pervasive Mob. Comput. **1**(4), 373–403 (2005)
28. Zadeh, L.A.: Fuzzy logic = computing with words. IEEE Trans. Fuzzy Syst. **4**(2), 103–111 (1996)

Chapter 6
User Interaction Adaptation Within Ambient Environments

Gaëtan Pruvost, Tobias Heinroth, Yacine Bellik, and Wolfgang Minker

Abstract Ambient environments introduce new user interaction issues. The interaction environment which was static and closed becomes open, heterogeneous and dynamic. The variety of users, devices and physical environments leads to a more complex interaction context. As a consequence, the user interface has to adapt itself to preserve its utility and usability. It is no longer reasonable to continue to propose static and rigid interfaces while users, systems and environments are more and more diversified. To the dynamic nature of the interaction context introduced by ambient environments, the user interface must also respond by a dynamic adaptation. Thanks to the interaction richness it can offer, multimodality represents an interesting solution to this adaptation problem. The objective is to exploit all the interaction capabilities available to the system at a given moment, to instantiate and evolve user interfaces. In this chapter, we start by presenting a survey of the state of the art on user interaction adaptation. After, discussing the limitations of the existing approaches, we present our proposals to achieve user interaction adaptation within ambient environments. Then we describe the derived software architecture and the user evaluation it led to. We conclude by some directions for future work.

6.1 Introduction

During the last years, the use of computers has largely popularized. From kids to seniors, from novices to experts, the ubiquity of computers has impacted a constantly growing variety of users. At the same time, advances in the miniaturization of electronic components have allowed the development of a large variety of portable devices [laptops, mobile phones, personnel digital assistants (PDA), etc.]. The devices are various but most of the tasks users want to realize are not

G. Pruvost • Y. Bellik (✉)
National Center for Scientific Research (LIMSI-CNRS) BP 133, 91403 Orsay cedex, France
e-mail: gaetan.pruvost@limsi.fr; yacine.bellik@limsi.fr

T. Heinroth • W. Minker
Institute of Information Technology, Ulm University, Ulm 89081, Germany
e-mail: tobias.heinroth@uni-ulm.de; wolfgang.minker@uni-ulm.de

© Springer International Publishing Switzerland 2016
S. Ultes et al. (eds.), *Next Generation Intelligent Environments*,
DOI 10.1007/978-3-319-23452-6_6

device specific. Thus, the *same* task can often be realized through *different* devices, depending on the situation. For instance, it is now common to see people read their e-mails in the bus with a smart phone whereas they would rather use the family computer when they are at home. Nowadays user interfaces (UI) are designed for the specific device they will run on. The same task can be executed on different devices, but manipulating user interfaces that don't have any common point. The users' mobility raises many new interaction situations and for each one, users must go through a new learning phase. Not being able to transfer the skills you have with computers to other interactive devices can be highly frustrating.

Furthermore, while the number of interactive devices is increasing, the capabilities and the richness of interaction should increase too. However, because each device lives in its own closed world, it is not trivial to combine the interaction capabilities of one another into a more synergetic user interface. Users are currently interacting with one independent device at a time. That behaviour is the opposite of the notion of ambient ecology described in Sect. 1.3. When it comes to interaction, the ecology should provide rich interaction with users by combining the capabilities of the different available interaction devices. In this chapter, we will discuss the adaptation of interactive systems in such environments.

In order to clearly explain what adaptation is and what its challenges are, we first need to define a few terms relative to interaction. Taxonomies of interaction generally involve three main concepts: **mode**, **modality** and **media**. The meaning of those terms might slightly change between authors [8, 11, 28, 47]. In our case we adopt user-oriented definitions [6, 69]. A mode refers to the human sensory system used to produce or perceive given information (visual, gestural, auditory, oral, tactile, etc.). A modality is defined by the information structure that is perceived by the user (text, ring, vibration, etc.) and not the structure used by the system. Finally, a medium is a physical device which supports the expression of a modality (screen, loudspeakers, etc.). These three interaction means are dependent. A set of modalities may be associated with a given mode and a set of media may be associated with a given modality.

Before situating our approach compared to similar work, we will define the concept of interaction adaptation—Sect. 6.1.1—and the specific challenges it introduces when it comes to Intelligent Environments (IEs)—Sect. 6.1.2.

6.1.1 Adaptation of User Interaction

The interaction adaptation [29] may have several meanings depending on the adopted point of view (user centred [13, 20], target-oriented adaptation [14, 36, 66], software architecture adaptation [71], etc.). Among the different points of view, it emerges that interaction adaptation involves the following main concepts:

- Actor: represents the entity responsible for the adaptation task. It can be the user, the designer, the system, etc.
- Components: represent the software entities that will be modified to achieve adaptation. It can be the help system, the kernel core, the task model, the dialog controller, the physical or logical interaction objects [5], etc.
- Time: represents the moment when the adaptation is performed. The adaptation can be *static* (performed during the design phase), *dynamic* (performed at runtime), and sometimes performed between sessions. Certain authors refer to static adaptation by the word *adaptability* while the dynamic adaptation is referred to by *adaptivity* [26, 65].
- Direction: represents the adaptation orientation. The system may adapt its outputs and/or adapt itself to inputs (artefacts adaptation).
- Target: represents the entity we want to adapt to. This is usually denoted by the *interaction context*.

Several definitions exist to describe the notion of *interaction context*. Within the HCI research community the most used definition is the triplet <User, System, Environment> [15, 60, 73]:

- User: the user is described by a profile which informs about their preferences, cultural characteristics, cognitive and sensory-motor capacities, etc. Those can be static (for instance a handicap) and/or dynamic (for instance, the user is not looking at the screen).
- System: the system represents the physical (devices) and logical (software) resources.
- Environment: it represents the physical environment where the interaction is done (luminosity, noise level, etc.)

Some authors include further information such as the current user activity [68] (which can be attached to the user profile), but what must be described in the interaction context depends on what you expect the system to adapt to. Next section explains what is specific in the interaction context of ambient systems.

6.1.2 *Ambient Specific Challenges*

Ambient systems are, by definition, integrated in a physical environment. Not only will they be physically located in those environments—like usual personal computers—but also they will offer users the opportunity to interact with this environment. The classical desktop interaction takes users into the virtual world, proposing them to interact only with virtual representations—for instance, files. On the contrary, the main application of ambient system is to enhance the real world with digital properties, to bridge the gap between the real world and the virtual world by proposing enhanced interaction with physical objects.

As a consequence, the main differences between classical interaction and ambient interaction are:

- Heterogeneity and Distributivity: The system is not composed of a static set of devices (Computing unit, screen, mouse, keyboard). It is the collaborative reunion of several screens, mice, keyboards, microphones and other devices, some of them offering interactive capabilities.
- Dynamic Media Mobility: Various media can enter and leave the ambient space during interaction (as users will move around taking some devices with them)
- User Mobility: User can move inside the ambient space. We can no longer assume that they are always behind the screen, holding the mouse and the keyboard.

The system needs to combine several media that are not known in advance and which can provide interaction through different modes and modalities. As a consequence, it must be able to provide interfaces that combine those different modalities of interaction. This property is called multimodality—see [34] for a survey of the domain. In general, when developing multimodal systems, the designer knows in advance which modalities it offers, and what is the structure of information that the media can send/receive. In ambient systems, however, no assumption can be made about the modalities and media that will be available. In fact, the system must adapt to what is available at runtime and may even have to change during execution if one of the media disappears from the ecology. Adapting to the mobility of devices requires to be able to discover the different media, to model their respective interaction capabilities, and to reason on those capabilities in order to combine them. Besides dynamically appearing/disappearing from the ecology, media are distributed across a network. In classical situations, all the interaction devices are connected to the same computer whereas in ambient systems, it is needed to have protocols that enable the dynamic discovery of those media and the routing of their events towards an entity that can interpret them. Interaction adaptation thus involves to dynamically re-route the flow of data between different media. Last but not least, when users move around in an ambient space, they can expect to seamlessly resume tasks they have started somewhere else. Adapting the interaction to the user mobility requires the possibility to migrate a user interface from one media to another while preserving the current state of the task. Of course, all the features of a task may not be well represented on every device. It is then the role of the system to provide the best compromise between the interaction capabilities of a media and the features that it can provide for a specific task.

In order to explain how such challenges can be addressed, we will first present a state of the art of this research field. We will discuss the different approaches and their respective limits. It will lead us, in Sect. 6.3, to introduce our vision of interaction adaptation and to present our focus within the ATRACO project. In Sects. 6.4 and 6.5, we will demonstrate how we can achieve adaptation on three different levels: Allocation, Instantiation and Evolution. We will first focus on a generic multimodal adaptation system that reasons on the interaction context to allocate and instantiate appropriate user interfaces. Then, we show how the developed concepts can be applied to provide adaptation within a single modality.

We will focus on the spoken modality, which offers rich interaction situations in ambient environments, and demonstrate how adaptation can be achieved during on-going dialogue—see Sect. 6.5.2. Based on this theoretical work, we show in Sects. 6.6.1 and 6.6.2 how a software agent responsible of interaction adaptation—called Interaction Agent—has been integrated and evaluated in the ATRACO project. We then conclude by giving some perspectives opened by our work.

6.2 Related Work

In this section, we establish an overview of the state of the art about the adaptation of the interaction to the context in ambient systems. This study of the state of the art gives us a global view of the different dimensions to which an interactive system can adapt. Based on this study, we define some limits of the current systems that we address in ATRACO.

6.2.1 Multimodality, Information Presentation and Model Driven Approaches

There are many interpretations of the term adaptation. In [37], Kolski and Le Strugeon present a typology of different categories of interaction adaptation. They demonstrate that interaction adaptation ranges from simple adaptation of presentation parameters to what they call intelligent agents. Intelligent agents have deep models of users, tasks and planification capabilities. The focus of this study is on how to mapped those adaptation mechanisms to existing interaction models. It results from this study that one research line for smart adaptation of interaction is in using multimodality and in breaking the "couple" composed of one human and one computer; replacing it by a group that includes several computer agents and users.

The use of multimodality for adaptation of interaction has been first explored through the issue of multimodal presentation [11] and the modelling of interaction context [64, 66]. Multimodal presentation focuses on selecting the best communication channels—i.e. the best modalities—to present information, given a certain context. User interface adaptation has been explored notably with the concept of plasticity [70] that tackles the issue of modifying the graphical user interface depending on the inputs, the outputs and the running platform. In ambient systems, two main issues complicate the adaptation of interaction. First, peripherals are heterogeneous and distributed over a network and second, the context of interaction, as defined in [19] is very dynamic and may even change during the interaction. In [17], the concept of Meta-UI is defined in order to represent the set of functions that are necessary and sufficient to control and evaluate the state of

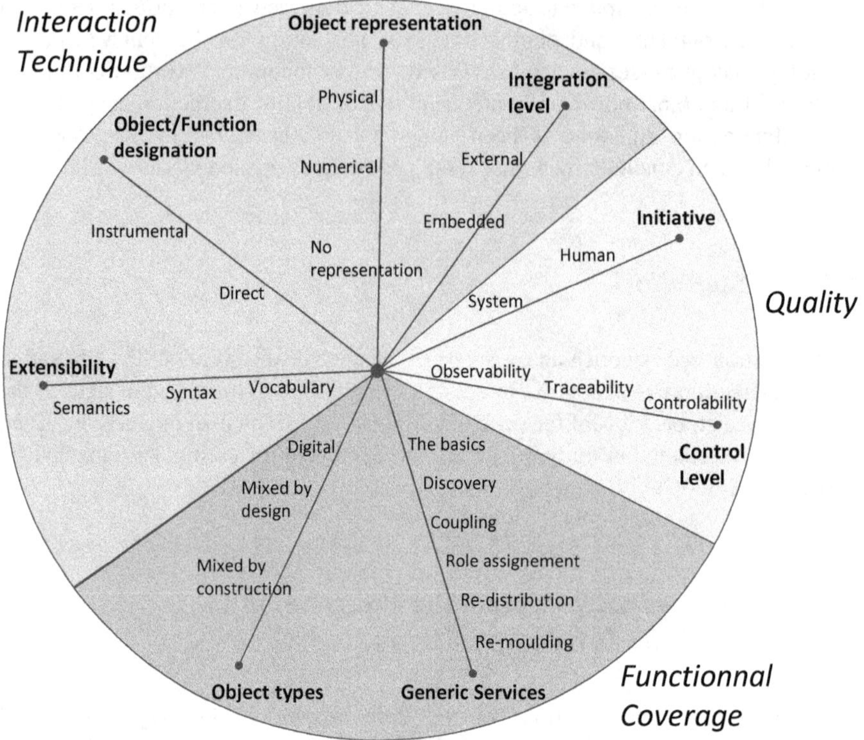

Fig. 6.1 Interaction adaptation taxonomy (courtesy of [17])

an interactive ambient space. A taxonomy is detailed—see Fig. 6.1—that shows 8 axes parted in three categories: Quality, Functional coverage and Interaction techniques. The Quality and Functional coverage are not specific to interactive systems, they are also reflected in other adaptation components of the ATRACO system. The category "Interaction techniques" provides a first classification of ambient interaction techniques.

Whereas previous works focused on output adaptation, plasticity inspired a more general framework called CAMELEON-RT [4]. This framework explores the different dimensions of interaction adaptation and describes a reference conceptual architecture to implement it both for inputs and outputs. From a technical point of view, this framework highlights the need for an architecture that deals with a multitude of interaction devices. From a conceptual point of view, CAMELEON-RT stresses the importance of using various modalities for improving communication with the user both on efficiency and intuitiveness levels. It provides concepts to further analyse the combination of modalities within a task-oriented approach [45].

As far as information presentation is concerned, a four-level model has been developed named WWHT [59]. This model defines four levels of interaction adaptation: What, Which, How, Then. It presents a rule-based system that allows selecting the best communication channels depending on a context model. Using the CARE properties—complementarity, assignation, redundancy and equivalence [18], richness of multimodality is used to improve robustness and expressivity of the system interaction. More details are given on this approach in Sect. 6.3.1.

The implementation of distributed interactive systems is still an open question though several architectures [3, 38, 57, 67] have been developed. More recently, the OpenInterface European project [62] has developed an architecture that overcomes some technical problems such as network communication and multiple input devices configurations. Next section presents the conceptual and technical issues that remains when it comes to move from a user interface localized on a specific platform to a user interface dispersed in an ambient environment.

6.2.2 Limitations of Those Approaches

Limits of adaptive interaction systems are strongly connected to the way context is modelled. The interaction context is usually defined as any information relative to a person, a place or an object considered as relevant for the interaction between the user and the system [19]. Research work in this field has analysed key-problems of those dynamic contexts [32] and proposed several solutions based on distributed systems [3] or on uncertainty models [49]. However, no general agreement has been reached on what should be modelled and how to use this knowledge. Those concerns are addressed in the WWHT framework [59]. Yet, another problem with this framework—and the others—is the adaptability of a specific model to a completely new context. Indeed, once the designer has overcome the first issue of choosing which information to represent, he/she needs to connect this information to an information source. This source can be a sensor, another program, or the user itself. No standard exists for those information sources and consequently, a model that has been specified for a specific environment might not be usable in another environment. This is the case, for instance, in the WWHT framework. In this framework, the context is described as a set of variables—i.e. a name associated to a native value: int or boolean. Reusing a specific model in a new environment would imply binding the new discovered sensors—and other information sources—to the variables of the model. Because no semantic description is associated to those variables, this process cannot be automated. Besides modelling context, the wide variety of channels of communication offered by ambient systems leads to a decidability issue for the interactive system. The interactive system must take a decision as to which modality and which language of interaction it should use to communicate with the user. Contrary to other decision-making systems, there is no such thing as an optimal solution with suboptimal versions of it. As explained by Rousseau in [59], a basic expert system cannot just apply rules to infer the best solutions because the "best" solution will be different from one designer to another. That is why Rousseau introduces an algorithm based on rules to evaluate solutions according to a given behavioural model. This means that behavioural rules

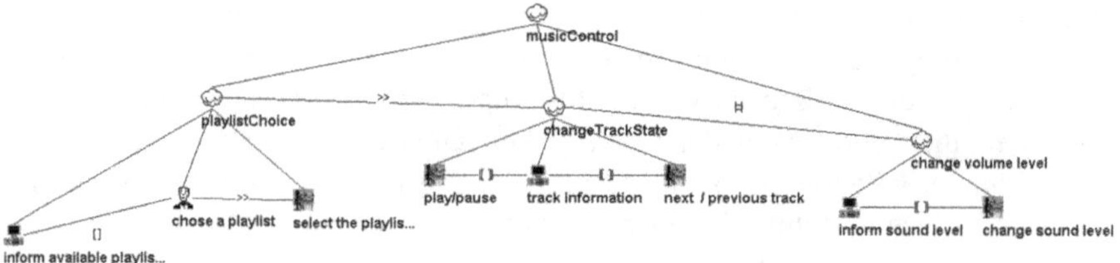

Fig. 6.2 An example of the CTT notation

are provided by the designer and by applying those rules, the different solutions are granted a mark that serves to rank them. Consequently, our aim is not to provide a tool to select the best modality. It is rather to provide a tool enabling designers—and possibly the user—to define in a generic fashion, what is the best way of adapting interaction. Of course, such a tool would have to apply those specifications at runtime and be able to instantiate the solution that it generated.

It leads us to our last problem: the automatic generation of a user interfaces. In order to automate the process of designing the interaction, it is first needed to formally describe interaction tasks. Some tools have been proposed for that, one of the most well-known is the ConcurTaskTree formalism [51]. Figure 6.2 shows an example of a task described in this formalism. It allows for a multi-level description of the task, its subtasks and the temporal relationships between them. Based on this description, the designer can have a platform independent view of the user interface he/she needs to implement. It helps him/her designing consistently for a same task across several platforms [45].

In the TERESA framework [45], the CTT notation is refined in an Abstract User Interface (AUI) that is composed of interactors which are abstract interaction objects identified by their semantics. This AUI is then refined in a Concrete User Interface (CUI) composed of concrete interaction objects that depends on the target platform. Those concrete interaction objects have attributes such as the size for instance. Finally, the attributes are selected and the CUI is translated into something usable—e.g. a XHTML description, Java code ...—called the Final User Interface (FUI). Similar approaches have designed with other formalisms like UsiXML [43, 63]. Normally, the designer should be necessary for at least refining the CTT into an AUI since this involves understanding the task's semantics. However, tools [44] have been defined in order to partially automate this generative procedure. This approach allows designing a user interface once and then seamlessly instantiate it on devices with different interaction capabilities such as a touch screen or a PDA. Those refinements are done once and for all at the instantiation of the task. Consequently, this approach does not allow adaptation during the task execution. If a modality suddenly becomes unavailable—for instance, if the user receives a phone call, we want to avoid using speech to interact with him/her—then the system needs to switch to another modality during the task, which is not possible with such an approach. Another counterpart for fully top-down approaches is that the user interface that is generated can only be very simple, it is limited to a form-based interaction for graphical user interfaces and to *multiple choices questions* for spoken dialogue.

Even though great progress has been done in the field of model-based HCI, there are still unexplored research directions. The modelling of interaction context, though being at the heart of this field, is still problematic. The dynamicity of the context model has been largely explored but not its reusability on different environments. A model defines a structure that enables the system to choose an appropriate communication channel. But the model instantiation might be foiled by the heterogeneity of the target platforms and by the wide variety of modalities and corresponding languages that are available. Consequently, the top-down generation of a user interface is an interesting approach, but it does not fill all of our needs, in particular the need for rich interaction techniques and for runtime adaptation.

6.3 The ATRACO Approach to User Interaction Adaptation

To resolve the problem of interaction adaptation, different kinds of approaches exist: translation approaches [24], retro-engineering and migrating approaches [9, 16, 50], Mark-up language-based interfaces and Model Driven Interfaces (MDI) [23, 63]. As seen in the previous section, most of these approaches try to design several interfaces for a same application, each one being adapted to a different interaction context. This kind of approaches can be used when the different possible interaction contexts are not very numerous and well identified, which is rarely the case in ambient systems. Another possibility consists in integrating adaptation mechanisms that allow the interface to dynamically modify its behaviour in order to stay pertinent with the interaction context [59, 70]. This approach is better suited to our needs as the interaction contexts in the case of ambient ecologies may be numerous and very variable. In Sect. 6.3.1, we present in more details the WWHT approach. Though this information presentation doesn't solve all our issues, it is a good starting point that guided the design of our interactive system. In particular, we apply it in Sect. 6.3.2 to specify the different levels of interaction adaptation that are relevant to ambient systems. Finally, in the last section, we detail the global vision of our interaction system and what we have focused on for implementation.

6.3.1 Breaking Down the Specifications of Interaction Adaptation

The WWHT approach models the problem of information presentation. It also provides guidelines to establish the needs and to specify the interaction system that supports information presentation. A complete presentation of this model is given in [30, 59].

The WWHT model relies on four main questions:

- What: what is the information to present?
- Which: which modality(ies) should we use to present this information?
- How: how to present the information using these modalities?
- Then: and then, how to continuously adapt the resulting presentation?

The first three questions (What, Which and How) refer to the initial building of a multimodal presentation while the last one (Then) refers to its future.

What

The starting point of the WWHT model is the semantic information that the system has to present to the user. To reduce the complexity of the problem, WWHT starts by decomposing the initial information into elementary information units. For instance, in the case of a phone call, the information "Call from X" may be decomposed into two elementary information units: the call event and the caller identity.

Which

When the decomposition is done, a presentation has to be allocated to each information unit. The allocation process consists in selecting, for each of them, a multimodal presentation adapted to the current state of interaction context. The resulting presentation is composed by a set of pairs (modality, medium) linked by redundancy/complementarity CARE relations. This process may be complex in particular when several communication modalities are available and/or the interaction context is highly variable. The selection process of presentation means is based on the use of a behavioural model. The representation of this behavioural model may vary depending on the considered system: rules [64], matrices [22], automata [35], Petri Nets [46], etc.

How

When the allocation is done, the resulting multimodal presentation has to be instantiated. Instantiation consists in determining the concrete lexico-syntactical content of the selected modalities and their morphological attributes depending on interaction context. First, a concrete content to be expressed through the presentation modality has to be chosen. Then, presentation attributes (modality-specific attributes like size spatial and temporal parameters) are set. This phase of the WWHT model deals with the complex problem of multimodal generation [1, 55]. Ideally, the content generation should be done automatically. However this is still an open problem for each considered modality such as text generation [56] or gesture generation [12].

Then

This phase addresses the problem of presentation expiration. Indeed, the presentation may be adapted when it is built but there is a risk that it becomes inadequate if the interaction context evolves. This constraint requires the use of mechanisms that allow the presentation to evolve according to the context.

6.3.2 *Our Vision of Adaptation*

The WWHT framework addresses the problem of information presentation but not of interaction. Indeed, it focuses on outputs whereas interaction requires both inputs and outputs. Moreover, strong links exist between inputs and outputs [7] which can influence one another. Thus, WWHT cannot be applied as it is to ambient systems. We propose to apply its philosophy and extend it to design an adaptive interaction system for ambient spaces: the Interaction Agent (IA). For this purpose, we assume that the system needs to interact with the user in two main cases:

- *Control tasks:* Those tasks imply the use of user interfaces to provide the user with persistent control of the environment.
- *Dialogue tasks:* Those tasks are used by the system to provide information to users (i.e. information presentation), collect information from them (what are their preferences, what behaviour the system should have) or collaborate/negotiate with them (helping users or finding a compromise between what they want and what the system can do).

We propose a knowledge-based system—the model will be described in Sect. 6.4—that addresses three levels of adaptation: *allocation* is the problem of selecting the modalities and devices through which the user interface will be expressed; *instantiation* is the issue of selecting the appropriate parameters for an allocated user interface and *evolution* addresses the evolution of the user interface during interaction. Evolution of a user interface is a very broad problem in itself. It can cover migration—re-allocation—in the case of a context change but also dynamically evolving the dialogue content—refinement. This is typically the case for negotiation or collaboration with the user via natural spoken dialogue.

The IA addresses these three levels of adaptation and answers the four WWHT questions. Both interaction tasks—control and dialogue—are triggered by the Sphere Manager (SM) (see Sect. 1.3.3) thereby partly answering the first WWHT question: *What?* In ATRACO, the *what* is not restricted to decomposable units of information, we extend it to complex interaction tasks. Like in WWHT, if adaptability is to be considered, the *what* needs to be described in a formalism that affords decomposing it and treating it at different levels of granularity. Our vision is that previously presented formalisms (like CTT) can be used to describe the interaction tasks at different levels—abstract task or sub task. Based on the system state, the SM can trigger such describing trees in the Interaction Agent. The IA should then analyse such a task description to generate a user interface by implementing tasks of different levels in the tree. For instance, the same task can be implemented as a complete graphical user interface on a big screen—i.e. root of the tree—whereas it would be decomposed into a succession of subtasks for the small screen of a PDA, i.e. the leaves of the tree. The *what* question is thus answered by the SM at the task level—in the activity sphere workflow presented in Sect. 1.4.

The *Which* and *How* stages are enforced by the IA, in an internal module called the Multimodal Manager (MM). It is responsible for deciding which modes and modalities should be instantiated by the use of which media and how—i.e. allocation and instantiation. The MM provides concepts and a rule engine to reason about the possible answers to those questions. It offers a framework for designers to design different behaviours by writing them as a set of rules in a simple language—described in Sect. 6.4.2.

The MM furthermore addresses these issues in a continuous way during an on-going task. In this manner it supports evolution, the *then* stage.

6.3.3 Our Focus

The range of research questions that we tackled in previous sections is quite large. They cannot be addressed all at the same time and require long-term efforts from the whole community in terms of plasticity of User Interfaces (UI), interaction context modelling, user modelling, automatic generation of UIs and multimodality. We showed in Sect. 6.2 that many efforts have already been done in the fields of automatic generation and plasticity. As a consequence, we assume that user interface generation and content generation techniques could be brought into our model later on. We thus chose to restrict ourselves to the problems of interaction context modelling, allocation/instantiation and evolution.

Because we don't know in advance which interaction capabilities the media will provide, user interfaces need to be generated and/or composed at runtime. As we discussed in Sect. 6.2.2, in order to provide rich user interfaces, we need to free ourselves from the fully top-down approach. Because a fully bottom-up approach would lead to inconsistencies in the flow of interaction tasks [61], we opted for a mixed approach in which a set of already existing interaction objects are combined to generate a user interface. We call those interaction objects Off-the-shelf Interaction Objects (OIO) because they are pre-implemented bundles of code that are reused and composed at runtime. Each of these OIOs can understand a specific interaction language and be manipulated by different interaction techniques. The crucial role of OIOs is further detailed in next sections. We discussed, in previous section, the role of automatic generation and composition of these elementary elements in the generation of a user interface. In our implementation though, we decided not to focus on this problem which has already been investigated—see Sect. 6.2.2. We simplify the problem by assuming there will always be at least one OIO available for each interaction task and focus our work on selecting and instantiating the OIO in a variable context of available devices.[1]

In ambient systems, the heterogeneity of available devices offers rich interaction possibilities. To get the most out of this richness, interaction will be multimodal [8] and rely on various peripherals. The problem of the IA is thus to instantiate each

[1]The model we propose in following sections was designed to take into account the global vision and will be expanded to include such plasticity in future versions.

of the interaction tasks through some modalities and choosing which devices to use for that. We do not address the issue of multimodality fusion and fission [40] for the moment and focus on the selection process of the right pairs <OIOs, devices> depending on the context. This approach is inspired by the refining process of Teresa [45]. However, the difference resides in the underlying model. In Teresa, the task model would be refined in abstract interaction objects that can be implemented on any platform. Those abstract interaction objects are elementary interaction objects. Since we wanted a mixed approach for the user interface generation, we designed OIOs to be of arbitrary complexity. An OIO can be a simple button, or it can be a whole widget for music playlist edition. This way, designers have a total freedom on the internal behaviour of the OIO they design, which lets them implement new and/or rich interaction techniques. On the other hand, low-level OIOs could still be composed at runtime to propose a "degraded"—less integrated—user interface.

Tasks are described as multi-level set of subtasks. An OIO can relate to any node of such a hierarchy whereas in Teresa, abstract interaction objects relate to leaves only. This recursive way of considering instantiation allows for instantiation with different granularity depending of the granularity of the task description. Indeed, one of the drawbacks of automatic generation—or computer assisted generation—of user interfaces is that the outcome is often monotonous and offers poor interaction capabilities whereas hand-crafted user interfaces may offer richer and more appealing interactions. Our approach allows for both hand-crafted user interfaces and automatically generated interfaces to coexist.

By taking this approach, we aim at including new hand-crafted interaction techniques right away. We did not implement automatic composition of OIOs since it was out of our focus. However, we designed this approach to create a system open and flexible enough to include automatic generation in later releases. Next section details the model that we developed to represent and reason about interaction context.

6.4 Adapting the Instance to the Context

To address the different levels of adaptation that we exposed, we decided to model the context as an ontology called the Interaction Ontology (IO). The content of this ontology is detailed in Sect. 6.4.1. It enables describing all the different elements of a user interface—i.e. from the user to the media—along with other information about the context. This ontology provides the basic vocabulary the IA will use to gather information about the environment via the Ontology Manager—defined in Sect. 1.4.3. It will support the description of a specific context in generic terms, which allows designing a rule system that can be seamlessly applied to different environments—see Sect. 6.4.2.

6.4.1 Modelling Distributed and Context-Dependent HCI

We developed a model that encompasses the description of the user interface along with two facets: the view of the user and the view of the system. It is important to understand and to take into account that whatever is theoretically possible may not be possible in reality because of physical/hardware constraints on the devices and platforms that are available. The system must take into account a grounded description of its capabilities while at the same time trying to serve the high-level needs of the user. Such high-level models have been introduced in Sect. 6.2. Our attempt at coupling high-level concepts to their more grounded counterparts is presented in the next section. This model is then encoded in an ontology that we describe in Sect. 6.4.1.2.

6.4.1.1 A Conceptual View

Cognitivist approaches conceptualized the problem of human–computer interaction with the notions of mode, modality and medium. Trying to model those concepts and their relationships with OIOs, we realized that they were not sufficient to describe the whole chain of perception/communication. In fact, whereas the relationships between modes and modalities are quite clear (see the notions of primary and secondary modes in [8]) the relations between modalities and media haven't been much investigated. A few decades ago, only one medium could be used for each modality (the screen for graphics or text output, the mouse for pointing input, the keyboard for text input, microphone for voice). Now that other technologies are available (touch screens, gesture recognition, accelerometers...), we do not only discover new modalities, but also new ways of using already existing modalities. A Wiimote,[2] for instance, offers different capabilities than a mouse, but it can also be used as a pointing device. The device is different but the language—and its structure—is the same, it still consists in moving a pointer across the screen and clicking. Thus, the user will be able to use and perceive the pointing input modality through a mouse, a touch screen or a Wiimote, with a low cognitive charge. Devices are also subject to variations. Even though the standard mouse has two buttons, some have more. The same observation can be made for keyboards. It is quite obvious to degrade the functionalities of such an extended device to the functionalities of the standard device it derives from. However, the extended capabilities should be used as much as possible. Finally, devices may be wrapped to emulate the behaviour of other devices. The device might not be really adapted but in the case where no suitable device is available, this would prevent the system from just failing and provide a degraded user interface. For instance, if the modality is typing and no keyboard is near the user, we can use a virtual keyboard controlled via the mouse.

[2]Wiimote is a sensor for video games that was designed by Nintendo in 2005. It is a remote control that includes vibrators, orientation sensors and screen pointing capabilities.

As we can see, there is no one to one relationship between the modality and the medium. One modality might need several media and one media can support several interaction languages, thus several modalities. By definition [6], modality refers to the structure of the information as it is perceived by the user. However it is also related to the structure of the interaction language used by the system. By modelling directly the relationship between modalities and media, we do not take into account the description of such an interaction language. That is why we developed the concept of interaction language that specifies the different communication acts within a modality. Devices provide inputs/outputs at a very low-level that is dependent on the hardware implementation of the device. This set of native signals is called interaction lexicon. As we have seen with the previous examples, there is a need to describe how the system can move from the interaction lexicon to the interaction syntax, or from the user point of view, how to make a device reflect/impact the interaction through a certain modality. This is the responsibility of modules that we call Mediators. Several kinds of mediators can be envisaged depending on the concerned modality. Indeed, the variety of inputs/outputs is such that there is no global approach to that problem. In graphical user interfaces, it is usually admitted that state machines and their extensions [2, 21] best suits the interpretation of device events. However, this approach is very bad for spoken dialogue, and even worse for natural spoken dialogues—we will see in Sect. 6.5.2 that spoken dialog is important in ambient environments and how to implement such a mediator. We thus allow different kinds of mediators to cohabit within our system.

Figure 6.3 summarizes the relationships between the different concepts that we have just presented. We can see that the initial concepts of Mode, Modality and Media represent the user—and designer—facet of the interaction components,

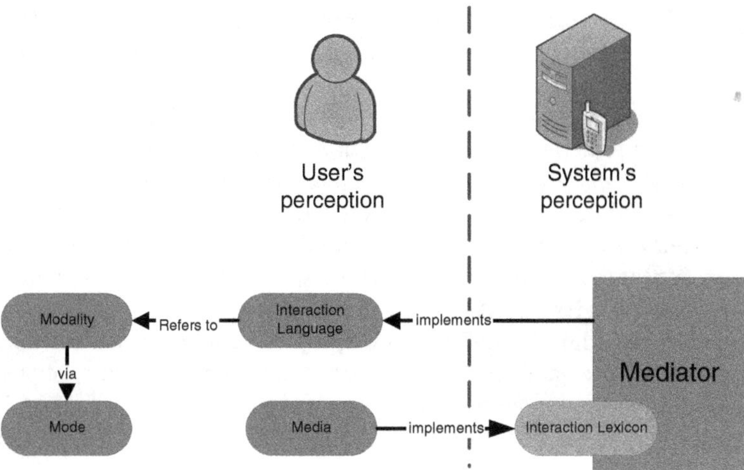

Fig. 6.3 The basic concepts and their relationships to the user or the system

whereas the Lexicon and the Mediators refer to a more formal specification that the system is able to interpret in order to instantiate the interaction. Our model thus unifies the system view of interaction with the users' and the designers' ones. Next section presents in details the modelling of those concepts.

6.4.1.2 Modelling Rich Interaction Context with Ontologies

The Interaction Ontology[3] is the knowledge base on which the Interaction Agent relies to describe the interaction context. In WWHT, the context was modelled as a simple set of variables, each having a certain range but with no connection between variables and thus no meaning associated to it. Moreover, no standard specification had been given concerning variables that could be found in different applications. For instance, if the noise level has to be part of the context, it could be represented by a variable named noise-level in application A and level-of-noise in application B. Consequently, the behavioural model of application A cannot be reused during the design of application B.

Our approach consists in modelling the context by structuring it in a generic way that includes semantics. This way, two different applications could exchange behavioural model seamlessly since they would rely on the same Interaction Ontology. Even if ontologies were slightly different, alignment methods could be used to try and automatically adapt a behavioural model to a new context of use (see Sect. 3.4). Moreover, standard relationships between concepts are reflected between instances, which allows reasoning. It enables us to make use of the richness of description logics inference in order to describe and apply the election rules. By using semantic web representation (OWL) of interaction context, behavioural models can be more complex and seamlessly adapted from one environment to another which was not the case in the WWHT framework.

Figure 6.4 shows the main concepts of the IO and their hierarchy. The *User* and the *EnvironmentalConditions* subconcepts serve as alignment points for other concepts of the Sphere Ontology. The *User* subconcepts, more particularly, should be aligned with User profile ontologies. This mechanism allows for semantically rich information gathering by the Interaction Agent that is able to ask queries such as *"What is the light level in the room where the user is?"*, *"where is the user?"*. Because those concepts are present in the Interaction Ontology, the IA is able to reason on them, even though the actual sensor value can only be found in some other concepts of the sphere ontology that are specific to a particular context. The alignment method hides the heterogeneity of the various information sources.

Figure 6.5 shows how the different concepts presented in previous section relate to each other in the ontology. In this figure, some concepts and relations have been hidden to make it more readable. The concepts are spread over three layers. The upper layer represents the designer's and user's view. This view encapsulates

[3]http://iroom.limsi.fr/atraco/InteractionOntology.owl.

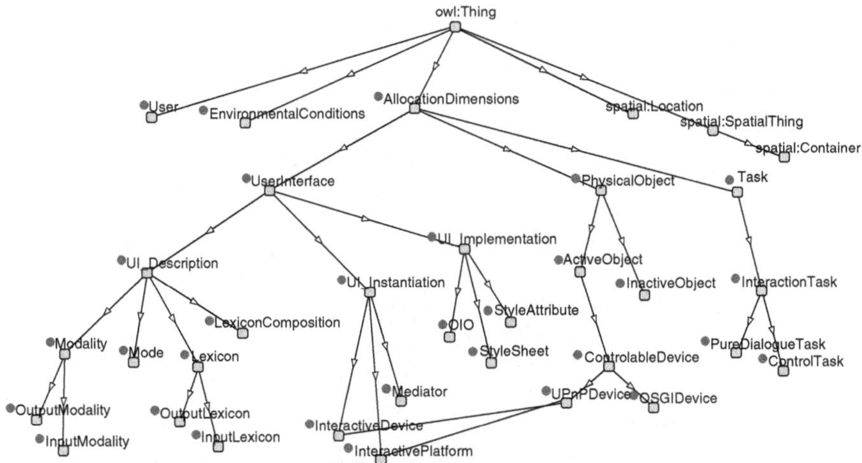

Fig. 6.4 The main categories of the Interaction Ontology

the concepts that are used by psychologists and ergonomists to describe the communication: *User*, *Mode* and *Modality*. The lowest layer represents the aim of the system: the *InteractionTask*. In between, the middle layer tries to establish a connection between the system's needs and the user's perception. The concepts appearing in this layer are hardware devices and software services that can be composed to generate a user interface. Among them, four concepts are mostly important: *OIO*, *Mediator*, *Lexicon* and *InteractiveDevice*. The *InteractiveDevice* is a device used for interaction. In this sense, it is a media. As shown in Fig. 6.3, the *InteractiveDevice* can send—for input devices—or receive—for output devices—a set of structured events. Such a set of events is called a lexicon. A device is said to "speak" a lexicon. Several lexicons can then be composed in a *Mediator*, a software entity responsible for mixing and interpreting events from a *LexiconComposition*. The *Mediator* can then host *OIOs* that have been designed to implement a specific *InteractionTask*.

At this stage of the design, we can draw a parallel between the usual interaction chain implemented in current graphical user interfaces and our model, which tries to abstract this chain of components. If we take the example of a classical computer, the *InteractiveDevices* are the mouse, the keyboard and the screen. They speak the following *Lexicons*: "Pointing" for the mouse, "Typing" for the keyboard and "Displaying" for the screen. We can informally refine the "Pointing" lexicon here by saying it is composed of three events:

- moveTo(x,y)
- right-click()
- left-click()

We could do the same refining for the other lexicons but this is not the point of this comparison—and the list would be too long to be of interest here. Now that we have defined our Lexicons, we would like to implement a language of interaction that we

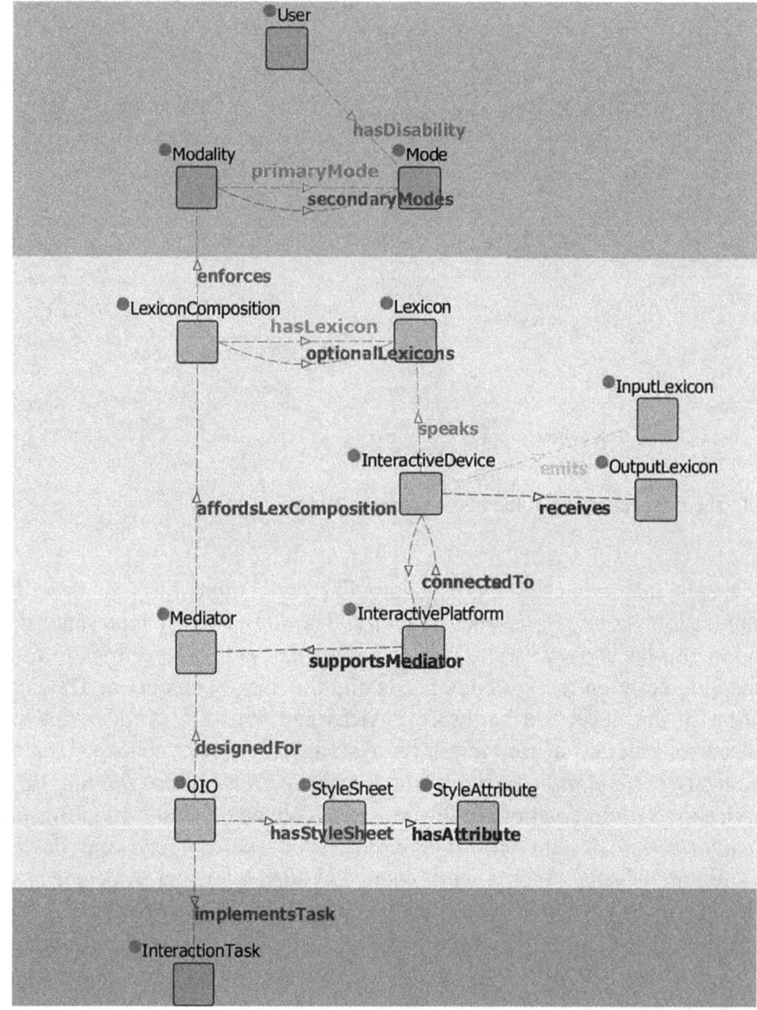

Fig. 6.5 User interface specification in the Interaction Ontology

are all used to: WIMP (Windows, Icon, Menu, Pointer). This graphical language is the one used in many mainstream operating systems. It results from the interpretation of mouse and keyboard events to generate drawing events on the screen. Many implementations have been proposed via libraries and protocols (X11 for Linux, GTK, Swing for Java…). In our model, each of those implementations would have been a specific *Mediator* that *affordsLexComposition* for the LexicalComposition that composes "Pointing", "Typing" and "Displaying".

The problem in classical computing is that this whole chain of components and their respective boundaries is not clearly defined and abstracted. This results in an ad-hoc implementation that prevents or makes more difficult the re-use of existing components and the implementation of new ones. The main asset of an architecture following this model is that by delimiting the boundaries and the role of each component, it allows for exposing their capabilities over the network using web-services protocols (SOAP, UPnP, ...). As a consequence, the system can reuse components that are located on different machines. The interaction is not centralized and isolated any more; it becomes distributed and opened to the world. Also, by disentangling the components of the user interface, the system is able to scale up to new devices with unknown *Lexicons*, as long as a specific language has been implemented as a *Mediator* somewhere over the network. Finally, because this model can handle distributivity of the hardware, and reusability of the code via a service-oriented approach, it allows for a certain plasticity of the communication. By composing the *InteractionDevices* with different *Mediators*, we can achieve different interaction styles relying on different modalities or that mix modalities.

Now, as with any distributed system, one question remains: where is the code running? Our vision is that embedded technologies have come to a point where each *InteractiveDevice* could come with an embedded chip that runs the service and the network connection needed to make it work. Said differently, it is technologically possible to create a mouse that would connect via a local network on Wi-Fi and propose on this network a web service that makes the mouse events available. Still, as we defined them, *OIOs* are pieces of program that realize specific interaction tasks. Those programs have to be run somewhere. For delay reasons and also implementation constraints, it is not trivial to run a graphical user interface on a machine and redirect its output to a screen that is not connected to it. In order to have better results and distribute the computing load, we envision that *OIOs* will be run on different computing units depending on the location of interaction. It will shorten delays and provide more reactivity. In order to provide such flexibility, we need that those computing units provide services to move *OIOs* from one place to another and to deal with several *OIOs* at a time. This set of services is called a meta-HCI [17]. In our model, an *InteractivePlatform* is a computing unit that offers such services. It can be dynamically—or statically—bound to *InteractionDevices*. We call an Interaction Space the pair mixing an *InteractionPlatform* with the set *InteractionDevices* it uses. Figure 6.6 shows the relationships between those concepts. It is interesting to see that the Interaction Space is a concept that matches both the system's view of how components are connected but also the user's view of which devices are "working together". This view helps users anticipate that a specific input device will have impact on a specific output device. Different OIOs can be run in the same Interaction Space, but in order to maintain consistency, an *InteractionDevice* should never be part of two different interaction spaces.

The knowledge base we presented in this section is of course tightly bound—at the implementation level—to the module of the IA that is responsible for reasoning on the context adaptation, the Multimodal Manager. Next section demonstrates how adaptation can be achieved by reasoning on such a model.

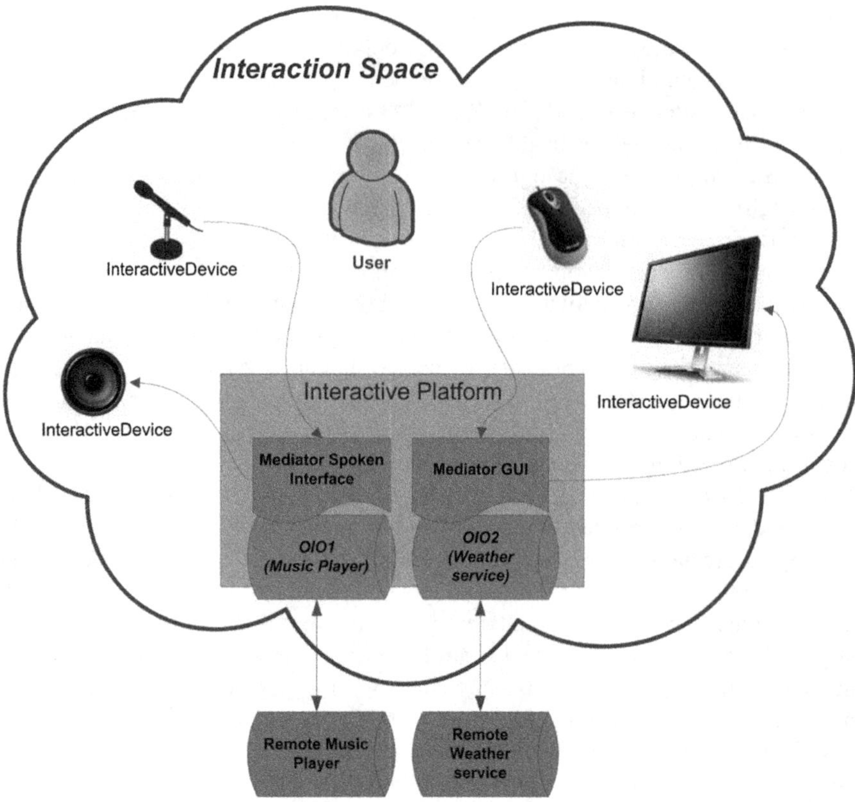

Fig. 6.6 Interaction spaces are formed by composing OIOs, Interactive Devices and Interactive Platforms

6.4.2 Reasoning for Contextual Allocation/Instantiation

In this section, we aim to describe how the model previously presented can be used to dynamically generate the Interaction Spaces—see Fig. 6.6. Having a model of interaction in ambient spaces is not enough. If we want to dynamically adapt, we need to reason over the concepts of this model. Among reasoning techniques, we chose to implement such a reasoner as a rule-based expert system. Rule-based expert systems such as Clips [54] provide the powerful expressivity of second order logic—i.e. logic with quantifiers. We first tried to implement our engine as a set of SWRL rules [33]. SWRL has the advantage of being integrated with OWL and other semantic web technologies. Our model being aligned with other knowledge bases from other ATRACO components, SWRL would have been a good way to express rules that could have been aligned with other concepts in the sphere ontology. However, SWRL has a limited expressivity (no support for quantifiers) and lacks flexibility—it is hard, for instance, to add new built-in predicates to the language.

Thus it didn't fit our needs for implementing a complex reasoning engine. As a consequence, we implemented a bridge from OWL to Jess [27], a Clips derivate that allows for more complex reasoning.

Our approach consists in three steps. First, contextual information from the ecology is gathered in the Interaction Ontology via the Ontology Manager. This information enables us to specify the interaction context within the Interaction Ontology terms previously described. The use of semantic web formats (OWL) at this phase is motivated by the integration in the more general context of ATRACO. This way, semantic web techniques such as alignment can be used by the Ontology Manager to interpret the heterogeneous knowledge bases from the ecology into terms that the Interaction Agent can understand. In the second phase, we use an expert system to apply rules that determine which are the possible associations of interactive devices. We call such solutions the candidates—see Sect. 6.4.2.2. Finally, a set of rules are applied to those candidates to elicit the best option. The behaviour of the interactive ambient system mainly relies on this set of rules. Thus, a consistent set of rules for eliciting the best candidate to interaction is named a behavioural model [59]. Different behavioural models can be described by designers of ambient spaces, the rules being fairly simple to write/modify—see Sect. 6.4.2.3. This section presents the main rules of our engine that enables eliciting the appropriate Interaction Space.

6.4.2.1 Introduction to Jess/CLips Syntax

Our examples relies on the Jess/Clips common syntax that is described in [27, 54].[4] This language uses a functional style to define three main structures:

- Functions
- Facts
- Rules

Facts are a set of predicates that are assumed to be true by the reasoning engine. Those predicates may have a complex structure involving several fields (called slots). Slots can be filled with one data or several (in this case, they are called multislots). The strict structure of Facts allows for fast pattern matching and unification.

A rule is composed of two parts separated by symbol "=>": a left-hand side (LHS) that describes a pattern that needs to be matched in order to execute the right-hand side (RHS). RHS is a set of functions that are called which can add new facts, modify/delete existing ones or modify global variables.

When the engine is run, the rules' LHS are matched against the facts using the Rete algorithm [25]. Each rule that has its LHS matched will be executed—or fired—which means that its RHS will be executed with the variables bound in the LHS.

[4]More details can also be found on the official jess website http://www.jessrules.com.

```
(defrule iplat-mediator-activate
    (InteractivePlatform (name ?ip))
    (supportsMediator ?ip ?mediator)
    (affordsLexComposition ?mediator ?lc)
    (forall (neededLexicons ?lc ?lexicon)
        (connects ?ip ?device)
        (InteractiveDevice (name ?device))
        (speaks ?device ?lexicon))
=>
    (assert (usesMediator (platform ?ip)
                (mediator ?mediator)
                (lexComposition ?lc))))
```
Listing 6.1 Rule that activate valid mediators on Interactive Platforms

6.4.2.2 Enumerating Candidates

In order to construct all the possible Interaction Spaces as described in Fig. 6.6, the system must first discover which are the eligible interactive platforms and which mediators it can run on them. Indeed, the use of a specific mediator requires that the interactive devices that are connected to the platform provide the appropriate lexicons—see Fig. 6.3. Having gathered information about the interactive spaces, the interactive devices and their lexicons via the Interaction Ontology, the system transforms this knowledge into facts in the reasoning engine.

Based on those facts, the inference engine will try to find which mediator can be instantiated for each interactive platform. For this, it looks at all the devices that are connected to the platform and the lexicons they provide. Each mediator needs a specific composition of lexicons (several compositions of lexicons can match the needs of one mediator). For instance, the GUIMediator needs either the displayOutput+mouseInput+KeyboardInput or the displayOutput+touchscreenInput to be able to run. Listing 6.1 demonstrates how mediators are activated on each Interactive Platforms.

After inferring which mediators can be used on which platform, the system will create every valid candidate. A candidate is the combination of an OIO,[5] a mediator, an Interactive Platform and its Interactive Devices. Each candidate is also associated with a style that allows for modifying the morphological attributes of the OIO (for instance, the text size, the voice to be used in speech synthesis...).[6]

Slots *valid*, *validity-reasons*, *salience*, *pros-reasons* and *cons-reasons* are used to grade the candidate. Their use is detailed in the next subsection.

Once mediators are activated, the creation of a candidate is quite simple, it relies on enumerating every possible associations. The LHS of Listing 6.3 declares the

[5] That is the bundle of code to be executed within a specific mediator.

[6] In our implementation, available OIOs and styles are statically specified in the Interaction Ontology, but they could also be specified and retrieved from a remote library if necessary.

```
(deftemplate Candidate
    "A candidate for task instantiation"
    (slot name)
    (slot targetTaskName) (slot oio) (slot platform)
    (slot mediator) (slot style)
    (slot valid) (slot validity-reasons)
    (slot salience (default 0))
    (slot pros-reasons) (slot cons-reasons))
```

Listing 6.2 The template defining Candidate facts

```
(defrule candidates-creation
    (TargetTask (name ?ttask) (interactionTask ?itask))
    (OIO (name ?oio))
    (InteractivePlatform (name ?ip))
    (Mediator (name ?med))
    (StyleSheet (name ?style))
    (hasStyleSheet ?oio ?style)
    (implementsTask ?oio ?itask)
    (designedFor ?oio ?med)
    (usesMediator (platform ?ip) (mediator ?med))
=>
    (bind ?cName (generateId "c"))
    (assert-candidate ?cName ?ttask ?oio ?ip ?med ?style))
```

Listing 6.3 Rule creating every possible candidates

constraints that a Candidate must fill in order to be valid. The TargetTask predicate is a placeholder for all the instances of interaction tasks that are required by the activity sphere. Interaction tasks are generic tasks like *controlling the lights* whereas a TargetTask represents the context-dependent instances of such tasks that are needed, like *controlling the lights in the living-room.*

Listing 6.3 represents how interaction tasks can be implemented via a unique OIO. However, as discussed in Sect. 6.3.3, it would be interesting to describe the whole interaction task as a CTT and use this task decomposition to derive a structure involving a composition of several OIOs. Work is still in progress to adapt our model to involve multiple-OIOs composition.

Once all possible candidates for a specific target task have been enumerated, we need to select the most appropriate ones.

6.4.2.3 The Behavioural Model

In order to implement the allocation level of adaptation, the most suitable candidates need to be identified. There is no unique way to solve this problem, it is rather a matter of design decisions to encourage a certain kind of behaviour from the system. As a consequence, we designed the model to be flexible and give designers the possibility to specify the intended behaviour of the system. Thus, we propose a

```
(defrule disableInvalidModes
    (User (name ?theUser))
    (hasDisability ?theUser ?disability)
    ?c <- (Candidate (name ?nc)
    (oio ?oio) (platform ?ip))
    (designedFor ?oio ?mediator)
    (forall (usesMediator (platform ?ip)
                (mediator ?mediator)
                (lexComposition ?lc))
        (enforces ?lc ?modality)
        (primaryMode ?modality ?disability))
    =>
    (forbid ?c "Implies the use of user's disability"))
```

Listing 6.4 Rule disabling candidates using a user's disability

```
(defrule preferTactileModality
    ?c <- (Candidate (name ?nc) (mediator ?mediator))
    (affordsLexComposition ?mediator ?lexComp)
    (enforces ?lexComp tactile)
    =>
    (approve ?c "User prefers the tactile mode"))
```

Listing 6.5 Rule improving the salience of candidates that use the tactile modality

flexible framework to design various behavioural models. This framework relies on the rule syntax presented earlier and a few set of predefined functions that allow granting marks to the different candidates.

Based on the facts encoded in the Multimodal Manager, the designers can apply rules to favour a certain candidate or forbid the use of some of them. The slot *validity*—see Listing 6.2—is used to forbid the use of some candidates. For instance, in Listing 6.4, a rule is written to disable candidates that make use of a mode that is a user's disability. Such a rule would prevent the system from using speech synthesis when the user is deaf for instance.

There are drawbacks in disabling a candidate altogether. If the candidate was the only one available, then even if it is bad, it might be better than nothing. Of course for deaf people, using speech synthesis will never make sense, whatever the situation is like. Thus it would be safe to use the forbid function. However, in other situations, designers might need to just discourage the use of some candidates and use them only as a last resort. To this extent, the *salience* slot has been added to hold a numeric value representing the appropriateness of a candidate regarding the context on a finer granularity. For instance, Listing 6.5 shows how designers can favour candidates that use the tactile modality over other candidates.

```
(defrule preferBigText
    (User (name ?user))
    ?c <- (Candidate (name ?nc) (platform ?ip)
    (mediator ?mediator) (style bigText))
    (usesMediator (platform ?ip)
    (mediator ?mediator)
    (lexComposition ?lc))
    (neededLexicons ?lc display)
    (connect ?ip ?device)
    (speaks ?device display)
    (not (isnear ?device ?user))
    =>
    (approve ?c "User far from screen, need big letters"))
```

Listing 6.6 Rule encouraging the use of big text when user is not near the screen

To sum up, like in the WWHT framework, four functions where designed to provide fine grained control of the voting process:

- forbid
- use
- approve
- disapprove

forbid and *use* make non reversible assertions on whether or not it is good to use this candidate. In case a conflict appears—i.e. a candidate is both declared usable and forbidden—different conflict resolution strategies can be envisioned. We chose the strategy to disable candidates for which such conflicts appeared. A candidate should be usable if all of the conditions that make it usable are met. Thus, it seems relevant to prefer disabling candidates if only one of the conditions that make them unusable would be met. *approve* and *disapprove* provide more fine grained tuning by increasing/descreasing the *salience* of the candidate. At the end of the process, candidates are ranked by validity: first, the candidates that are declared valid, then the ones for which no assumption has been done about validity and finally the ones that are forbidden are discarded. Within those three categories, candidates are ranked by decreasing salience.

The slots *validity-reasons*, *pro-reasons* and *cons-reasons* are used to hold *string* explanations about the rules that were applied. It is then easier to trace back which rule was applied to explain the Multimodal Manager's decision in a test case or to the user in real conditions.

The instantiation level of adaptation that involves selecting the morphological attributes of the user interface is addressed via the slot *style* of the candidate. Rules can also be written to select the best style. For instance, Listing 6.6 represents a rule that encourage choosing big text style when the user is not in the proximity of the screen.

It is important to note that any arbitrary contextual information that would have been aligned with the Sphere Ontology can be used in rules, as long as they are defined in the Interaction Ontology. Alignment is impossible if the concept is not sufficiently described in both ontologies. Also, the designer needs a concept name as reference to designate such information in the rules. For that reason, the Interaction Ontology acts as a set of reference terms that can be extended. For instance, in the environmental conditions part of the Interaction Ontology, basic concepts are defined to represent the current level of sound and the current luminosity. It allows for writing rules about generic environmental conditions as if they were generally available. If the sensors from the ecology provides such information and the Ontology Manager is able to align it to these concepts in the Interaction Ontology, then the rule would apply with the sensor value. Otherwise, it would be discarded. The behavioural model is thus adaptative to interaction context and environmental context, and, thanks to the alignment technique,[7] the switching from one environment to another is seamless for the designer who writes the rules.

Of course, the allocation and instantiation adaptation that we present here are adapted to the situation when the engine is run, which means at the beginning of the interaction task. We will see in next section how evolution of the interaction task can be achieved during the interaction with the user.

6.5 Continuous Adaptation During Execution

There are different ways adaptation of a task over time can be seen. Several factors might require evolution of the user interface [58]:

- *Information factor:* Whenever the kernel application status change, the change must be reflected in the user interface. For instance, on a multimedia task such as music control, when the current track in the playlist changes, the new track title has to be reflected by the user interface.
- *User's actions:* The user interface must adapt to the user's actions on the user interface itself. For instance, when the user uses the pointing modality, tool tips can be displayed when the user's pointer hovers an icon.
- *Interaction context:* Because of the natural evolution of the ecology, a candidate might become invalid when for instance a device enters or leaves the space.
- *Time factor:* For some applications, like agendas, the system must pro-actively deliver a message to the user depending on a scheduled event.
- *Spatial factor:* The location of different objects and users in the ecology might change and influence the validity of a candidate.

Those different factors require the Multimodal Manager to be aware of the current status of the activity sphere. It is needed that the list of valid candidates and their salience continuously evolves during interaction. Sect. 6.5.1 details how events need to be exchanged with other ATRACO components to adapt during interaction. We can also remark that among these factors, the *user's actions* is the only one

[7]We assume here that such techniques exist. We showed in Sect. 3.4 that even though it is the case, the reliability of those techniques is variable. Our model is thus flexible to the extent that third-party ontologies that describe ecology devices are of a sufficient quality to allow alignment.

that involves only the user and that does not need input from the system. In fact, adaptation to user's actions depends only on the structure of the language that is used to communicate with users, i.e. the modality. Such a modality dependent adaptation cannot be handled at the Multimodal Manager level because it does not imply re-evaluating the salience of a candidate. Such kind of modality dependent adaptation is particularly relevant and complex when it comes to use linguistic modalities [8] such as text interaction or natural spoken interaction. Section 6.5.2 introduces an example of such adaptation, the continuous evolution of spoken dialogue.

6.5.1 Revisiting Allocation/Instantiation Over Time

Information factor and Time factor are both application dependent factors. They occur as a spontaneous change in the status of the kernel application. In ambient spaces, the kernel application is a composition of web services that expose their status over the network. Thus, the Sphere Manager, which is responsible for the handling the task, is able to monitor those changes by subscribing to the corresponding web services. Every time such a change occur, it will send a message to the Interaction Agent to inform it of the new status. The IA then transmits the new status to the OIO so that it updates its internal representation of the task status and reflects the change.

Other factors—Interaction context and Spatial factor—do not have direct impact on the content of the OIO but they can change the salience of a specific candidate. They might require that the candidate be replaced by a similar candidate with another style—this is called refinement—or by a candidate relying on different interactive devices to interact with the user—this is called re-allocation. An example of refinement required by a spatial factor is the size of a text on "text" modality that must continuously adapt to the user's location. Re-allocation is more frequently used to respond to a change in the set of available interactive devices. For instance, when a candidate makes use of a wireless mouse and this one runs out of battery, the interaction task could be re-allocated onto another screen or using a different modality.

The Multimodal Manager handles allocation and instantiation at the start of the interaction task. For that purpose, it is fed with information gathered by the OM. In order to provide refinement and re-allocation, the candidate's saliences must be re-evaluated as soon as the ecology status get deprecated by a dynamic change. The sphere manager, by managing the ecology of the activity sphere, is the component responsible for being aware of such dynamic changes. It will thus advertise those changes to the OM and the Interaction Agent. There are two sorts of events, each representing the structure of the corresponding factor of change:

- activity sphere change event : Represents the apparition/removal of an interactive device in/from the ecology
- Location event : Represents a change in the location of objects and users

Every time such an event is received by the Interaction Agent, the state of the Multimodal Manager is updated and the Allocation/Instantiation process is run again. At the end of it, the status of a candidate might have undergone the following changes:

- *Birth/Death* The candidate is removed if it made no sense any more. New candidates can also be created.
- *Validity change* The candidate's validity is changed.
- *Salience change* The candidate's salience changes.

In case the current candidate becomes outperformed by another candidate, the system is in a non-optimal situation and the conflict needs to be resolved. Different strategies can be envisaged. For instance, designers might prefer sticking to the consistency principle [61] and select the best candidate that uses same devices or the same modalities. Another strategy consists in taking always the best candidate as the current candidate which leads to a much more mobile and adaptive behaviour.

Note that the evolution requirements described until now do not take into account the *User's action* factor. As explained in the introduction, dealing with such kind of context changes requires fine control of the OIO at language level. Such adaptation does not impact the election process but only the dialogue status within one modality. This can be achieved only from within the entity controlling the OIO: the mediator. There is no general solution to that issue and for each modality that is implemented, a different approach must be devised. In next section, we give an example of such kind of adaptation in the complex situation of spoken dialogue.

6.5.2 Ongoing Interaction Evolution: The Example of Spoken Dialogue

Within the scientific area of IEs, Spoken Dialogue System (SDS) technologies offer one of the most natural interfaces. In this context for many tasks such as command-and-control of devices or services, proactive behaviour (warning, information, etc.), and negotiative dialogues, speech is a promising modality. However, to realize spoken dialogue adaptation it is not only necessary to provide advanced voice recognition and speech synthesis capabilities but also furthermore to provide a Spoken Dialogue Manager (SDM) residing at the core of an SDS able to manage adaptation. Nowadays one of the most widespread technologies to implement an SDM is the W3C standardized VoiceXML description language [48]. The idea behind this approach is to simplify the development of dialogues that would allow even non-experts to implement speech applications in a similar manner as websites are implemented. VoiceXML has undoubtedly many advantages but the structure of such dialogue systems is limited to system-initiative and mixed-initiative layouts. More complex structures such as negotiative or task-oriented dialogue flows still lack of a sufficient approach that combines the ease-of-use of VoiceXML with the more powerful expressiveness of scientific approaches such as [10, 41, 74].

In this section, we introduce our approach regarding the third level of adaptation—i.e. evolution—within the spoken dialogue modality. After classifying the situations that require adaptation of the ongoing dialogue, we will present the spoken dialogue manager that we implemented and the ontological model that is associated to it and that affords such flexibility of the dialogue content.

6.5.2.1 A Classification of Evolution Triggers

In [31] we have defined the behaviour of the SDM to be adaptive regarding the following changes that may happen during (spoken) human–computer interaction in the context of IEs and distinguished between three levels of adaptation:

- Environmental changes that relate to *Device Adaptation*: This relates to the continuous modifications of devices and services that are active and accessible within the user's context. Depending on the surrounding and the situation of the user (kitchen, living room, car, etc.), the availability of devices and services may vary. We assume that users would usually "talk" to devices (i.e., control them by speech) that are within sight and correspond to the current situation of the user. This requires the capability to continuously change grammars, utterances, and system commands to a changing device population and changing user focus within the IE. Furthermore since it seems not to be possible to always be aware of the user's situation (doing housework, relax, prepare a dinner, etc.) it is necessary to provide an option to activate the control of devices or services that even are not within the context of the current location and/or situation. Thus the context of the user may also change depending on the time of day and even on the actual level of trust in nearby entities (guests, newly added technical devices). It is obvious that the spoken commands and/or utterances that can be understood / that are said by an SDM will therefore change continuously as well. In the case of spoken dialogue an important influence on the availability of spoken dialogue is the level of noise. In practice it will hardly be possible to control an environment via speech while music is playing loudly.
- Environmental changes that relate to *Event Adaptation*: Since various tasks within IEs are to be accomplished it is necessary to move the current focus from an on-going dialogue to other (contingently more urgent) dialogues. These dialogues may consist of informative system utterances, alerts, and questions. Afterwards the on-going dialogue would be resumed. We have recognized two kinds of events: external and internal. While external events always need an entity that throws the specific event, internal events can be initiated by the dialogue manager itself. Reasons for initiating an internal event can be various and sundry: fixed priorities, dynamic priorities (i.e. changing over time), semantics (i.e. semantically similar dialogues can extend a dialogue) and depending on the progress done within an on-going dialogue.
- Task-driven changes that relate to *Task Adaptation*: During more complex tasks that might come up during a conversation the task itself may vary.

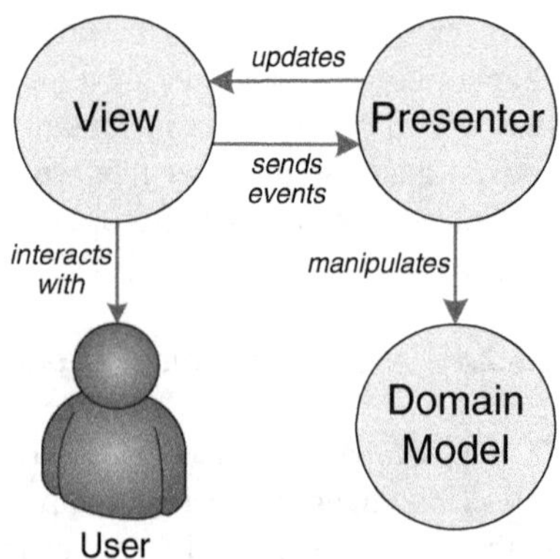

Fig. 6.7 The MVP Passive View pattern as used to implement the ATRACO SDM

Especially during a spoken human–computer negotiation the requirements and/or constraints may change. The effects of task adaptation may result in task cancellation or in slight changes such as an extension of the initially planned dialogue. Obviously the task adaptation level corresponds to the most complex level of adaption as it relates to automatic learning and/or direct user-system collaboration.

6.5.2.2 The Implementation of the SDM

In order to meet the requirements listed in Sect. 6.5.2.1 the ATRACO SDM is implemented considering the Passive View variation of the Model-View-Presenter (MVP) [52] pattern, itself a model-derivative of the popular Model-View-Controller (MVC) paradigm [53]. The underlying idea of MVP is that an application or system should be divided into three logical parts, the Model, the View and the Presenter. As shown in Fig. 6.7 the user only interacts with the View layer. Contrary to MVC the Presenter mediates between Model and View—the Model conveys no functionality, i.e. it is not an application but solely encodes the knowledge that is used by the Presenter. The term Model in this case refers to a *Domain Model*. Therefore especially for systems that directly interface to the user (i.e. user interfaces) MVP is perfectly suitable. To be able to communicate with a user or with other external entities the application needs a knowledge base that describes facts and the relations between such facts. A fact could, for example, be the name of a person or an ID number. A relation could be “has”, which could be used for person “A” has ID number “4711”.

Without this knowledge the system wouldn’t be able to generate useful output, i.e. act as *knowledge source* or to understand input that is provided by external entities, thus acting as *knowledge sink*. The term Domain Model could therefore be specified as the knowledge a system needs in order to be able to interact with the context in a meaningful way. There are many ways to establish such a knowledge base: to name but a few, SQL databases or XML files could be utilized. A more sophisticated option is to make use of ontologies to provide a common understandable knowledge base.

The ATRACO SDM makes use of a specific number of dialogue representations. These representations serve as Domain Models. Each representation provides knowledge about both dialogue flow and state of a specific spoken conversation. Depending on contextual information various sets of spoken dialogues can be activated or deactivated. It is furthermore possible to add new representations for dialogues during runtime and therefore extend the knowledge base, i.e. the Model. For the prototype we use OWL ontologies to implement the Model. In the next paragraphs we provide a detailed look on the capabilities of this structure and the way we use it to describe spoken dialogues.

The underlying knowledge base of the SDM is modelled using OWL ontologies, so called Spoken Dialogue Ontologies (SDOs). We have implemented this tree shaped structure to arrange the data-bearing individuals using a defined set of classes. The root of each knowledge base is DialogueDomain, which has the two subclasses Speech and State. We divide the ontology into these two main branches since we want to distinguish between knowledge that corresponds to the static structure and knowledge that corresponds to the dynamic state of the current dialogue. Figure 6.8 shows an overview of all classes populating the SDO together with the relations interlinking them.

The main purpose of separating the Domain Model from the actual intelligence—the Presenter—and the View is that the model can be modified, (partly) removed, extended, and exchanged. Furthermore by separating static from dynamic knowledge the data (grammar, utterances, and semantics) that might be needed for

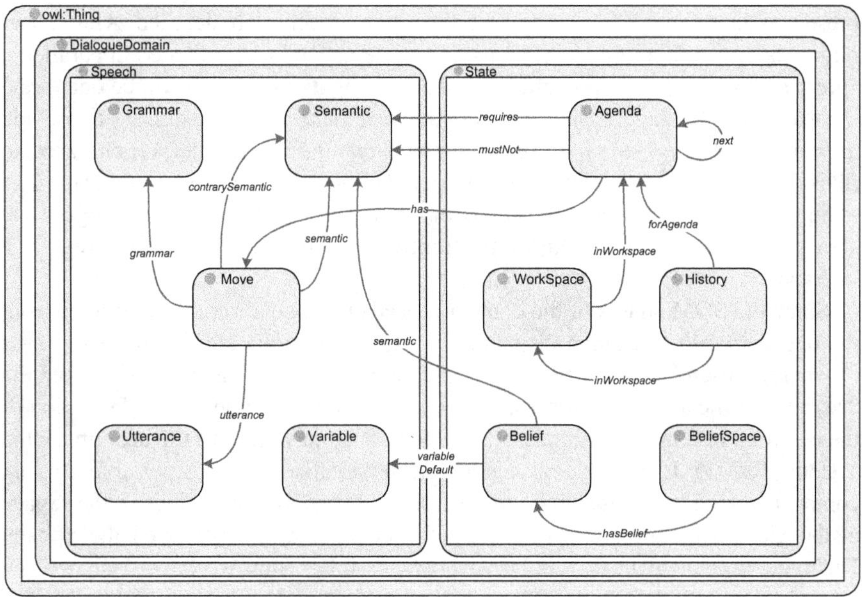

Fig. 6.8 Overview of the classes and main relations of the Spoken Dialogue Ontology

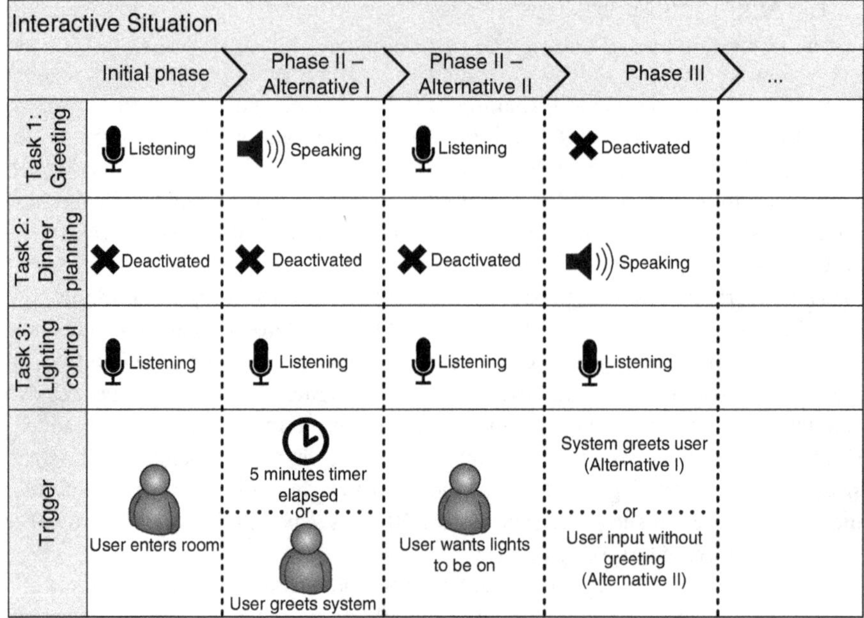

Fig. 6.9 An interactive situation that may occur with two alternatives

various dialogues can be exchanged and reused as well. In the following we present an interactive situation to describe our approach more detailed. After a user returns home there might be several spoken interactive tasks activated in parallel: a "greeting" task, a dinner preparation task, and a light control task. Since one of the main duties an IE should handle is to control specific tasks, it is necessary to provide a (probably varying) set of spoken commands that the IE can interpret and execute (thus device and event adaptation). An example for such a behaviour could be a user telling the system to switch the lights on after entering the apartment. Figure 6.9 shows the three interactive tasks mentioned above that may form the exemplary interactive situation.

Since the SDM adapts to the context it needs to be able to receive triggers from the outside world to change its state. The initial phase therefore is triggered by a "user enters room" event. This event might happen only once a day and/or when the user has left the apartment for a specified period depending on the configuration of the environment. In our example the SDM is set up to wait until the user greets the system (Task 1). It further activates a control task that listens to possible lighting control commands the user may utter (Task 3). Initial system studies in the iSpace at the University of Essex [72] revealed that the subjects preferred the Spoken Dialogue System to be as unobtrusive as possible. Thus we have designed the system to behave rather passively and not to proactively initiate a conversation if this can be avoided. A control task such as Task 3 waits for user input by default. However, if the user starts talking to the system by uttering a spoken command the

Table 6.1 A dialogue snipped that might occur during an interactive situation

Speaker	Utterance
Suki	Hello Julia!
Julia	Hi Suki!
Suki	Switch the lights on!
Julia	Do you want to start preparing the dinner for your friends now?

system could take this opportunity to start dialogues that otherwise would have to be proactively initiated and therefore would have been more obtrusive. Figure 6.9 presents two alternatives showing how the situation could proceed in Phase II: Alternative I contains two triggers that might allow the system to perform Task 1; the five-minutes-timer elapsed since the user entered the room or—probably the more usual case—the user greets the system. As mentioned above the reason for such a five-minutes-timer is that the system should act as unobtrusively as possible.

Note that Task 3 is still active since the system is meant for handling more than one interactive task in parallel. If one of the two triggers is actuated the system would greet the user and would add a semantic value such as "userInitiatedConversation" to the knowledge base. This would allow Phase III to start. Table 6.1 shows a possible conversation that might occur using the proposed set of SDOs.

Alternatively Phase II could be activated by the user telling the system to switch the lights on. This would make Task I obsolete—the system shouldn't greet the user in response to a spoken command. It would be more natural that the system skips the greeting task and activates the proactive Task 2 "Dinner planning" instead. Figure 6.9 shows Phase III constituted by the additionally activated Task 2 and the still-running Task 3. The preceding greeting task has either become obsolete or has already being processed. Since the user can dynamically activate or deactivate the tasks, the SDM may perform, it is possible at any time to terminate a conversation with the system or to start a dialogue the system hasn't been aware of.

The proposed Spoken Dialogue Ontology refers to the informational components and their formal representation as originally described in the Information State theory [41]. It reflects both static and dynamic knowledge and therefore facilitates a generic representation of a model that can be processed by the Presenter layer. Since the SDO is not only interpreted before the dialogue starts but is also involved dynamically in the on-going dialogue our approach allows for a great flexibility and provides many opportunities for adaptation. It is possible to learn new grammars, semantics, or utterances during an on-going dialogue by simply adding new individuals to the SDO. Furthermore by extending or reducing the set of activated SDOs various interactive situations for a changing device and task population can be generated. Thus we provide a fertile ground to cope with the challenges adaptation brings along within Ambient Environments.

This concludes our presentation of the different layers of adaptation regarding user interaction. We explored the evolution through the structure of the user interface, using a model-driven approach to allocate and instantiate user interfaces on existing interaction devices. Though the set of available interaction devices is dynamic, we showed how to evolve the composition during runtime. Finally, we showed that the structural changes are not enough, adaptation is required within an instantiated modality and we demonstrated in this section how such adaptation could be achieved for the most complex modality: the spoken dialogue. Next section will develop our approach regarding the evaluation of this adaptive system.

6.6 Experiments and Results

As a result of the ATRACO project, a prototype of the Interaction Agent was setup in the iSpace at the University of Essex, United Kingdom. We focus in the next sections on the implementation of that prototype and on the validation results we gathered from it.

6.6.1 ATRACO Integration

The integration of our approach can be distinguished into two major parts: the implementation of the IA itself and its purpose to adapt the interaction to the context.

6.6.1.1 Role of the IA in the Activity Sphere

Within the ATRACO architecture, the role of the IA is to interface the system with the user. As such, it needs to provide some functionalities to other components of the system. Table 6.2 shows the features that the Interaction Agent exposes to other components. The main component with which the IA interacts is the SM. The SM will trigger the IA to create interaction tasks on-demand during the execution of the BPEL workflow—defined in Sect. 1.4. TM and PM will require the IA to disable or hide crucial commands and information when trust or privacy rules are violated—see Sects. 4.3.4.2 and 4.2.5.2.

Table 6.2 The features the IA provides to other ATRACO components

Target component	Features
Sphere manager	Start/Stop Interaction tasks Send *user will* as predefined events Display system status updates
Trust manager	Inform on where interaction occurs Disable a user interface (because of trust breach)
Privacy manager	Inform on where interaction occurs Hide a user interface (because of privacy breach)

6.6.1.2 Controlling Remote OIOs

We have highlighted, throughout this chapter, the main role of the IA: to adapt the interaction to the context. A part of this role that we have not focused on is the routing of events. This role, crucial though it may be, doesn't really represent a research topic but rather an architectural and technical challenge. In ATRACO, to keep things simple, we have assimilated events to a pair of strings: one for the event name, and one for the content of the message. Events are divided into two categories: *System Events* represent updates from the system—for instance, the value of the luminosity; *User Will Events* are events generated by the user interface that represent an order from the user to the system—for instance, when the user wants to set the light level, the event < "*light_level*", "30 %" > will be sent.

The IA is composed of three internal modules:

- Multimodal Manager: It is responsible for the allocation, instantiation and evolution of the interaction.
- Spoken Dialogue Manager: It is a mediator responsible for handling OIOs for spoken dialogue.
- Graphical Dialogue Manager: It is a mediator responsible for handling OIOs for graphical user interfaces.

Figure 6.10 shows how system events and user will events are propagated between the User and the System through the IA.

6.6.2 Model Validation

The validation of our model for interaction adaptation was made in two steps. After having implemented the Multimodal Manager that reasons over the Interaction Ontology information, we implemented a simulator for interaction. This simulator helped us design a behavioural model for the IA and test it. After this simulation phase, a social evaluation was driven in the iSpace to evaluate the impact of the whole system in real users' life.

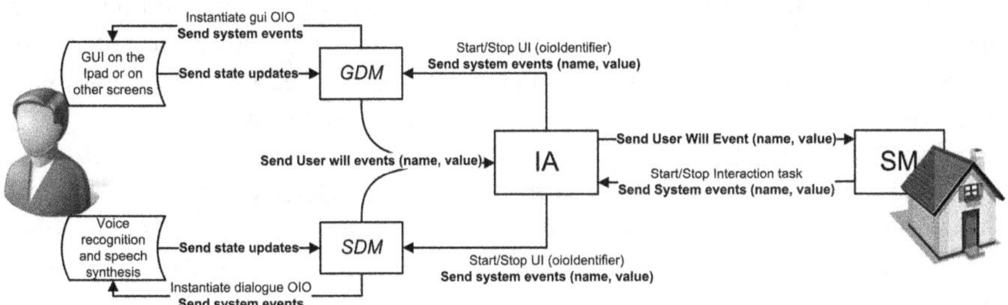

Fig. 6.10 Flow of events within the IA

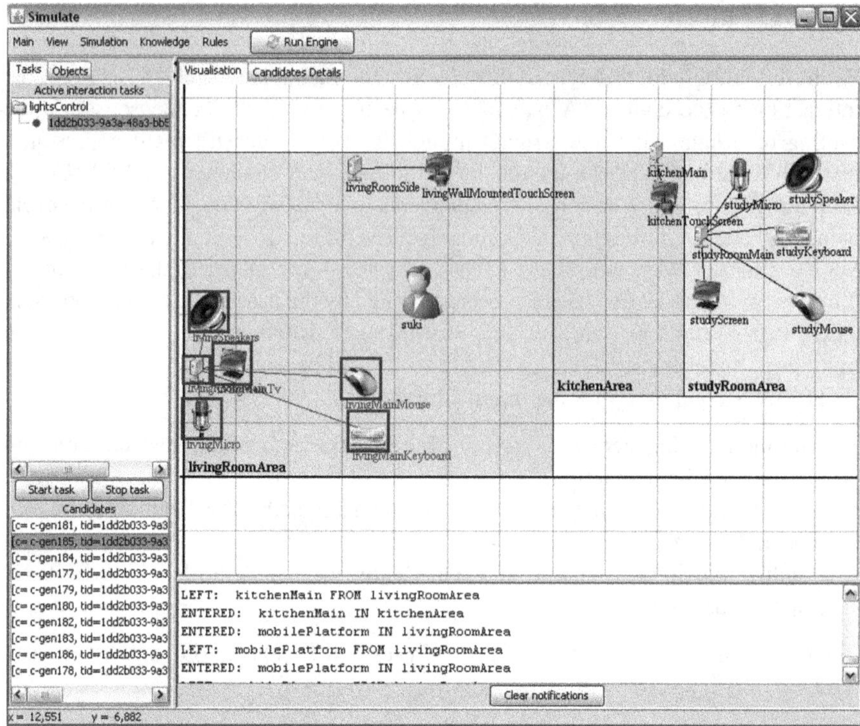

Fig. 6.11 Simulator implemented to test behavioural models—shows the model

6.6.2.1 Simulations

We designed a tool named Simteraction—for SIMulation of inTERACTION. This tool enables designers to write behavioural rules and immediately test them. It offers the possibility to add any interaction component to a simulated system, to give them a location and to evaluate the result of the allocation process in a specific situation.

Figure 6.11 shows a screenshot of the application in a simulation of the iSpace. We can see the different rooms of the apartment, the devices that have been deployed in it, and the user—here, his name is Suki. The application also draws the connections between available devices, thus showing the active Interaction Spaces. On the left side, an interaction task has been started for controlling lights. A list of candidates is proposed. By selecting one of them—here the second—the corresponding Interaction Space is highlighted in the main view. More details about a specific candidate can be found in the "Candidates Details" tab—see Fig. 6.12. This view is particularly useful for the designer as it gives explanation concerning the salience and validity of each candidate. For instance, in Fig. 6.12, we can see that the candidate $c - gen177$ has a salience of $+1$ because it uses the tactile mode.

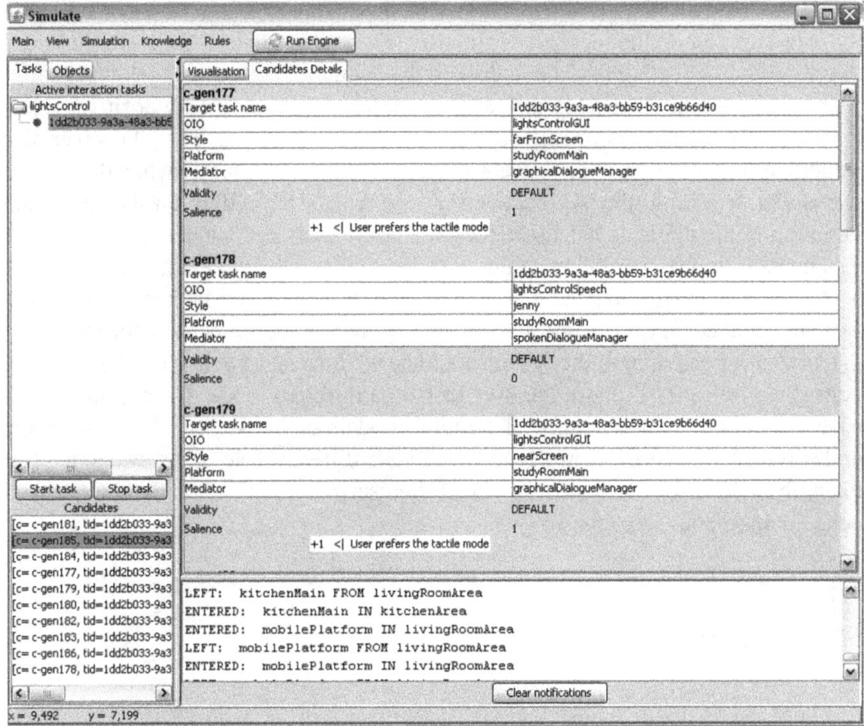

Fig. 6.12 Simulator implemented to test behavioural models—shows the candidates

This simulator was very useful to explore the expressivity of our rule language and of the overall model. It also helped us in imagining and testing the reactions of the system to new devices or interaction languages that do not exist yet—or for which we don't have any implementation. Finally, it has been used—and that was the main reason for its implementation—during the implementation process to design and debug the behavioural model. This first phase of local evaluation by simulation was then followed by a real world application in the iSpace.

6.6.2.2 User Evaluation

The social evaluation process and its results are detailed in Chap. 9 and in [72]. Such evaluations gives us an interesting view of how users appreciated the behavioural models we proposed them. However, they do not provide direct feedback on the architecture and the interaction itself, since they are hidden from the user. However, having been able to run such an evaluation is a proof in itself, that the system is consistent.

Moreover, it enabled us to test a first behavioural model that focuses on adapting the user interface to the location of the user. This behavioural model was designed to give users the feeling that the interfaces follow them.

The outcome of this evaluation regarding interaction was mainly positive. Users reacted well on the following behaviour of the interfaces. They however felt sometimes that their personal space was quite invaded, particularly when the system was doing or asking things proactively. It looks like, depending on the cases, the initiation of the dialogue has its importance in the user acceptance. This represents a significant improvement that could be added to the interaction model we propose.

Another interesting outcome was that users felt that their interaction with the system would change with time and depending on their mood. We had not envisaged that the same behaviour of the system could be well-accepted at first and be irritating after some time—or because the user feels in a different mood. Users proposed to have the possibility to change behavioural models of the system. Such a change is theoretically possible, it just requires to have different rule bases available. The main obstacle is for the system to detect the user mood and adapt by changing its behavioural model accordingly.

6.7 Conclusion

Ambient interactive systems create new challenges in the field of user interaction adaptation. Though this adaptation is made complex by the distributed and dynamic aspects of ambient systems, achieving some kind of adaptation in the user interfaces is crucial as it will be a determining factor in user acceptance of these systems.

In this chapter we have presented some directions regarding the adaptation of interaction. We have focused on two main aspects: the structure of user interfaces and the adaptation of a running dialogue. The structural adaptation of user interfaces requires three steps: allocation, instantiation and evolution. We introduced those three steps and detailed how they could be implemented. Reasoning techniques are of first importance when it comes to dynamically choose the best option in the scope of interaction capabilities that are proposed by a system. This selection is complex and relies on two key points: a structured representation of the context—that we designed as an ontology for interaction, and a reasoning engine that can apply some rules on this context description to make a decision. The set of rules used to make a decision is called a behavioural model. We presented a language for writing those rules and a complete framework for helping designers writing, testing and simulating/evaluating different behavioural models.

After exploring structural adaptation of the user interface, we explored the structural adaptation of the language itself. We focused this analysis on adapting the on-going interaction within one modality, the spoken dialogue. We presented how ontology technologies could be used again, to perform evolution of an on-going dialogue depending on contextual events.

Finally, we showed how we integrated all those concepts in a unique software architecture, the Interaction Agent. We demonstrated how this IA relate to other ATRACO components and how it was perceived by users during the social evaluation.

While our approach proved to be functional, a remaining challenge consists in integrating automatic user interface generation techniques to provide an even more flexible structural adaptation. We imagined a system where components implemented by different designers could be mixed in at runtime to provide just the features that are needed in a unified user interface. Our model was designed in that direction but theoretical aspects of computer assisted generation needs to be solved before our vision can be realized. Furthermore, the realization of such dynamic composition would be even more powerful if it could mix components implementing different modalities. The problem of synergetic multimodality fusion/fission has been investigated but solutions always involve decision taking from a designer. We can only hope that future results will concern its realization as an automated process.

6.8 Further Readings

As far as multimodality is concerned in pervasive systems, the book *Multimodality in Mobile Computing and Mobile Devices: Methods for Adaptable Usability* [39] compiles an interesting set of visions ranging from theoretical visions to real-world applications.

Concerning the adaptation of natural spoken dialogue, the book *Spoken Dialogue Systems for Ambient Environments* [42] groups major publications of that field for year 2010.

References

1. André, E.: The generation of multimedia presentations. In: Handbook of Natural Language Processing, pp. 305–327. Marcel Dekker Inc, Basel (2000)
2. Appert, C., Huot, S., Dragicevic, P., Beaudouin-Lafon, M.: Flowstates : prototypage d'application interactives avec des flots de donnees et des machines a etats. In: Proceedings of 21eme Conference Francophone sur l'Interaction Homme-Machine, IHM 2009, pp. 119–128. ACM Press, New York (2009)
3. Athanasopoulos, D., Zarras, A., Issarny, V., Pitoura, E., Vassiliadis, P.: CoWSAMI: interface-aware context gathering in ambient intelligence environments. Pervasive Mob. Comput. **4**(3), 360–389 (2008)
4. Balme, L., Demeure, A., Barralon, N., Coutaz, J., Calvary, G., et al.: Cameleon-rt: a software architecture reference model for distributed, migratable, and plastic user interfaces. In: Ambient Intelligence. Lecture Notes in Computer Science, vol. 3295, pp. 291–302, Springer, Berlin (2004)
5. Bass, L., Faneuf, R., Little, R., Mayer, N., Pellegrino, B., Reed, S., Seacord, R., Sheppard, S., Szczur, M.: A metamodel for the runtime architecture of an interactive system. ACM SIGCHI Bull. **24**(1), 32–37 (1992)

6. Bellik, Y.: Interfaces multimodales: concepts, modèles et architectures. Ph.D. Thesis, Université d'Orsay Paris-Sud (1995)
7. Bellik, Y., Rebai, I., Machrouh, E., Barzaj, Y., Jacquet, C., Pruvost, G., Sansonnet, J.: Multimodal interaction within ambient environments: an exploratory study. In: Human-Computer Interaction, INTERACT'09: IFIP International Conference on Human-Computer Interaction; Uppsala, Sweden (2009)
8. Bernsen, N.: Modality Theory in support of multimodal interface design. In: Proceedings of Intelligent Multi-Media Multi-Modal Systems, pp. 37–44 (1994)
9. Berti, S., Paternó, F.: Migratory multimodal interfaces in multidevice environments. In: Proceedings of the International Conference on Multimodal Interfaces, pp. 92–99. ACM Press, New York (2005)
10. Bohus, D., Rudnicky, A.I.: The ravenclaw dialog management framework: architecture and systems. Comput. Speech Lang. **23**, 332–361 (2009)
11. Bordegoni, M., Faconti, G., Maybury, M.T., Rist, T., Ruggieri, S., Trahanias, P., Wilson, M.: A standard reference model for intelligent multimedia presentation systems. Comput. Stand. Interfaces **18**(6), 477–496 (1997)
12. Braffort, A., Choisier, A., Collet, C., Dalle, P., Gianni, F., Lenseigne, B., Segouat, J.: Toward an annotation software for video of Sign Language, including image processing tools and signing space modelling. In: 4th International Conference on Language Resources and Evaluation, held in Lisbon, Portugal (2004)
13. Browne, D., Totterdell, P., Norman, M.: Adaptive User Interfaces. Academic Press, London (1990)
14. Brusilovsky, P.: Adaptive hypermedia. User Model. User-Adap. Inter. **11**, 87–110 (2001)
15. Calvary, G., Coutaz, J., Thevenin, D., Limbourg, Q., Bouillon, L., Vanderdonckt, J.: A unifying reference framework for multi-target user interfaces. J. Interacting with computers **15/3**, 289–308 (2003). Elsevier Science B.V
16. Chikofsky, E., Cross II, J.: Reverse engineering and design recovery: a taxonomy. IEEE Softw. **7(1)**, 13–17 (1990)
17. Coutaz, J.: Meta-user interfaces for ambient spaces. In: Task Models and Diagrams for Users Interface Design. Lecture Notes in Computer Science, vol. 4385, pp. 1–15. Springer, Berlin (2007)
18. Coutaz, J., Nigay, L., Salber, D., Blandford, A., May, J., Young, R.: Four easy pieces for assessing the usability of multimodal interaction: the CARE properties. In: Proceedings of INTERACT, vol. 95, pp. 115–120 (1995)
19. Dey, A.K.: Providing architectural support for building context-aware applications. Ph.D. Thesis, Georgia Institute of Technology (2000)
20. Dieterich, H., Malinowski, U., Kühme, T., Schneider-Hufschmidt, M.: State of the art in adaptive user interfaces. Hum. Factors Inform. Technol. **10**, 13–13 (1993)
21. Dragicevic, P., Fekete, J.: Input device selection and interaction configuration with ICON. In: People and Computers, pp. 543–558. Springer, London (2001)
22. Duarte, C., Carriço, L.: A conceptual framework for developing adaptive multimodal applications. In: Proceedings of the 11th International Conference on Intelligent User Interfaces, pp. 132–139. ACM Press, New York (2006)
23. Eisenstein, J., Vanderdonckt, J., Puerta, A.: Applying model-based techniques to the development of UIs for mobile computers. In: Proceedings of the 6th International Conference on Intelligent User Interfaces, pp. 69–76. ACM Press Publication, Santa Fe, New Mexico, USA (2001)
24. Florins, M., Trevisan, D., Vanderdonckt, J.: The continuity property in mixed reality and multiplatform systems: a comparative study. In: Proceedings of CADUI'04, pp. 13–16, Madeira Island (2004)
25. Forgy, C.: Rete: a fast algorithm for the many pattern/many object pattern match problem* 1. Artif. Intell. **19**(1), 17–37 (1982)
26. Frasincar, F., Houben, G.J.: Hypermedia presentation adaptation on the semantic web. In: Proceedings of the 2nd International Conference on Adaptive Hypermedia and Adaptive Web-Based Systems, LNCS 2347, pp. 133–142. Malaga, Spain (2002)

27. Friedman, E.: Jess in action: rule-based systems in java. Manning Publications Co., Greenwich, CT, USA (2003)
28. Frohlich, D.M.: The design space of interfaces, multimedia systems, interaction and applications. In: Proceedings of 1st Eurographics Workshop, held in Stockholm, Sweden (1991)
29. Gram, C., Cockton, G.: Design Principles for Interactive Software (IFIP International Federation for Information Processing). Chapman & Hall Ltd, London (1996)
30. Hagras, H., Goumopoulos, C., Bellik, Y., Minker, W., Meliones, A.: Research Results on Adaptation & Evolution. Technical Report, 7th Framework program, ATRACO Project Report - Deliverable D13 (2011)
31. Heinroth, T., Denich, D., Schmitt, A.: Owlspeak - adaptive spoken dialogue within intelligent environments. In: 8th IEEE International Conference on Pervasive Computing and Communications Workshops (PERCOM Workshops), pp. 666–671. Mannheim (2010). http://ieeexplore.ieee.org/stamp/stamp.jsp?tp=&arnumber=5470518
32. Henricksen, K., Indulska, J., Rakotonirainy, A.: Modeling context information in pervasive computing systems. Pervasive Comput. **2414**, 79–117 (2002)
33. Horrocks, I., Patel-Schneider, P., Boley, H., Tabet, S., Grosof, B., Dean, M.: SWRL: a semantic web rule language combining OWL and RuleML. W3C Member submission **21** (2004). http://www.w3.org/Submission/2004/03/
34. Jaimes, A., Sebe, N.: Multimodal human-computer interaction: a survey. Comput. Vis. Image Underst. **108**(1–2), 116–134 (2007)
35. Johnston, M., Bangalore, S.: Finite-state multimodal integration and understanding. Nat. Lang. Eng. **11**(02), 159–187 (2005)
36. Kobsa, A., Koenemann, J., Pohl, W.: Personalised hypermedia presentation techniques for improving online customer relationships. Knowl. Eng. Rev. **16**(2), 111–155 (2001)
37. Kolski, C., Le Strugeon, E.: A review of intelligent human-machine interfaces in the lights of the arch model. Int. J. Hum. Comput. Interact. **10**(3), 193–231 (1998)
38. Kolski, C., Forbrig, P., David, B., Girard, P., Tran, C., Ezzedine, H.: Agent-based architecture for interactive system design: current approaches, perspectives and evaluation. In: Jacko, J.A. (ed.) Human-Computer Interaction. New Trends, Lecture Notes in Computer Science, vol. 5610, pp. 624–633. Springer, Berlin (2009). doi:10.1007/978-3-642-02574-7_70. http://dx.doi.org/10.1007/978-3-642-02574-7_70
39. Kurkovsky, S.: Multimodality in Mobile Computing and Mobile Devices: Methods for Adaptable Usability. IGI Global, 701 E. Chocolate Avenue Hershey, PA 17033, USA (2010)
40. Landragin, F.: Physical, semantic and pragmatic levels for multimodal fusion and fission (2007). Accessible on HAL server : http://hal.archives-ouvertes.fr
41. Larsson, S., Traum, D.: Information state and dialogue management in the TRINDI Dialogue Move Engine Toolkit. Nat. Lang. Eng. (Special Issue) **6**(3), 323–340 (2000)
42. Lee, G.G., Mariani, J., Minker, W. (eds.): Spoken Dialogue Systems for Ambient Environments. Lecture Notes in Computer Science, vol. 6392. Springer, Heidelberg (2010)
43. Limbourg, Q., Vanderdonckt, J., Michotte, B., Bouillon, L., López-Jaquero, V.: Usixml: a language supporting multi-path development of user interfaces. In: Engineering Human Computer Interaction and Interactive Systems, pp. 200–220. Springer, Berlin (2005)
44. Mori, G., Paternò, F., Santoro, C.: Tool support for designing nomadic applications. In: Proceedings of the 8th International Conference on Intelligent User Interfaces, pp. 141–148. ACM Press, New York (2003)
45. Mori, G., Paternó, F., Santoro, C.: Design and development of multidevice user interfaces through multiple logical descriptions. IEEE Trans. Softw. Eng. **30**(8), 507–520 (2004)
46. Navarre, D., Palanque, P., Bastide, R., Schyn, A., Winckler, M., Nedel, L., Freitas, C.: A formal description of multimodal interaction techniques for immersive virtual reality applications. In: Human-Computer Interaction-INTERACT 2005, pp. 170–183, Springer, Berlin (2005)
47. Nigay, L., Coutaz, J.: A generic platform for addressing the multimodal challenge. In: Proceedings of the Conference on Human Factors in Computing Systems, CHI'95 held in Denver, Colorado, USA (1995)

48. Oshry, M., Auburn, R., Baggia, P., Bodell, M., Burke, D., Burnett, D., Candell, E., Carter, J., Mcglashan, S., Lee, A., Porter, B., Rehor, K.: Voice extensible markup language (voicexml) version 2.1. Technical Report, W3C – Voice Browser Working Group (2007)
49. Padovitz, A., Loke, S., Zaslavsky, A.: The ECORA framework: a hybrid architecture for context-oriented pervasive computing. Pervasive Mob. Comput. **4**(2), 182–215 (2008)
50. Paganelli, L., Pernó, F.: Automatic reconstruction of the underlying interaction design of web applications. In: Proceedings of SEKE 2002. Ischia, Italy (2002)
51. Pernó, F., Mancini, C., Meniconi, S.: Concurtasktrees: A diagrammatic notation for specifying task models. In: INTERACT '97: Proceedings of the IFIP TC13 International Conference on Human-Computer Interaction, pp. 362–369. Chapman & Hall, Ltd (1997)
52. Potel, M.: MVP: Model-View-Presenter The Taligent Programming Model for C++ and Java. Technical Report, Taligent Inc (1996). Available on http://www.wildcrest.com/Potel/Portfolio/mvp.pdf
53. Reenskaug, T.: Models - views - controllers. Technical Report, Xerox PARC (1979). http://heim.ifi.uio.no/~trygver/1979/mvc-2/1979-12-MVC.pdf
54. Riley, G.: Clips: an expert system building tool. In: NASA, Washington, Technology 2001: The Second National Technology Transfer Conference and Exposition, vol. 2 (1991)
55. Rist, T.: Supporting mobile users through adaptive information presentation. Multimodal Intell. Inf. Presentation **27**, 113–139 (2005)
56. Rist, T., Andre, E.: Building smart embodied virtual characters. In: Smart Graphics. Lecture Notes in Computer Science, vol. 2733, pp. 123–130. Springer, Berlin (2003)
57. Román, M., Hess, C., Cerqueira, R., Ranganathan, A., Campbell, R., Nahrstedt, K.: A middleware infrastructure for active spaces. IEEE Pervasive Comput. **1**(4), 74–83 (2002)
58. Rousseau, C.: Présentation multimodale et contextuelle de l'information. Ph.D. Thesis, Université d'Orsay Paris-Sud (2006)
59. Rousseau, C., Bellik, Y., Vernier, F., Bazalgette, D.: A framework for the intelligent multimodal presentation of information. Signal Process. **86**(12), 3696–3713 (2006)
60. Samaan, K., Tarpin-Bernard, F.: Task models and interaction models in a multiple user interfaces generation process. In: In Proceedings of TAMODIA 2004, vol. 86, pp. 137–144. ACM Press, Czech Republic (2004)
61. Scapin, D., Bastien, J.: Ergonomic criteria for evaluating the ergonomic quality of interactive systems. Behav. Inform. Technol. **16**(4), 220–231 (1997)
62. Serrano, M., Nigay, L., Lawson, J.Y.L., Ramsay, A., Murray-Smith, R., Denef, S.: The openinterface framework: a tool for multimodal interaction. In: CHI '08 Extended Abstracts on Human Factors in Computing Systems, pp. 3501–3506. ACM, New York, NY (2008). doi:http://doi.acm.org/10.1145/1358628.1358881
63. Stanciulescu, A., Limbourg, Q., Vanderdonckt, J., Michotte, B., Montero, F.: A transformational approach for multimodal web user interfaces based on UsiXML. In: Proceedings of the 7th International Conference on Multimodal Interfaces, pp. 259–266. ACM, New York, NY (2005)
64. Stephanidis, C., Karagiannidis, C., Koumpis, A.: Decision making in intelligent user interfaces. In: Intelligent User Interfaces, IUI'97 held in Orlando, Florida, USA, pp. 195–202 (1997)
65. Stephanidis, C., Paramythis, A., Sfyrakis, M., Stergiou, A., Maou, N., Leventis, A., Paparoulis, G., Karagiandidis, C.: Adaptable and adaptive user interfaces for disabled users in avanti project. In: Intelligence in Services and Networks: Technology for Ubiquitous Telecom Services. Lecture Notes In Computer Science. vol. 1430, pp. 153–16. Springer, Berlin (1998)
66. Stephanidis, C., Savidis, A.: Universal access in the information society: methods, tools, and interaction technologies. UAIS J. **1**(1), 40–55 (2001)
67. Tandler, P.: The BEACH Application Model and Software Framework for Synchronous Collaboration in Ubiquitous Computing Environments. J. Syst. Softw. **69**(3), 267–296 (2004)
68. Tarpin-Bernard, F.: Interaction Homme-Machine Adaptative. Habilitation a diriger des recherches (hdr), Université Claude Bernard de Lyon (2006)

69. Teil, D., Bellik, Y.: The Structure of Multimodal Dialog II. Multimodal Interaction Interface Using Voice and Gesture, chap. 19, pp. 349–366. John Benjamins Publishing Co., P.O.Box 75577, 1070 AN Amsterdam, The Netherlands, John Benjamins Noth America, P.O.Box 27519, Philadelphia PA 19118-0519, USA (2000)
70. Thevenin, D., Coutaz, J.: Plasticity of user interfaces: framework and research agenda. In: Human-computer Interaction, INTERACT'99: IFIP TC. 13 International Conference on Human-Computer Interaction, 30th August-3rd September 1999, Edinburgh, UK, p. 110. IOS Press (1999)
71. Thevenin, D., Coutaz, J.: Adaptation des IHM: taxonomies et archi. logicielle. In: Proceedings of the 14th French-Speaking Conference on Human-Computer Interaction, pp. 207–210. ACM, New York, NY (2002)
72. van Helvert, J., Hagras, H., Kameas, A.: D27 - prototype testing and validation. Restricted deliverable, The ATRACO Project (FP7/2007-2013 grant agreement 216837) (2009)
73. Vanderdonckt, J., Grolaux, D., Van Roy, P., Limbourg, Q., Macq, B., Michel, B.: A design space for context-sensitive user interfaces. In: Proceedings of IASSE 2005, held in Toronto, Canada (2005)
74. Young, S., Williams, J., Schatzmann, J., Stuttle, M., Weilhammer, K.: D4.3: Bayes net prototype - the hidden information state dialogue manager. Technical Report, TALK - Talk and Look: Tools for Ambient Linguistic Knowledge, IST-507802, 6th FP (2006)

Chapter 7
User-Centred Spoken Dialogue Management

Florian Nothdurft, Stefan Ultes, and Wolfgang Minker

Abstract Adaptivity of intelligent environments to their surroundings provided by the ATRACO Spoken Dialogue Manager is only one means of adaptation. Recent work in Spoken Dialogue Systems focuses on the integration of user-centred adaptation means to alter the content, flow and structure of the ongoing dialogue. In this chapter, we introduce a general user-centred adaptation cycle, accompanied by two implemented adaptation approaches focusing respectively on short-term and long-term goals in human–computer interaction. After motivating the need for short-term and long-term goals to entail different adaptation mechanisms, we provide exemplary adaptation entities for each case with corresponding experiments and implementations. The short-term goal user satisfaction allows for detecting whether the user is not satisfied with the interaction and for triggering counter measures to improve the interaction. As a long-term goal, maintaining human–computer trust attempts to keep users still willing to use the system even if the interaction was confusing.

7.1 Introduction

Providing speech interfaces for intelligent environments requires all different kinds of adaptability of the corresponding Spoken Dialogue System (SDS). While the adaptivity described in previous chapters deals with a changing environment, e.g. adding or removing control devices during run-time, a recent research field has emerged: adapting the system behaviour to the user. Here, not only adaptivity to statically changing information, i.e. integrating user models into the interaction, is of interest. Adaptivity to dynamically changing information plays a key role in user-friendly speech interfaces influencing soft measures like user satisfaction or human–computer trust (HCT).

While the subjective user experience should clearly be one of the most important measures in a dialogue system, it might seem odd to take HCT into account as well. However, trust has been shown to be a crucial part in the interaction between

F. Nothdurft (✉) • S. Ultes • W. Minker
Institute Communications Engineering, Ulm University, Ulm 89081, Germany
e-mail: florian.nothdurft@uni-ulm.de; stefan.ultes@uni-ulm.de; wolfgang.minker@uni-ulm.de

© Springer International Publishing Switzerland 2016
S. Ultes et al. (eds.), *Next Generation Intelligent Environments*,
DOI 10.1007/978-3-319-23452-6_7

humans and machines. If the user does not trust the system and its actions, advice or instructions, this may lead to the user avoiding any further interaction with the system [28]. Those situations in particular where the user does not understand the system's behaviour or finds that the system behaves in an unexpected way are likely to impact negatively on the HCT relationship [23] and more generally on human–computer interaction (HCI).

Hence, it seems reasonable to incorporate adaptivity to user satisfaction and HCT into an SDS to foster a highly qualitative and trustworthy interaction. However, enabling an SDS to incorporate further dynamic information into the decision-making process is a twofold problem: the user information must be collected (e.g., estimated using statistical classifiers) and the course of the dialogue must be altered to match the extracted user information. Thus, a general model to dynamic user-adaptive dialogue management is described in Sect. 7.3, preceded by an introduction into the corresponding related work in the next section.

Finally, the implementation of several adaptation concepts is described. These concepts have been implemented within experiments with real and simulated users showing the general benefit of extending the speech interface by user-adaptivity capabilities. Roughly speaking, the adaptation approaches can be grouped into adaptation to dynamically changing information with short-term and long-term goals. While the former is aimed at the improvement of the current interaction step, the latter focuses on reaching a long-term goal, which does not necessarily result in an effective current interaction step.

7.2 Significant Related Work

The field of adaptive dialogue spans over different types of adaptation. While some systems adapt to their environment, the focus in this chapter lies on systems which are capable of adapting to the user and their characteristics. More specifically, an emphasis is put on adaptation to dynamically changing information, i.e. the dynamic adaptation to the user during the ongoing dialogue. In the following, we will present significant work on dynamic adaptation grouped by adaptation targeting a short-term or a long-term goal.

7.2.1 Short-Term Goal

For short-term goal adaptation, i.e. reaching the goal within the same interaction, very prominent work has been presented by Litman and Shimei [18]. They identify problematic situations in dialogues by analyzing the performance of the speech recognizer (ASR) and use this information to adapt the dialogue strategy. Each dialogue starts off with a user initiated strategy without confirmations. Depending on the ASR performance, a system-directed strategy with explicit confirmations may eventually be employed. Applied to TOOT [27], a system for getting information about train schedules, they achieved significant improvement in task success compared to a non-adaptive system.

Further work on user-adaptive dialogue with a short-term goal has been presented by Gnjatović and Rösner [6]. For solving the Tower-of-Hanoi puzzle with an SDS, they identify the emotional state of the user in order to recognize if the user is frustrated or discouraged. The dialogue is adapted by answering the questions "When to provide support to the user?", "What kind of support to provide?" and "How to provide support?" depending on the emotional state of the user. By that, the system is capable of providing well-adapted support for the user which helps to solve the task.

Nothdurft et al. [25] created a dialogue system which is adaptive to the user's knowledge with the short-term goal of increasing the user knowledge after the interaction has been performed. For the task of connecting a home cinema system, the multimodal system provides explanations on how to solve the task presenting text, spoken text, or pictures. The system makes assumptions about the user's knowledge by observing critical as well as successful events within the dialogue (e.g., failed tries, accomplished tasks). Based on the user's knowledge model, the system selects the appropriate explanation type and generates explanations so that the user can be expected to be capable of solving the upcoming task.

7.2.2 Long-Term Goal

Previous relevant work on adaptive dialogue systems with a long-term goal, i.e. maintaining a goal over more than one interaction, mostly involves trust in technical systems. Glass et al. [5] investigated factors that may change the level of trust users are willing to place in adaptive agents. Among these verified findings were statements like "provide the user with the information provenance for sources used by the system", "intelligently modulating the granularity of feedback based on context- and user-modeling" or "supply the user with access to information about the internal workings of the system". However, what is missing in Glass et al.'s work is the idea of rating the different methods to uphold HCT in general and the use of a complex HCT model.

Other related work was for example done by Lim et al. [16] on how different kinds of explanations can improve the intelligibility of context-aware intelligent systems. They concentrate on the effect of Why, Why-not, How-to and What-if explanations on trust and understanding system's actions or reactions. The results showed that Why and Why-not explanations were the best kind of explanation to increase the user's understanding of the system, though trust was only increased by providing Why explanations. Drawbacks of this study were that they only concentrated on understanding the system and trusting the system in general and did not consider that HCT is on the one hand not only influenced by the user's understanding of the system and on the other hand that if one component of trust is flawed, the HCT in general will be damaged [21].

7.3 User-Adaptive Dialogue Management

Realizing user-adaptive dialogue management represents the main contribution of this chapter. Here, two general adaptation types exist: adaptation to statically and dynamically changing information. For the latter, system behaviour is statically influenced, e.g. by user preferences stored in a user model. However, we focus on adaptation to dynamically changing information where the course of the ongoing dialogue is influenced dynamically by some adaptation entity (AE). For this, the general dialogue management concept has to be altered. Furthermore, for adapting the course of the dialogue to the user, two different types of adaptation have been identified:

Adaptation to dynamically changing information with Short-term Goal
This means adapting the dialogue to an AE derived from the ongoing dialogue and modifying the course of the dialogue to improve the AE for the current interaction, e.g. user satisfaction.

Adaptation to dynamically changing information with Long-term Goal
This means adapting the dialogue to an AE derived from the ongoing dialogue and modifying the ongoing dialogue to reach a long-term goal, e.g. establish HCT.

Both types have in common that an AE is derived from the interaction and used to influence the action selection. The difference is—for the first—the interaction is adapted to reflect an increase in the target AE directly for the same interaction. For instance, if the user is not satisfied with the interaction, the goal is to increase the satisfaction of the user for the same interaction. This is in contrast to a long-term goal. For example, if the long-term goal is to maintain HCT, it is not necessarily important that the current interaction is going well but that the users feel they can trust the system nonetheless.

For adapting the dialogue, several adaptation modes exist. A straight-forward mode is to use the AE to influence the selection of the next system action out of the pool of existing system actions (cf. [41, 47]). Thus, the dialogue strategy may depend on the AE. Here, all different kinds of dialogue strategy aspects are possible, e.g. the grounding strategy, the dialogue initiative or the prompt design. Of course, there are many more options.

Another way of adaptation mode is to add extra system actions only triggered by the adaptation mechanism. A help action or an error recovery strategy might be activated depending on the AE.

To implement any type of adaptation to dynamically changing information, the processing sequence of a SDS has to be extended. It may be viewed as a cyclic process—involving the human as one part. For the extension of the dialogue cycle to allow for this kind of adaptation, a new module has to be introduced (see Fig. 7.1). Without loss of generality, the cycle may be regarded to start with the system selecting the first system action. This can be seen as valid for all situations if the set of system actions also includes the action of only waiting for user input without

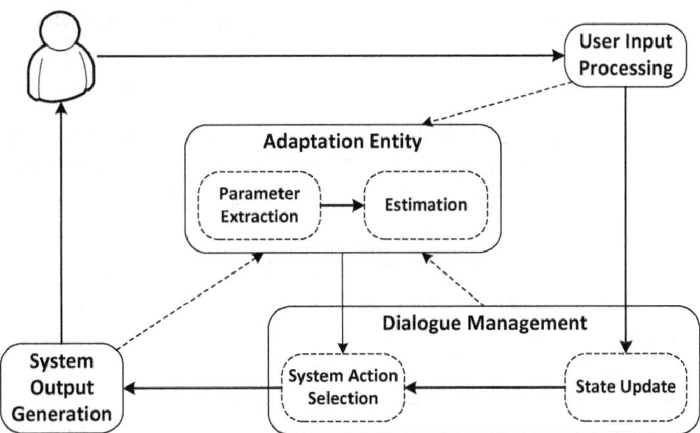

Fig. 7.1 The adaptive dialogue processing cycle. For adapting to additional user values, the modules Parameter Extraction and Value Estimation are integrated producing the estimation of the adaptation value

producing any output. Based on the selected system action, output is created and presented to the user. After the user turn, the created output of the user is processed as user input to the system. Usually, this involves automatic speech recognition and a semantic interpretation.

For enabling the dialogue system to react adaptively, the AE must be determined. As this has to be done without human intervention, often, an automatic estimation approach, e.g. statistical classification, is used. For this, parameters used as input to the estimator must be derived from the interaction taking into account information from all dialogue system components, i.e. speech recognition, semantic interpretation, dialogue management, and output generation. Based on these input parameters, the AE may then be determined and fed into the action selection module of the dialogue management. Based on the AE and the updated system state, the next system action is selected and the cycle starts anew.

Having this type of adaption to dynamically changing information also encompasses several issues. Adding an AE to the action selection may be regarded as an increase in dimensionality and complexity of the problem. This also results in an added uncertainty to the system which should be handled adequately, e.g. using Partially Observable Markov Decision Processes (POMDPs). Furthermore, having several adaptation modes, the question arises which mode is suitable, taking into account the type of adaptation to dynamically changing information as well as the dialogue situation. Finally, the concepts of dialogue management presented in this section represent a system where the only modifications of the system state are due to user input. However, in an intelligent environment, there are multiple entities which are able to modify the state. Hence, the mechanisms have to be integrated into such a framework. The following two sections will give more insight into the aforementioned issues for adaptation to dynamically changing information with short-term and with long-term goals.

7.4 Adaptation to Dynamically Changing Information with Short-Term Goal

Adapting the course of the ongoing dialogue to a dynamically derived adaptation entity with a short-term goal means that the goal should be reached within the same interaction. For this, adaptation entities which may be used for this are, for instance, the intoxication level [42], the emotional state [35, 45] or the user satisfaction [32, 33]. For the latter, if the system detects that the user is not sufficiently satisfied with the interaction, it may take measures to increase the user's satisfaction level. As user satisfaction is a domain-independent entity which may occur in almost all types of dialogue, this section focuses on adaptation to user satisfaction.

Today, there are multiple approaches and metrics which model user satisfaction. To be useful for adaptation to dynamically changing information, the user satisfaction metric must meet certain criteria [44]. The most important criterion is that it must be derivable automatically for each system-user exchange without human intervention. While many approaches exist to automatically determine user satisfaction on the exchange level [3, 7, 9, 10, 33], the Interaction Quality paradigm presented by Schmitt et al. [32] seems to be most suitable.

Consequently, in this section, an adaptation mechanism will be described adapting the course of the dialogue to the Interaction Quality (IQ) in order to increase the IQ of the ongoing dialogue. The Interaction Quality paradigm will be described in the following (Sect. 7.4.1). As adding an extra dimension to the action selection problem also results in adding more uncertainty, the dialogue system should be cast as a POMDP which are specially designed to handle uncertainty. Hence, theoretical aspects of mapping a POMDP to SDS and extending the ATRACO SDM presented in Sect. 6.5.2 to incorporate the POMDP will be described in Sect. 7.4.2. Finally, experiments incorporating quality-adaptivity using real users and a user simulator are also presented showing the viability of this approach in Sect. 7.4.3.

7.4.1 Interaction Quality

Interaction Quality (IQ) was originally proposed by Schmitt et al. [32] as an alternative and more objective measure of user satisfaction. For the authors, the main aspect of user satisfaction is that it is assigned by real users. However, this is impractical in many real world scenarios. Therefore, the usage of expert raters is proposed. Further studies have also shown that ratings applied by experts and users have a high correlation [46].

Furthermore, IQ fulfills all requirements identified by Ultes et al. [44] which are needed for a quality metric to be employable for dialogue adaptation:

- exchange level quality measurement
- automatically derivable features

- domain-independent features
- consistent labeling process
- reproducible labels
- unbiased labels
- sufficient estimation performance

The performance of a SDS may be evaluated either on the dialogue level or on the exchange level. As dialogue management is performed after each system-user exchange, dynamic adaption of the dialogue strategy to the dialogue performance requires exchange level performance measures. Therefore, dialogue-level approaches are of no use.

Features serving as input variables for a classification algorithm must be automatically derivable from the dialogue system modules. This is important because manually annotated features produce high costs and are also not available immediately during run-time in order to use them as additional input to the Dialogue Manager. Furthermore, for creating a *general* quality metric, features have to be domain-independent, i.e. not dependent on the task domain of the dialogue system.

Another important issue is the consistency of the labels. Labels applied by the users themselves are subject to large fluctuations among the different users [17]. As this results in inconsistent labels, which do not suffice for creating a generally valid quality model, ratings applied by expert raters yield more consistent labels. The experts are asked to estimate the user's satisfaction following previously established rating guidelines. Furthermore, expert labelers are also not prone to be influenced by certain aspects of the SDS, which are not of interest in this context, e.g. the character of the synthesized voice. Therefore, they create less biased labels.

Finally, the process of deriving the measure must perform adequately. Otherwise, the quality value is not reliable and hence no reasonable adaptation may be applied.

The IQ paradigm describes the Interaction Quality on a scale from five to one: 5 ("satisfied"), 4 ("slightly unsatisfied"), 3 ("unsatisfied"), 2 ("very unsatisfied"), and 1 ("extremely unsatisfied"). The paradigm is based on automatically deriving interaction parameters from the SDS and feeding these parameters into a statistical classification module which predicts the IQ level of the ongoing interaction at the current system-user-exchange. The interaction parameters are rendered on three levels (see Fig. 7.2): the exchange level the window level, and the dialogue level. The exchange level comprises parameters derived from the SDS modules Automatic Speech Recognizer, Spoken Language Understanding, and Dialogue Management directly. Parameters on the window and the dialogue level are sums, means, frequencies or counts of exchange level parameters. While dialogue level parameters are computed from all exchanges of the dialogue up to the current exchange, window level parameters are only computed from the last three exchanges.

These interaction parameters are used as input variables to a statistical classification module. The statistical model is trained based on annotated dialogues of the Lets Go Bus Information System in Pittsburgh, USA [30].[1] The annotated

[1] The Lets Go domain will be introduced in more detail in Sect. 7.4.3.2.

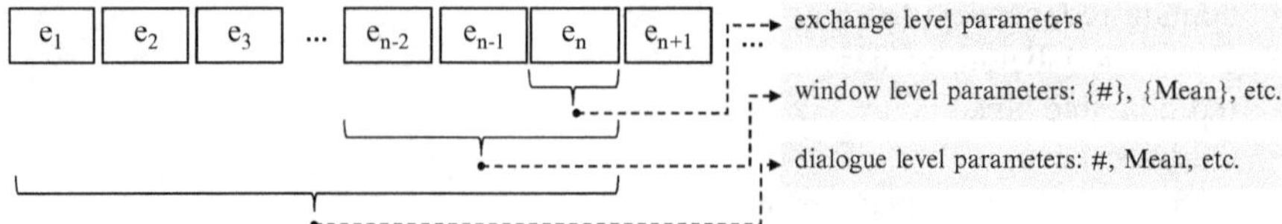

Fig. 7.2 The interaction parameters consist of three levels [34]: the exchange level containing information about the current exchange, the window level, containing information about the last three exchanges, and the dialogue level containing information about the complete dialogue up to the current exchange

data is packaged in the LEGOv2 corpus [34, 50]. Each of the 9638 exchanges (401 calls) has been annotated by three different raters resulting in a rating agreement of $\kappa = 0.52$. Furthermore, the raters had to follow labeling guidelines to enable a consistent labeling process [34]. Applying a Support Vector Machine [51] (SVM) for estimating the Interaction Quality achieved an unweighted average recall of 0.59 when including domain information [32] and 0.49 without domain information [39] both using only automatically derivable features. For the latter, Ultes et al. were able to improve the performance by applying an hierarchical approach introducing error correction [38] to a UAR of 0.53. While modeling IQ estimation as a sequential problem using regular Hidden Markov Models was not successful [43], the performance could be improved by applying a Hybrid Hidden Markov Model [39] achieving a UAR of 0.51.

7.4.2 Probabilistic Dialogue Management for Intelligent Environments

In order to enable the Dialogue Manager to deal with the added uncertainty inherent in adaptation to dynamically changing information when adding an automatic AE estimation module, a POMDP [12] is utilized which has been shown to work well with SPDs [54]. Formally, a POMDP consists of a set S of state variables, a set A of system actions, and a set O of all possible observations of the system. Furthermore, transition probabilities $P(s'|s, a)$ and observation probabilities $P(o'|s')$ are included. As the state of the underlying process cannot be determined exactly, a probability distribution over all possible states, called the *belief state* $b(s)$, is used instead. It is updated with the following equation:

$$b'(s') = p(o'|s') \cdot \sum_{s} P(s'|s, a) \cdot b(s) . \quad (7.1)$$

However, casting an SDS as a POMDP yields the problem that computing the probability distribution over all dialogue states is intractable. A promising methodology of handling multiple state hypotheses is the Hidden Information State

(HIS) introduced by Young et al. [55] which will be described in Sect. 7.4.2.1. To integrate the HIS approach into the ATRACO SDM (see Sect. 6.5.2), several alterations have to be made which are described in Sect. 7.4.2.2.

7.4.2.1 The Hidden Information State Approach

To cast an SDS as a POMDP within the Hidden Information State (HIS) approach proposed by Young et al. [55], the state s is decomposed into (u, g, h) representing user action u, user goal g, and dialogue history h as proposed by Williams and Young [54], who also introduce reasonable independence assumptions. The user goal space is further partitioned into equivalence classes, or partitions p, according to the possible values a slot can take. Introducing further simplification, one slot in the partition may only take all values or one single value, or it may exclude a set of values. For user input, the partitions are first split and the probability mass of the originating partition is distributed to the resulting partitions. In the second phase, the belief $b(s)$ of state $s = (u, p, h)$ is updated according to equation

$$b'(u', p', h') = k \cdot P(o'|u')P(u'|p', a) \sum_h P(h'|u', p', h, a) \sum_u P(p'|p)b(u, p, h) , \tag{7.2}$$

where o' is the current observation and a the last system action. $P(p'|p)$ denotes the probability of partition p' originating from partition p or, in other words, the fraction of probability mass which is transferred from p to p' if p is split into p' and $p - p'$.

According to Williams [53], the splitting probability

$$P(p'|p) = \frac{b_0(p')}{b_0(p)} \tag{7.3}$$

is computed as the ratio of the prior probability of the new partition $b_0(p')$ to the prior probability of the originating partition $b_0(p)$.

In order to make optimization of the policy $\pi(b)$ that determines the next system action more tractable, the resulting belief is transformed to a summary belief point $\hat{b}$ containing only information about the two most likely partitions. Based on this, a summary system action $\hat{a}$ is selected according to the trained policy $\pi(\hat{b})$. This summary system action then has to be refined using heuristics. The resulting action a is then executed by the system.

For better illustration of the partitioning approach, an example in a flight booking domain is shown in Fig. 7.3. Initially, there is only one partition containing all values for each slot, namely `Origin` and `Destination`. If new user input arrives, the partition is split according to the slot the user input belongs to. In this example, the n-best list belongs to the slot `Destination` and contains two entries. Therefore, the root partition is split and two new partitions are created, each one containing one of the two values in the slot `Destination`. The range

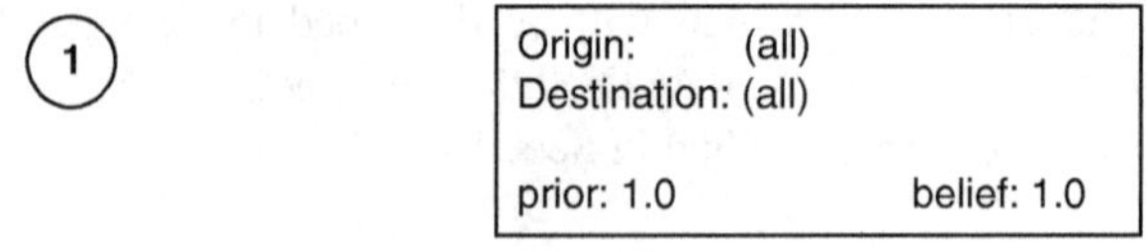

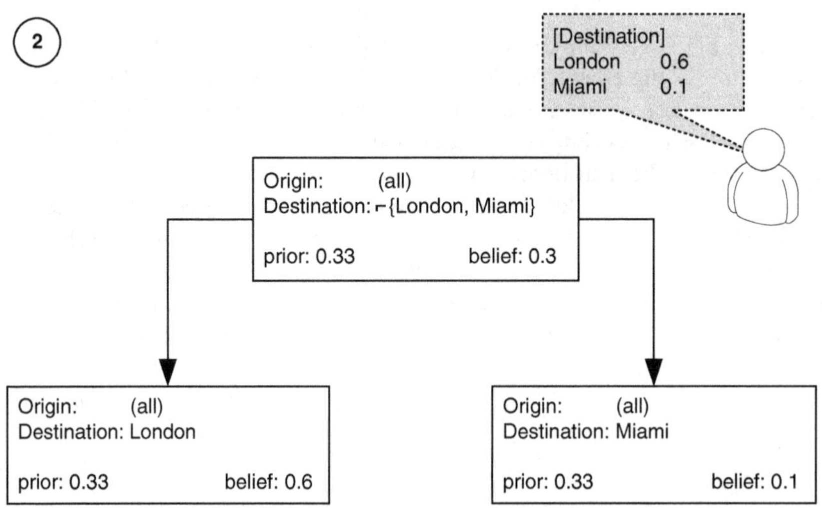

Fig. 7.3 An example of partition splitting [40] with three possible destinations: "London", "Miami", and "Paris". First, there is only one partition subsuming all values for the two slots. After splitting the partition on the user input, two new partitions are created each representing all goals containing "London" or "Miami", respectively, as destination, while the original partition excludes both values

of slot-values the original partition is representing has been reduced to exclude the two destinations provided by the user. Following that, the new belief values are determined. In order to select the next system action, the summary belief is computed. Based on this, the policy is applied and the resulting system action is refined and executed.

7.4.2.2 ATRACO Spoken Dialogue Management and the Hidden Information State Approach

For modeling spoken dialogue interaction between the user and the IE, the ATRACO SDM has been developed. As it is based on the Information State (IS) approach [13], applying the Hidden Information State approach—an extension of the IS—for introducing probabilistic dialogue into the ATRACO SDM seems natural. The resulting system contains the capability of having different operation modes concerning the number of state hypotheses and the policy. A more detailed description may be found in [40].

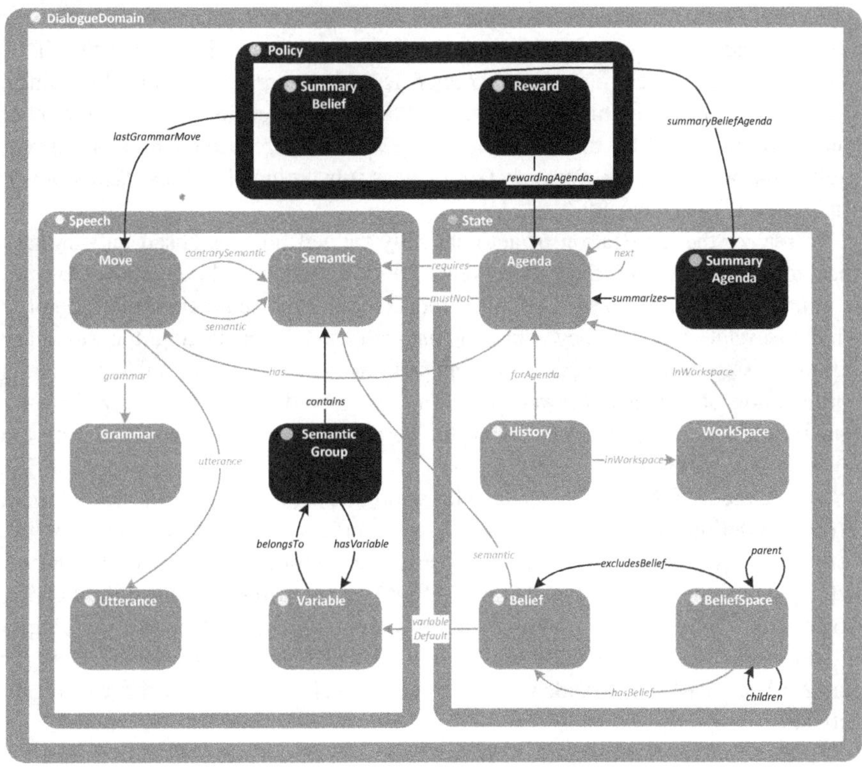

Fig. 7.4 A scheme of the Extended Spoken Dialogue Ontology (SDO) [40] based on Heinroth et al. [8]. Concepts belonging to the original SDO are *light grey* while concepts and relations introduced for the HIS implementation are *dark grey*. The static dialogue description is shown on the *left side* of the picture within the *Speech* class while the concepts belonging to the dynamic *State* of the system are shown on the *right side*. Additionally, concepts belonging to the *Policy* are shown at the *top*

The core of the ATRACO SDM, which has been designed following the *model-view-presenter* design pattern [29] allowing for a strict separation of data management, dialogue logic and dialogue interface, depicts the Domain Model. This SDO is internally structured—using the Web Ontology Language (OWL) [1]—into a static and a dynamic part. To integrate HIS functionality into the model, some relations and classes have to be altered and added. The new ontology is shown in Fig. 7.4 with all additional classes and relations colored in dark grey.

The main difference between the original SDO—implementing the IS approach—and the HIS approach is the state model. While for the IS, only one single state exists (represented by the *BeliefSpace* class), the HIS includes a tree of states associated with a probability. To introduce this hierarchical structure of the HIS partition state into the SDO, the relations *parent* and *children* have been added.

As each HIS partition either represents all values (for each slot), exactly one value, or excludes a set of slot values this has also be modeled within the SDO. For the latter, the relation *excludesBelief* is added representing all slot values, which are excluded by this partition. A slot taking one specific value, on the other hand, is represented by the already existing *hasBelief* relation. The slot concept itself is introduced by the *SemanticGroup* class subsuming all values (*Semantics* or *Variables*) of the given slot.

To realize the application of automatically trained policies based on summary space belief points [56], the general class *Policy* is added to the *DialogueDomain*. It contains the *SummaryBelief* class representing the summary belief point $\hat{b}$ which is related to the new class *SummaryAgenda* representing the summary action $\hat{a}$ (summarizing "actual" system actions *Agendas*). To enable automatic optimization of policies, a reward function has to be defined. It is realized by the *Reward* class defining a reward value for all agendas which are connected by the *rewardingAgendas*.

Basing the ATRACO SDM on the HIS approach also entails a probability model $P(o|u)$ modeling the probability of the user input. It is usually approximated with $P(o|u) \approx P(u|o)$ and thus modeled using the confidence scores or the n-best list input. As the view of the ATRACO SDM is based on automatically creating VoiceXML [26] documents for each turn, the created documents have to be altered to add n-best list functionality. Fortunately, VoiceXML provides mechanisms for using n-best list with confidences inherently which only have to be added and activated.

Spoken dialogue management, in general, has two major tasks: updating the internal state representation and, based on this updated state, selecting the next system action. For the ATRACO SDM, this is handled within the presenter which hence contains the dialogue logic. In order to incorporate HIS functionality, i.e. handling multiple state hypotheses and applying an optimized ontology, the AT&T Statistical Dialog Toolkit (ASDT) [52] is integrated into the SDM. Thus, only certain probability models have to be designed within the ATRACO SDM: the partition splitting model $P(p'|p)$, the user model $P(u'|p', a)$, and the history model $P(h'|u', p', h, a)$.

While the latter two are modeled rather simply by checking if the user input matches the current partition and history state, the partition splitting is more complex. First, the decision if the partition is split or solely updated has to be made. The latter happens if the user action does not represent new slot information but only a confirmation. The partition is split if the user action contains new slot information.

While the new ATRACO SDM offers HIS functionality, the original control modes and system state representations are still usable. Hence, the new ATRACO SDM incorporating the HIS approach offers multiple operation modes:

- Rule-Based Control + Single State Hypothesis
- Rule-Based Control + Multiple State Hypotheses
- Trained Policy-Control + One State Hypothesis
- Trained Policy-Control + Multiple State Hypotheses

The original ATRACO SDM conveys rule-based control with a single state hypothesis, i.e. the dialogue state is modeled within the ontology class *Beliefspace*. By extending the SDM with the HIS approach, both the capability of training an optimized policy automatically as well as handling multiple parallel state hypotheses, i.e. the HIS partitions, is introduced.

7.4.3 Experiments

To evaluate the impact of adding IQ-adaptivity to the dialogue, two studies have been conducted using the ATRACO SDM presented in the previous section. First, a pilot user study employing IQ-adaptation techniques adapting the grounding strategy in a limited train booking domain has been conducted [48]. The general aim of the study was to gain an initial insight into the capabilities of IQ-adaptive dialogue. A second study with a simulated user has been conducted in a more complex domain adapting the initiative [49]. The design of both studies along with the results will be described in the following.

7.4.3.1 Pilot User Study in the Train-Booking Domain

To gain an initial insight into the capabilities and opportunities of IQ-adaptive dialogue, a study within a simple train booking dialogue with real users was conducted. Depending on the current IQ value, the grounding strategy was adapted, i.e. each time the system requests a confirmation about a certain slot value from the user, the IQ value is used to decide whether the system uses an explicit or implicit confirmation prompt. In the following, the design and setup of the study will be presented before giving details about the results.

Design and Setup

For conducting a pilot study for IQ-adaptive dialogue, the grounding strategy was selected as it is an easily adaptable concept which occurs in almost every dialogue. A dialogue in the train booking domain was created asking the user for information about the origin, the destination, the day of the week, and the time of travel. The user could choose out of 22 cities which were used as origin and destination alike. Furthermore, the time of travel was restricted to every full hour (1 pm, 2 pm, 3 pm, etc.). Three different dialogues were created: one only applying explicit confirmation (all-explicit), one applying only implicit confirmation (all-implicit), and one adapting the confirmation type to the current IQ value (adapted). Besides these differences, the dialogues were the same. The complete dialogue was system initiated and the course of the dialogue was predetermined, i.e. the order of information the user was asked to provide was given. As only two different options for adapting the dialogue exist, i.e. either selecting implicit or explicit confirmation, the IQ value has been limited to only two values: two representing a satisfied user and one representing an unsatisfied user. If the user was recognized as being satisfied with the dialogue (high IQ value), slot values were confirmed implicitly while explicit confirmation was applied for unsatisfied users (low IQ value). In the end of the dialogue, the user was provided with a dummy message stating that the reservation has been made.

The IQ estimation module was based on the LibSVM implementation [2] using a linear kernel.

Before the experiment, each participant was presented with a sheet of paper stating all options they could say during the dialogue. This also included a list of all cities. Furthermore, each user participated in three runs of the dialogue—one for each type of confirmation strategy. During the experiment, the order of these dialogues has been alternated to get an equal distribution over all combinations so that learning effects are taken account of. However, the user was not aware of the different dialogue types. After each dialogue, the participants were asked to fill out a questionnaire based on the SASSI questionnaire [11] to evaluate their overall experience with the dialogue. Each item was rated on a seven-point scale.

In total, there were 24 participants (eight female, 16 male) creating 72 dialogues with an average number of turns of 33.58. The participants, who were students from multiple disciplines, were between 19 and 38 years old with an average age of 26.42.

Experimental Results

The results for all questions from the questionnaires are depicted in Table 7.1. Each row shows the average score for one of the three different strategies. It is a well-known fact that, for simple tasks like this, an all-implicit strategy is usually preferred over an all-explicit strategy (cf. [4]). Hence, as expected, the all-implicit strategy performed best outperforming the all-explicit strategy clearly: it achieved a better score for almost all questions. The difference is even significant for 16 out of 25 values ($\alpha < 0.05$ applying the Mann–Whitney U test [20]). Comparing the all-explicit to the adapted strategy gives a similar impression: The scores for almost all questions are better for the adapted strategy. However, this is not as significant having only seven significant different values. More revealing is the conclusion drawn from comparing the all-implicit with the adapted strategy. While the all-implicit strategy again governs the scores, almost all results are not significantly different. Hence and in contrast to the expectations, the adapted strategy did not perform significantly worse despite the dialogue being very simple.

This result is underpinned by looking at the users' overall satisfaction score with the dialogue as an emphasis was put on the question which strategy people liked best. A bar graph showing the average outcome of the user ratings grouped by the respective dialogue strategy is depicted in Fig. 7.5. While the adapted strategy resulted in 45.6 % explicit and 54.4 % implicit confirmations, it is very interesting that it was not rated significantly different compared to the all-implicit strategy.

Table 7.1 The average results of the user questionnaires

	Fun with interaction	System reaction	Hard to lose track	Natural communication	User is in control	Overall impression	User felt calm	**user felt tense**	Dialogue length appropriate	User knew what to say	High level of concentration necessary	System understands user input	Functionality	Interaction was pleasant
All-implicit	5.3^{e}	5.9^{e}	6.2	4.8^{e}	3.0	5.4^{e}	5.0	$6.3^{a,e}$	5.4^{e}	6.0	2.7	5.9^{e}	5.5	5.6^{e}
All-explicit	$3.7^{a,i}$	4.8^{i}	5.8	3.5^{i}	2.7	4.3^{i}	4.5	5.0^{i}	$3.5^{a,i}$	6.1	3.3	4.2^{i}	4.6	4.3^{i}
Adapted	4.8^{e}	5.3	6.1	4.5	3.0	5.0	4.8	5.4^{i}	5.0^{e}	6.2	3.2	5.2	5.3	5.0

	Interaction was not exhausting	Interaction was not boring	Interaction was diversified	Interaction was not frustrating	System made no errors	Error due to user behaviour	System is friendly	System is easy to use	System is flexible	System is understandable	Reaction time	Would use system again	Human operator preferred	Overall satisfaction
All-implicit	5.6^{e}	4.7^{e}	3.9^{e}	5.9^{e}	5.9^{e}	3.3	5.8	6.3^{e}	4.1	5.8	5.0^{e}	5.2	4.7	5.5^{e}
All-explicit	4.3^{i}	$3.3^{a,i}$	$3.0^{a,i}$	$4.4^{a,i}$	4.5^{i}	2.1	5.8	5.6^{i}	3.8	5.5	$3.9^{a,i}$	4.2	5.1	$4.4^{a,i}$
Adapted	5.3	4.3^{e}	3.9^{e}	5.5^{e}	5.3	2.8	5.6	6.0	4.2	5.8	4.9^{e}	5.0	4.7	5.3^{e}

Each question could be answered by a seven-point scale being translated to scores from one to seven. Significant differences are marked with a, e, and i marking significance with the adaptive, explicit, and implicit strategy respectively. (Please note: the original questionnaire was in German)

That is even although the ASR component made almost no errors (due to the limited number of options). Moreover, calculating Spearman's Rho [37] shows significant correlation ($\alpha < 0.01$) with $\rho = 0.6$ between the users' overall satisfaction of the all-implicit and adapted strategy. Additionally, the dialogue length, which is one main indicator for user satisfaction in simple dialogues like this, is significantly higher for the adapted strategy compared to the all-implicit strategy.

In other words, although the task was quite simple, there was no difference between the all-implicit and adapted strategies encouraging the hope that for more complex dialogues, quality-adaption will perform best.

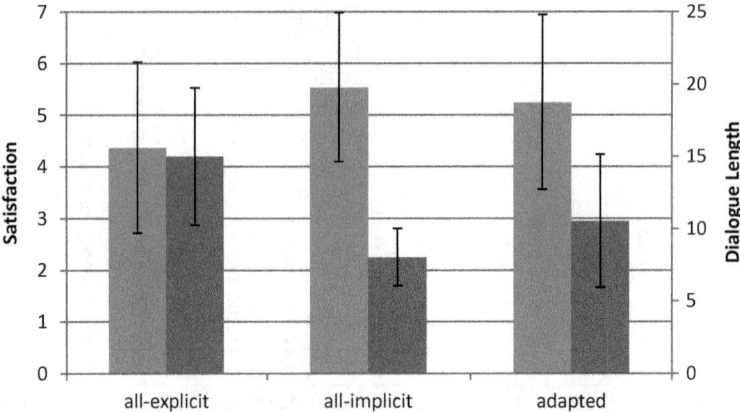

Fig. 7.5 The overall satisfaction with the dialogue (*left bar*, left *y*-axis) and the average dialogue length in number of turns (*right bar*, right *y*-axis) according to questionnaire evaluation. Satisfaction for implicit and adapted do not differ significantly while all other differences are significant

7.4.3.2 User Simulator Study in the Bus Schedule Information Domain

For a second experiment with a system providing bus schedule information, the dialogue initiative was adapted. Conventional dialogue initiative categories are *user initiative*, *system initiative* and *mixed initiative* [22]. As there are different interpretations of what these initiative categories mean, we stick to the understanding of initiative as used by Litman and Pan [18]: the initiative influences the openness of the system question and the set of allowed user responses. The latter is realized by defining which slot values provided by the user are processed by the system and which are discarded. Hence, for *user initiative*, the system asks an open question allowing the user to respond with information for any slot. For *mixed initiative*, the system poses a question directly addressing a slot. However, the user may still provide information for any slot. This is in contrast to the *system initiative*, where the user may only respond with the slot addressed by the system. For instance, if the system asks for the arrival place and the user responds with a destination place, this information may either be used (*mixed initiative*) or discarded (*system initiative*).

Design and Setup

In order to evaluate the dialogue strategies, the adaptive ATRACO SDM is used interacting with a user simulator having rule-based control with a single state hypothesis. For creating dialogues, the Lets' Go Domain is chosen as it represents a domain of suitable complexity. The Let's Go Bus Information System [30] is a live system in Pittsburgh, USA providing bus schedule information to the user. It consists of four slots: bus number, departure place, arrival place, and travel time.

However, the bus number is not mandatory. The original system contains more than 300,000 arrival or departure places, respectively. The Let's Go User Simulator (LGUS) by Lee and Eskenazi [14] is used for evaluation to replace the need for human evaluators.

The IQ estimation module was based on the LibSVM implementation [2] using a linear kernel. The trained model achieves an accuracy of 54.1 % on the training data using tenfold cross-validation. All exchanges of the LEGO corpus have been used for training.

For evaluation, a total of 5000 simulated dialogues for each strategy have been created. In accordance to Raux et al. [30], short dialogues (less than 5 exchanges[2]) which are considered "not [to] be genuine attempts at using the system" are excluded from all statistics in this paper.

Three objective metrics are used to evaluate the dialogue performance: the average dialogue length (ADL), the dialogue completion rate (DCR) and task success rate (TSR). The ADL is modeled by the average number of exchanges per completed dialogue. A dialogue is regarded as being completed if the system provides a result—whether correct or not—to the user. Hence, DCR represents the ratio of dialogues for which the system was able to provide a result, i.e. provide schedule information:

$$\mathrm{DCR} = \frac{\#\mathrm{completed}}{\#\mathrm{all}} .$$

TSR is the ratio of completed dialogues where the user goal matches the information the system acquired during the interaction:

$$\mathrm{TSR} = \frac{\#\mathrm{correctResult}}{\#\mathrm{completed}} .$$

Here, only destination place, arrival place, and travel time are considered as the bus number is not a mandatory slot and hence not necessary for providing information to the user. Furthermore, the average IQ value (AIQ) is calculated for each strategy based on the IQ values of the last exchanges of each dialogue.

Experimental Results

Figure 7.6 shows the ration of complete, incomplete, and omitted dialogues for each strategy with respect to the total 5000 dialogues. As can be seen, about the same ratio of dialogues is omitted due to being too short. The DCR clearly varies more strongly for the five strategies.

[2]The minimum number of exchanges to successfully complete the dialogue is 5.

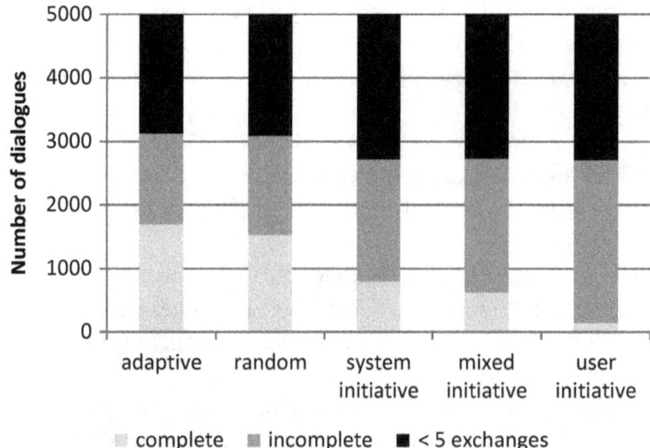

Fig. 7.6 The ratio of omitted dialogues due to their length (<5 exchanges), the completed dialogues (complete), and the dialogues which have been aborted by the user (incomplete) with respect to the dialogue strategy. While the amount of short dialogues is similar for each strategy, the number of completed dialogues varies strongly

The results for DCR, TSR, ADL, and AIQ are presented in Fig. 7.7. TSR is almost the same for all strategies, meaning that, if a dialogue completes, the system almost always found the correct user goal. Hence, TSR is not further regarded. DCR, ADL and AIQ on the other hand vary strongly. They strongly correlate with a Pearson's correlation of $\rho = -0.953$ (level of significance $\alpha < 0.05$) for DCR and ADL, $\rho = 0.960$ ($\alpha < 0.01$) for DCR and AIQ, and $\rho = -0.997$ ($\alpha < 0.01$) for ADL and AIQ.

Comparing the performance of the adaptive strategy to the three non-adaptive strategy clearly shows that the adaptive strategy performs significantly best for all metrics achieving a DCR of 54.27 % (which is comparable to the rate achieved on the training data of LGUS (cf. [14]).

Furthermore, the adaptive strategy has a significant higher average IQ (AIQ) value calculated from the IQ value for the whole dialogues, i.e. the IQ value of the last system-user-exchange, than all other non-adaptive strategies.

7.4.4 Conclusion

Using the short-term goal Interaction Quality for adapting to dynamically changing information seems to be a promising approach to increase the overall dialogue performance for both user experience and objective metrics. Adapting the grounding strategy as well as the dialogue initiative in a rule-based setting are both reasonable measures. Moreover, casting the dialogue system as a POMDP—resulting in an

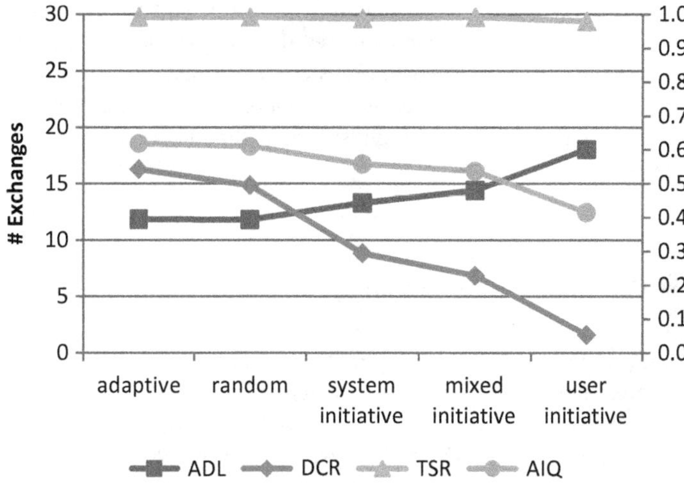

Fig. 7.7 The average dialogue length (ADL), task success rate (TSR), the dialogue completion rate (DCR), and the average Interaction Quality (AIQ) for all for dialogue strategies[3]. With decreasing DCR, also AIQ decreases and ADL increases. (AIQ values are normalized to the interval [0–1])

extension of the ATRACO SDM—allows not only for an improvement in dialogue performance but also for a better handling of the added uncertainty inherent in IQ estimation.

7.5 Adaptation to Dynamically Changing Information with Long-Term Goal

While the adaptation to IQ is focused on achieving short-term goals, the HCT relationship between human and dialogue system represents a long-term goal. A user's HCT model can not be measured directly during run-time, but only by means of a questionnaire. What can be assessed are though the dialogue history and the user's affective state (at least a hypothesis on that can be used). This means, that symptoms like user frustration or confusion, which may indicate the possibility of a decreasing HCT relationship, have to recognized and then the resulting change in HCT must be estimated. Now the question might arise why we should bother with modeling HCT in the first place instead of reacting only to affective user states?

[3]All results for DCR and TSR are significantly different (chi-squared test). Significant differences in ADL (unpaired t-test) and AIQ (Mann–Whitney U test) with the respective strategy to the right are on the level of $\alpha < 0.01$ for system initiative (ADL) and mixed initiative (ADL) and on the level of $\alpha < 0.05$ for adaptive (AIQ), random (ADL, AIQ), and mixed initiative (AIQ). All other comparisons between non-neighbours are significant with $\alpha < 0.01$.

The most important difference is that human–computer trust is a model which evolves long-term, but affective states model dynamic events. Though we want to use dynamic events as well to determine system behaviour, we also want to develop a healthy HCT relationship between human and computer. This way, despite, for example, being confused, the user might still be willing to continue interacting with a dialogue system.

7.5.1 Human–Computer Trust

Trust has shown to be a crucial part of the interaction between human and machines. If the user does not trust the system and its actions, advice or instructions, the interaction with the machine may change up to a complete abortion of future interaction [28]. Situations, where the system's actions do not match the user's expectations, are likely to have a negative impact on the HCT relationship [23]. Those situations occur as a consequence of incongruent models of the system: During the interaction the user builds a mental model of the system and its underlying processes that determine system actions and output. However, if this perceived mental model and the actual system model do not match, the HCT relationship may be influenced negatively [23].

Mayer et al. [21] define trust in human–human interaction to be "the extent to which one party is willing to depend on somebody or something, in a given situation with a feeling of relative security, even though negative consequences are possible". For HCI, trust can be defined as "the attitude that an agent will help achieve an individual's goals in a situation characterized by uncertainty and vulnerability" [15]. Machines that serve as intelligent assistants with the purpose of helping the user, in complex as well as in critical situations, seem to be very dependent on an intact HCT relationship. However, trust is multi-dimensional and consists of several components. For human relationships, Mayer et al. defined three levels that build trust: ability, integrity, and benevolence. The same holds for HCI, where HCT is a composite of several components. For human–computer trust Madsen and Gregor [19] constructed a hierarchical model (see Fig. 7.8) resulting in five basic components of trust, which can be divided into two general categories, namely cognitive-based and affect-based ones. In short-term HCI, cognition-based HCT components seem to be more important because it will be easier to influence those. Perceived understandability can be seen in the sense that the human supervisor or observer can form a mental model and predict future system behaviour. Perceived reliability in the usual sense of repeated, consistent functioning. Furthermore, technical competence in the sense that the system is perceived to perform the tasks accurately and correctly based on the input information. In this context it is important to mention, that as Mayer already stated, the components of trust are separable, yet related to one another. All components must be perceived highly for the trustee to be deemed trustworthy. If any of the components does not fulfill this requirement, the overall trustworthiness can suffer [19]. Hence, a dialogue system

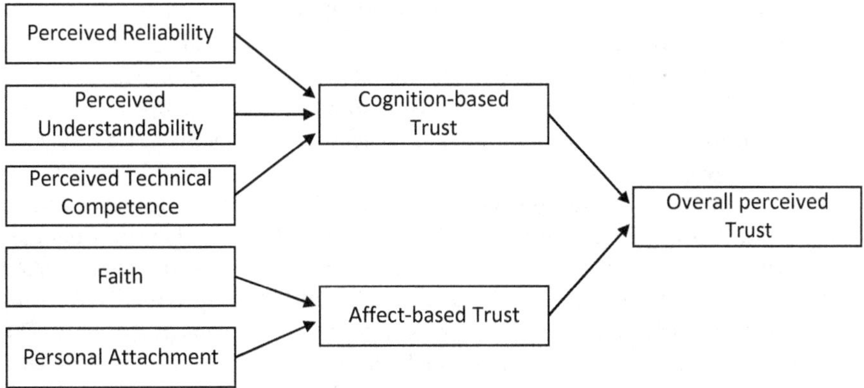

Fig. 7.8 Human–computer trust model: personal attachment and faith are the components of affect-based trust. Perceived understandability, technical competence and reliability for cognition-based trust

should not only adapt to the estimated general trust, but to the single components. In the following we will introduce our approach of adapting the dialogue to single HCT components to foster the long-term relationship between user and system.

7.5.2 Integrating HCT Adaptation

While the adaptation of the task-oriented dialogue flow using a probabilistic model described previously is a good way to foster a high Interaction Quality, it still has its drawbacks. The integration of other *Adaptation Entities* like single HCT components (e.g., the user's perceived understandability of the system, the perceived system's technical competence, or the system's reliability) will result in a highly complex dialogue structure. Dialogue moves might only be suitable for a certain combination of AE values and, therefore, several permutations of AE values might be required, resulting in a highly complex dialogue flow. However, the most important dialogue strategy for coping with HCT issues is to provide explanations. Explanations are, however, related to the task-oriented dialogue at hand, not directly connected with them. This means that the flow of a task-oriented dialogue is not altered by including or augmenting the ongoing task-oriented dialogue with additional explanations. Therefore, we developed a dialogue system which incorporates a dedicated decision-making component dealing with domain-independent situations in HCI, which are also independent of the task-oriented dialogue. For example, the decrease of the user's perceived understandability of system can be estimated when observing user confusion. This type of situation is not domain-dependent and may be therefore handled by a domain-independent probabilistic decision model. Though the situation and the resulting decision may be domain-independent, it is important to note that the resulting dialogue strategy does not necessarily have to be domain-independent. In our example of observed user confusion, the dialogue strategy is to provide some explanation corresponding to the ongoing dialogue. Though the decision to provide explanations is domain-independent, the explanation itself should, at least in the optimal case, integrate domain knowledge.

In a nutshell, our goal was to integrate a dedicated probabilistic decision-making component, modeling domain-independent characteristics or adaptation entities, which can be plugged into existent dialogue systems. Hence, as the task-oriented dialogue should remain untouched, the kind of adaptation strategy has to be domain-independent and not conflicting but supplementary to the planned dialogue. The task of the dedicated component described here is to estimate the user's human–computer trust model and to augment the IQ-adaptive dialogue. Therefore, we perform an adaptation to dynamically changing information with the long-term goal of HCT adaptation.

7.5.2.1 Probabilistic HCT Model

A probabilistic model of the HCT relationship between user and dialogue system is used to determine strategies that lead in the long run to a trustworthy HCI. The AE, which are in this case the cognition-based components of HCT (i. e. perceived reliability, perceived understandability and perceived technical competence) are estimated by the observation of affective user states along with the dialogue history. This is described using a POMDP (cf. Sect. 7.4.2) and formalized in the Relational Dynamic Influence Diagram Language (RDDL) [31]. RDDL is a uniform language that allows an efficient description of POMDPs by representing its constituents (actions, observations, belief state) with variables. Figure 7.9 shows a simplified

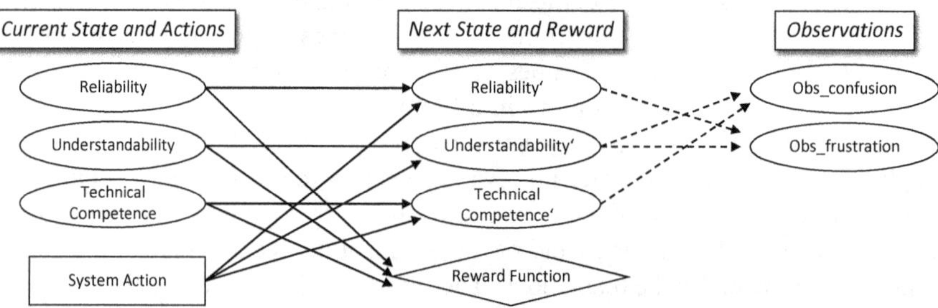

Fig. 7.9 This simplified figure of the POMDP model incorporates exemplary observations of the affective states *confusion* and *frustration* which influence the current state. These observations combined with the cognition-based components of trust (perceived reliability, perceived understandability and perceived technical competence), which are also part of the system state, and the current system action determine whether the next system action should be for example a transparency explanation

model of the in RDDL defined POMDP model. The system actions A are the dialogues presented to the user. These are the different goals of explanations (justification, transparency, conceptualization, relevance and learning). The POMDP model is a probabilistic representation of the domain, which determines when and how to augment the dialogue with explanations at run-time. Now, the quest is to define the reward function $R(s, a)$ in a way that it leads to an optimal flow of actions. I. e. the system should receive a penalty when the dimensions of HCT do not remain intact, and actions should incur a cost so that the system only executes them when the human–computer trust is endangered. For example, following the conducted experiment, the reward is defined in a way that providing transparency explanations is beneficial for increasing the state variables perceived understandability and reliability, though it also inflicts a cost for providing extra information dialogues.

The POMDP is then used by a planner [24, 36] to search for a policy that determines the system's behaviour. This policy is, e. g. represented as a decision tree that recommends the most suitable action based on the system's previous actions and observations. For example, a policy for a POMDP that models HCI with respect to HCT, can thus represent a decision tree which represents a guideline for a dialogue flow that ensures an intact HCT relationship.

7.5.2.2 Dialogue Augmentation Process

Integrating the probabilistic HCT model is done by plugging the component into the existent pipeline depicted earlier, by integrating the component in the system selection (see Fig. 7.10). The HCI is started using the task-oriented dialogue approach. The POMDP checks during the ongoing dialogue whether the user's trust or components of it are endangered. If this is the case, the proposed explanation has to be integrated into the ongoing task-oriented dialogue. Hence, the POMDP is used only for the augmentation of the task-oriented part of the dialogue with explanations and serves two purposes. First, it proposes the integration of domain-independent dialogue strategies into the task-oriented dialogue. Second, it selects what kind of explanation has to be selected or generated.

Though we know that explanations can help in keeping a system trustworthy (cf. Sect. 5.2), the question remains what kind of explanation is the best for which situation in HCI. Since the components of HCT are impaired differently in different situations, the system reaction to those situations should be directed as well. Hence, we conducted an experiment to test which explanations work best to deal with impairments of specific HCT components, to be able to generate directed explanation dialogue strategies.

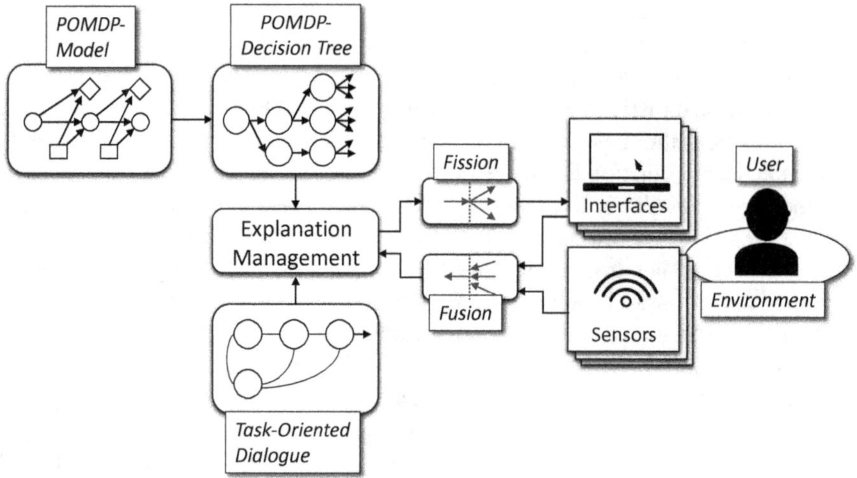

Fig. 7.10 This figure shows the architecture used for the augmentation of the task-oriented dialogue with domain-independent dialogue strategies

7.5.3 Experiments

The experiment was a web-based study inducing events to create unclear or not anticipated situations and to compare the effects of different explanations on the components of HCT. For our experiment, we concentrated on justification and transparency explanations. Justifications are the most obvious goal an explanation can pursue. The main idea of this explanation is to provide support for and increase confidence in given system advice or actions. The goal of transparency is to increase the user's understanding of how the system works. It may help to change the user's perception of the system from a black-box to a system the user can comprehend (i. e. a white box). Thereby, the user can build a mental model of the system and its underlying reasoning processes. Therefore, our hypothesis was that transparency explanations will perform best to recover the user's perceived understandability. The user's perceived reliability, measuring the impression of consistent functioning, was also expected to be recovered best by transparency explanations, because they explain how the system works, and thus explain why the system reacted inconsistently.

Design and Setup

The main objective for the participants was to organize four parties for friends or relatives in a web-based environment. They had to use the browser at home or the university to organize, for example, the music, select the type and amount of food or order drinks. The first two rounds were meant to go smoothly and were supposed to get the subject used to the system and in this way build a mental model of it. After the first two rounds, a HCT questionnaire [19] was presented to the user. As expected the users had built a relationship with the system by gaining an understanding of the systems processes. The next two rounds were meant to influence the HCT-relationship negative with incomprehensible, unexpected external events. These unexpected and incongruous system events in terms of the user's mental model would pro-actively influence the decisions and solutions the user could make to solve the task. Without warning, the user was overruled by the system and either simply informed by this change or was presented an additional justification or transparency explanation.

Results

One hundred and thirty-nine starting participants were distributed among the three test groups (no explanation, transparency, justifications). Ninety eight accomplished round 2, reaching the point when the external events were induced and 59 participants completed the experiment. The first main result was that 47 % from the group receiving no explanations quit during the critical rounds 3 and 4. However, if explanations were presented only 33 % (justifications) and 35 % (transparency) did quit. This means that even though the participants would encounter negative consequences of losing the reward money, they dropped out of the experiment. Therefore, we can state that the use of explanations in incomprehensible and unexpected situations can help to keep the HCI running.

The main results from the HCT questionnaires can be seen in Fig. 7.11. The data states that providing no explanations in rounds three and four resulted in a decrease in several components of trust. Therefore, we can conclude that the external events did indeed result in our planned negative change in trust. *Perceived understandability* diminished on average over the people questioned by 1.2 on a Likert scale with a range from 1 to 5 when providing no explanation at all

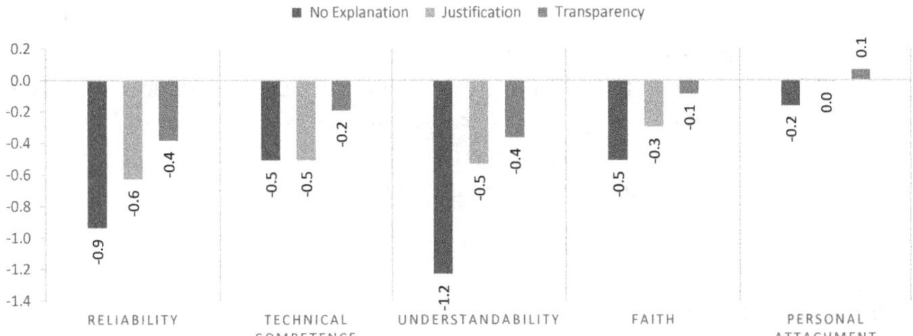

Fig. 7.11 This figure shows the changes of HCT components from round 2 to round 4. The scale was a 5 point Likert scale with e. g., 1 the system being not understandable at all and 5 the opposite

compared to only 0.4 when providing *transparency* explanations [no explanation vs. transparency $t(34) = -3.557, p = 0.001$], and on average by 0.5 with *justifications* [no explanation vs. justifications $t(36) = -2.023, p = 0.045$]. Omitting explanations resulted in an average decrease of 0.9 for the *perceived reliability*, with transparency explanations in a decrease of 0.4 and for *justifications* in a decrease of 0.6 [no explanation vs. transparency $t(34) = -2.55, p = 0.015$].

Discussion

These results support our hypothesis that transparency explanations can help to reduce the adverse effects of trust loss regarding the user's perceived understandability and reliability of the system in incomprehensible and unexpected situations. Particularly for the perceived understandability, meaning the prediction of future outcomes, transparency explanations fulfill their purpose in a good way. Additionally, they seem to help with the perception of a reliable, consistent system. Though justifications do not perform best in any situation, they are still helpful to keep an intact HCT relationship. Justifications have the significant advantage that, compared to transparency explanations, they do not require reasoning about system processes, but can be predefined. Hence, despite being less efficient in keeping HCT, they might still be a valuable option if the system's reasoning capabilities about inner processes are not available in the dialogue system.

In our dialogue system for transparency explanations, this includes the imparting of domain-dependent knowledge (i. e. reasoning about system processes and functionalities) to foster a deeper understanding of the system. For the other explanations predefined content suitable for the present domain can be selected and presented to the user. For example, if the user is frustrated during an instruction, a justification explanation for the present dialogue may be chosen from the domain, without the need to generate it at run-time.

In general the results show that it is worthwhile to augment ongoing dialogues with explanations to maintain HCT. However, HCT can only be estimated by the use of additional factors like affective user states. Hence, to estimate impairment in the HCT relationship, we need to incorporate observations prone to uncertainty, resulting in the necessity to model HCT in a probabilistic way.

Apart from adaptation to HCT, the present approach can be used for any domain-independent adaptation entity with a long-term goal. Probably it is even more advantageous for domain-independent AE which do not require the integration of domain-dependent content in the dialogue strategy.

7.6 Conclusion

While conventional non-adaptive Spoken Dialogue Systems are rigid and inflexible regarding the user's needs, introducing user-centred dynamic adaptation mechanisms into the spoken dialogue results in an improvement in user experience.

For this, a general model of adaptive dialogue management has been presented. This approach yields the possibility to adapt not only to short-term but also to long-term goals of adaptation to dynamically changing information. For distinguishing between long-term and short-term goals, two example implementations and corresponding experiments have been presented showing the general benefit of this approach. Furthermore, POMDP structures have been introduced into the dialogue management to handle additional uncertainty introduced by the entities of the adaptation mechanism.

References

1. Antoniou, G., van Harmelen, F.: Web ontology language: owl. In: Staab, S. (ed.) Handbook on Ontologies in Information Systems, pp. 76–92. Springer, Berlin (2003)
2. Chang, C.C., Lin, C.J.: LIBSVM: a library for support vector machines. ACM Trans. Intell. Syst. Technol. **2**, 27:1–27:27 (2011). Software available at http://www.csie.ntu.edu.tw/~cjlin/libsvm
3. Engelbrecht, K.P., Gödde, F., Hartard, F., Ketabdar, H., Möller, S.: Modeling user satisfaction with hidden markov model. In: SIGDIAL '09: Proceedings of the SIGDIAL 2009 Conference, pp. 170–177. Association for Computational Linguistics, Morristown, NJ, USA (2009)
4. Fraser, N.M.: The sundial speech understanding and dialogue project: results and implications for translation. In: Aslib Proceedings, vol. 46, pp. 141–148. MCB UP Ltd (1994)
5. Glass, A., McGuinness, D.L., Wolverton, M.: Toward establishing trust in adaptive agents. In: IUI '08: Proceedings of the 13th International Conference on Intelligent User Interfaces, pp. 227–236. ACM, NY, USA (2008)
6. Gnjatović, M., Rösner, D.: Adaptive dialogue management in the nimitek prototype system. In: PIT '08: Proceedings of the 4th IEEE Tutorial and Research Workshop on Perception and Interactive Technologies for Speech-Based Systems, pp. 14–25. Springer, Berlin (2008). doi:10.1007/978-3-540-69369-7_3
7. Hara, S., Kitaoka, N., Takeda, K.: Estimation method of user satisfaction using n-gram-based dialog history model for spoken dialog system. In: Calzolari, N., Choukri, K., Maegaard, B., Mariani, J., Odijk, J., Piperidis, S., Rosner, M., Tapias, D. (eds.) Proceedings of the Seventh conference on International Language Resources and Evaluation (LREC'10). European Language Resources Association (ELRA), Valletta, Malta (2010)
8. Heinroth, T., Denich, D.: Spoken interaction within the computed world: evaluation of a multitasking adaption spoken dialogue system. In: 35th Annual IEEE Computer Software and Applications Conference (COMPSAC), Munich, Germany, pp. 134–143. IEEE (2011)
9. Higashinaka, R., Minami, Y., Dohsaka, K., Meguro, T.: Issues in predicting user satisfaction transitions in dialogues: Individual differences, evaluation criteria, and prediction models. In: Lee, G., Mariani, J., Minker, W., Nakamura, S. (eds.) Spoken Dialogue Systems for Ambient Environments. Lecture Notes in Computer Science, vol. 6392, pp. 48–60. Springer, Berlin (2010). doi:10.1007/978-3-642-16202-2_5
10. Higashinaka, R., Minami, Y., Dohsaka, K., Meguro, T.: Modeling user satisfaction transitions in dialogues from overall ratings. In: Proceedings of the SIGDIAL 2010 Conference, pp. 18–27. Association for Computational Linguistics, Tokyo, Japan (2010)
11. Hone, K.S., Graham, R.: Towards a tool for the subjective assessment of speech system interfaces (sassi). Nat. Lang. Eng. **6**(3–4), 287–303 (2000). doi:10.1017/s1351324900002497
12. Kaelbling, L.P., Littman, M.L., Cassandra, A.R.: Planning and acting in partially observable stochastic domains. Artif. Intell. **101**(1–2), 99–134 (1998)

13. Larsson, S., Traum, D.: Information state and dialogue management in the TRINDI dialogue move engine toolkit. Nat. Lang. Eng. **6**, 323–340 (2000)
14. Lee, S., Eskenazi, M.: An unsupervised approach to user simulation: toward self-improving dialog systems. In: Proceedings of the 13th Annual Meeting of the Special Interest Group on Discourse and Dialogue, pp. 50–59. Association for Computational Linguistics (2012)
15. Lee, J.D., See, K.A.: Trust in automation: designing for appropriate reliance. Hum. Factors J. Hum. Factors Ergon. Soc. **46**(1), 50–80 (2004)
16. Lim, B.Y., Dey, A.K., Avrahami, D.: Why and why not explanations improve the intelligibility of context-aware intelligent systems. In: Proceedings of the SIGCHI Conference on Human Factors in Computing Systems, CHI '09, pp. 2119–2128. ACM, NY (2009)
17. Lindgaard, G., Dudek, C.: What is this evasive beast we call user satisfaction? Interact. Comput. **15**(3), 429–452 (2003)
18. Litman, D., Pan, S.: Designing and evaluating an adaptive spoken dialogue system. User Model. User-Adap. Inter. **12**(2–3), 111–137 (2002). doi:10.1023/a:1015036910358
19. Madsen, M., Gregor, S.: Measuring human-computer trust. In: Proceedings of the 11th Australasian Conference on Information Systems, pp. 6–8 (2000)
20. Mann, H.B., Whitney, D.R.: On a test of whether one of two random variables is stochastically larger than the other. Ann. Math. Stat. **18**(1), 50–60 (1947)
21. Mayer, R.C., Davis, J.H., Schoorman, F.D.: An integrative model of organizational trust. Acad. Manag. Rev. **20**(3), 709–734 (1995)
22. McTear, M.F.: Spoken Dialogue Technology – Toward the Conversational User Interface. Springer, London (2004)
23. Muir, B.M.: Trust in automation: part i. theoretical issues in the study of trust and human intervention in automated systems. In: Ergonomics, pp. 1905–1922 (1992)
24. Müller, F., Späth, C., Geier, T., Biundo, S.: Exploiting expert knowledge in factored POMDPs. In: Proceedings of the 20th European Conference on Artificial Intelligence (ECAI 2012), pp. 606–611 (2012)
25. Nothdurft, F., Honold, F., Kurzok, P.: Using explanations for runtime dialogue adaptation. In: Proceedings of the 14th ACM International Conference on Multimodal Interaction, pp. 63–64. ACM, New York (2012)
26. Oshry, M., Auburn, R., Baggia, P., Bodell, M., Burke, D., Burnett, D., Candell, E., Carter, J., Mcglashan, S., Lee, A., Porter, B., Rehor, K.: Voice extensible markup language (voicexml) version 2.1. Technical Report, W3C – Voice Browser Working Group (2007)
27. Pan, A.S., Litman, D.J., Pan, S.: Empirically evaluating an adaptable spoken dialogue system diane j. litman. In: Proceedings of the 7th International Conference on User Modeling, pp. 55–64 (1999)
28. Parasuraman, R., Riley, V.: Humans and automation: use, misuse, disuse, abuse. Hum. Factors J. Hum. Factors Ergon. Soc. **39**(2), 230–253 (1997)
29. Potel, M.: MVP: model-view-presenter the taligent programming model for C++ and java. Technical Report, Taligent Inc (1996). http://www.wildcrest.com/Potel/Portfolio/mvp.pdf
30. Raux, A., Bohus, D., Langner, B., Black, A.W., Eskenazi, M.: Doing research on a deployed spoken dialogue system: one year of let's go! experience. In: Proceedings of the International Conference on Speech and Language Processing (ICSLP) (2006)
31. Sanner, S.: Relational dynamic influence diagram language (rddl): language description (2010). Http://users.cecs.anu.edu.au/ ssanner/IPPC2011/RDDL.pdf
32. Schmitt, A., Schatz, B., Minker, W.: Modeling and predicting quality in spoken human-computer interaction. In: Proceedings of the SIGDIAL 2011 Conference, pp. 173–184. Association for Computational Linguistics, Portland, Oregon, USA (2011)
33. Schmitt, A., Schatz, B., Minker, W.: A statistical approach for estimating user satisfaction in spoken human-machine interaction. In: Proceedings of the IEEE Jordan Conference on Applied Electrical Engineering and Computing Technologies (AEECT), pp. 1–6. IEEE, Amman, Jordan (2011)
34. Schmitt, A., Ultes, S., Minker, W.: A parameterized and annotated spoken dialog corpus of the cmu let's go bus information system. In: International Conference on Language Resources and Evaluation (LREC), pp. 3369–337 (2012)

35. Sidorov, M., Ultes, S., Schmitt, A.: Emotions are a personal thing: Towards speaker-adaptive emotion recognition. In: IEEE International Conference on Acoustics, Speech, and Signal Processing (ICASSP), pp. 4836–4840 (2014)
36. Silver, D., Veness, J.: Monte-carlo planning in large POMDPs. In: NIPS, pp. 2164–2172 (2010)
37. Spearman, C.E.: The proof and measurement of association between two things. Am. J. Psychol. **15**, 88–103 (1904)
38. Ultes, S., Minker, W.: Improving interaction quality recognition using error correction. In: Proceedings of the 14th Annual Meeting of the Special Interest Group on Discourse and Dialogue, pp. 122–126. Association for Computational Linguistics (2013). http://www.aclweb.org/anthology/W/W13/W13-4018
39. Ultes, S., Minker, W.: Interaction quality estimation in spoken dialogue systems using hybrid-hmms. In: Proceedings of the 15th Annual Meeting of the Special Interest Group on Discourse and Dialogue (SIGDIAL), pp. 208–217. Association for Computational Linguistics (2014). http://www.aclweb.org/anthology/W14-4328
40. Ultes, S., Minker, W.: Managing adaptive spoken dialogue for intelligent environments. J. Ambient Intell. Smart Environ. **6**(5), 523–539 (2014). doi:10.3233/ais-140275
41. Ultes, S., Heinroth, T., Schmitt, A., Minker, W.: A theoretical framework for a user-centered spoken dialog manager. In: Proceedings of the Paralinguistic Information and its Integration in Spoken Dialogue Systems Workshop, pp. 241–246. Springer, New York (2011)
42. Ultes, S., Schmitt, A., Minker, W.: Attention, sobriety checkpoint! can humans determine by means of voice, if someone is drunk... and can automatic classifiers compete? In: Proceedings of the 12th Annual Conference of the International Speech Communication Association (INTERSPEECH 2011), pp. 3221–3224 (2011)
43. Ultes, S., ElChabb, R., Minker, W.: Application and evaluation of a conditioned hidden markov model for estimating interaction quality of spoken dialogue systems. In: Mariani, J., Devillers, L., Garnier-Rizet, M., Rosset, S. (eds.) Proceedings of the 4th International Workshop on Spoken Language Dialog System (IWSDS), pp. 141–150. Springer, New York (2012)
44. Ultes, S., Schmitt, A., Minker, W.: Towards quality-adaptive spoken dialogue management. In: NAACL-HLT Workshop on Future directions and needs in the Spoken Dialog Community: Tools and Data (SDCTD 2012), pp. 49–52. Association for Computational Linguistics, Montréal, Canada (2012). http://www.aclweb.org/anthology/W12-1819
45. Ultes, S., ElChabb, R., Schmitt, A., Minker, W.: Jachmm: a java-based conditioned hidden markov model library. In: IEEE International Conference on Acoustics, Speech and Signal Processing (ICASSP), 2013, Vancouver, BC, Canada, pp. 3213–3217. IEEE (2013)
46. Ultes, S., Schmitt, A., Minker, W.: On quality ratings for spoken dialogue systems – experts vs. users. In: Proceedings of the 2013 Conference of the North American Chapter of the Association for Computational Linguistics: Human Language Technologies, pp. 569–578. Association for Computational Linguistics (2013)
47. Ultes, S., Dikme, H., Minker, W.: Dialogue management for user-centered adaptive dialogue. In: Proceedings of the 5th International Workshop On Spoken Dialogue Systems (IWSDS) (2014). http://www.uni-ulm.de/fileadmin/website_uni_ulm/allgemein/2014_iwsds/iwsds2014_lp_ultes.pdf
48. Ultes, S., Dikme, H., Minker, W.: First insight into quality-adaptive dialogue. In: International Conference on Language Resources and Evaluation (LREC), pp. 246–251 (2014)
49. Ultes, S., Kraus, M., Schmitt, A., Minker, W.: On objective performance measures of quality-adaptive spoken dialogue—and their correlation with interaction quality. In: Proceedings of the 2015 Conference of the North American Chapter of the Association for Computational Linguistics: Human Language Technologies (2015, submitted)
50. Ultes, S., Platero Sánchez, M.J., Schmitt, A., Minker, W.: Analysis of an extended interaction quality corpus. In: Proceedings of the 6th International Workshop On Spoken Dialogue Systems (IWSDS) (2015). Accepted for publication
51. Vapnik, V.N.: The Nature of Statistical Learning Theory. Springer, New York (1995)
52. Williams, J.D.: At&t statistical dialog toolkit (2010). http://www2.research.att.com/sw/tools/asdt/

53. Williams, J.D.: Incremental partition recombination for efficient tracking of multiple dialog states. In: Acoustics Speech and Signal Processing (ICASSP), 2010 IEEE International Conference on, Dallas, Texas, pp. 5382–5385. IEEE (2010)
54. Williams, J., Young, S.: Partially observable markov decision processes for spoken dialog systems. Comput. Speech Lang. **21**(2), 231–422 (2007)
55. Young, S., Schatzmann, J., Weilhammer, K., Ye, H.: The hidden information state approach to dialog management. In: IEEE International Conference on Acoustics, Speech and Signal Processing, 2007 (ICASSP 2007), vol. 4 (2007)
56. Young, S.J., Gačić, M., Keizer, S., Mairesse, F., Schatzmann, J., Thomson, B., Yu, K.: The hidden information state model: a practical framework for POMDP-based spoken dialogue management. Comput. Speech Lang. **24**(2), 150–174 (2010)

Chapter 8
Artificial Intelligence Planning for Ambient Environments

Julien Bidot and Susanne Biundo

Abstract In this chapter, we describe how Artificial Intelligence planning techniques are used in The Adapted and TRusted Ambient eCOlogies (ATRACO) in order to provide Sphere Adaptation. We introduce the Planning Agent (PA) which plays a central role in the realization and the structural adaptation of activity spheres. Based on the particular information included in the ontology of the execution environment, the PA delivers workflows that consist of the basic activities to be executed in order to achieve a user's goals. The PA encapsulates a search engine for hybrid planning—the combination of hierarchical task network (HTN) planning and partial-order causal-link (POCL) planning. In this chapter, we describe a formal framework and a development platform for hybrid planning, PANDA. This platform allows for the implementation of many search strategies, and we explain how we realize the search engine of the PA by adapting and configuring PANDA specifically for addressing planning problems that are part of the ATRACO service composition. We describe how the PA interacts with the Sphere Manager and the Ontology Manager in order to create planning problems dynamically and generate workflows in the ATRACO-BPEL language. In addition, an excerpt of a planning domain for ATRACO is provided.

8.1 Introduction

Intelligent environments (IEs), such as smart homes, offices, and public spaces, are featured with a large number of devices and services that help users in performing efficiently various kinds of tasks. In the scope of The Adapted and TRusted Ambient eCOlogies (ATRACO), we use workflows to model how a large number of services should interact with one another as well as with the user in IEs based on available resources, environment characteristics, user's tasks and profile.

Sphere Adaptation (SA) is one of the dimensions of adaptation which is realized within the ATRACO project. In ATRACO, we use the notions of Ambient Ecology

J. Bidot • S. Biundo (✉)
Institute of Artificial Intelligence, Ulm University, Ulm 89069, Germany
e-mail: julien.bidot@uni-ulm.de; susanne.biundo@uni-ulm.de

© Springer International Publishing Switzerland 2016
S. Ultes et al. (eds.), *Next Generation Intelligent Environments*,
DOI 10.1007/978-3-319-23452-6_8

to describe the resources of an Ambient Intelligence (AmI) environment and activity spheres (ASs) to depict the specific Ambient Ecology resources, data, and knowledge required to support a user in realizing a particular goal. In our approach, a ubiquitous computing system supports the execution of overlapping or disjoint ASs using the resources provided by the AmI space. An AS which is a temporary entity at run time is set up in order to enable a specific user goal: once the user goal is no longer relevant, the AS is dissolved. As long as this goal persists, AS is adaptive, in the sense that it can be instantiated within different environments containing similar resources and adaptively pursue its goal, whenever it remains possible. An AS is represented by a workflow that consists of activities to be executed and that are described in terms of services required or produced by resources of the Ambient Ecology.

Our approach to service composition consists in two steps: (1) making strategic decisions by considering abstract services; i.e., identifying what basic activities are to be executed and in which order they are to be executed; (2) taking operational decisions; i.e., determining what resources should execute the activities. Describing IEs with ontologies allows us to decompose the problem of orchestrating the services into two parts: in this chapter, we focus on the Planning Agent (PA), an agent integrating Artificial Intelligence (AI) planning techniques, that is responsible for making strategic decisions in order to create workflows with abstract services at design time, while Sect. 1.4.2 introduces the Sphere Manager, an agent in charge of taking the operational decisions at run time in order to generate executable workflows with concrete services.

In ATRACO, the planning problem can be stated as "discover an execution path of tasks to achieve a user goal given some state of the world." A plan that is solution to the problem is called a task model in the ATRACO terminology. We use a library of planning domains, each of them corresponding to a user goal. AI planning techniques are used for the realization and the structural adaptation of ASs in ATRACO. Based on the available services in the AmI environment, we apply these techniques in order to determine the basic activities to be executed for attaining user goals. The structural adaptation of ASs refers to the persistent achievement of the goals when the type of the available resources changes (as agents and users may come and go, devices and services may appear and disappear over time) and when the cardinality of the available resources varies (as the number of devices or users that participate in the realization of an AS may differ over time).

The formation of a system that realizes adaptive ASs is based on a service-oriented architecture (see Sect. 1.2.3), which integrates the PA to allow for adaptive planning.

8.2 Related Work

The composition of services has been a hot research topic in the last years, and various AI planning approaches have been used to address this issue [9].

Zhao and Doshi proposed a framework that can handle the uncertainty inherent in Web services [20]. This framework is based on semi-Markov decision processes, and it implements symbolic techniques that operate directly on first-order logic based representations of the state space to obtain the compositions. In addition, time constraints of the services are explicitly represented and taken into account. Although this approach is applicable to Web processes that are nested to an arbitrary depth, it is not possible to express complex causal relations between them in the abstraction hierarchy. Since we do not have any information about the uncertainty relative to the services in ATRACO, we use a deterministic AI planning approach to addressing the service composition issue.

Pistore, Traverso, and Bertoli presented an approach to the automated composition of Web services that integrates symbolic model checking [10]. They modeled BPEL4WS Web services with non-deterministic and partially observable behaviors, and they expressed composition requirements with extended goals. Unlike our work, they dealt with AI planning problems with uncertainty.

The work of R-Moreno et al. [13] integrated AI planning and scheduling techniques to automatically generate business process models, avoiding going through all the drawing process, and making sure that the established connections among activities conform to a valid sequence of activities. After the models have been generated, a user can simulate and optimize the process. The authors use partial-order causal-link (POCL) planning and constraint propagation techniques to address this problem. Unlike our work, they implemented a system that can deal with explicit resource and time constraints.

El Falou et al. [3] addressed the problem of automated composition of Web services by using AI planning techniques. They proposed a multiagent framework and an algorithm for guiding the planner of each agent towards the best local plan. Contrary to their work, the search for plans performed by the PA is centralized; i.e. we do not address the issue of merging several local plans generated by a number of planners.

Like the work of Sirin et al. [17], we use hierarchical task network (HTN) planning techniques to model and solve planning problems. However, our implementation is more flexible than theirs, and our planning domain models are purely declarative and do not contain any control structure for guiding search in contrast to theirs. In their application, Web services are specified with OWL-S.

Marquardt and Uhrmacher focus on using AI planning to solve the problem of service composition in smart environments [7]. They compare the runtime performance of four different planning systems using an abstract simulation model, and the evaluation results show that some of these planners are suitable for composing services in time. Their modeling of the problem is different from ours, and none of the evaluated planners implements HTN planning. In addition, the service composition is completely done at design time, which makes replanning from scratch necessary each time a new device appears or disappears in the smart environment.

In the Gaia project, Ranganathan and Campbell presented a paradigm for the operation of pervasive computing environment that is based on AI planning [12].

The first difference with our system lies in the modeling of planning problems, since they use PDDL, the initial world state aggregates the states of all entities (i.e.s̃ervices, devices, and applications) of the environment along with the context of the environment, and the goal world state is determined given a user goal, a template world state, and a utility function to be maximized, but there are no abstract tasks and expansion methods. In addition, unlike our approach, their planning component is responsible for the binding of concrete services, devices, and applications, which means that scalability problems inevitably appear due to the very large world states for realistic testbeds (no abstraction mechanism) and due to the impossibility to declare abstract tasks and procedures in the planning domains (no HTN planning). Their solving procedure becomes very prohibitive, if it has to restart search for solution plans from scratch several times, each time with a different goal world state. Finally, their planning component is also in charge of executing plans, and it may re-execute an action or replan when an execution failure occurs.

Amigoni et al. proposed an AI planning system, D-HTN, that performs a centralized plan-building activity which is tailored to use the capabilities of the devices currently available in smart environments [1]. In their experimental system, each device is associated with one agent, and another single agent is responsible for building plans. D-HTN implements an HTN planning approach that differs from ours, although their modeling of the problem is similar to ours. D-HTN is not formally grounded and is much more rigid than our search engine. In consequence, D-HTN does not allow for a general backtracking mechanism during the planning process, which is inefficient for addressing complex and large-scale planning problems. An important problem with using D-HTN, also encountered in [17], resides in including search control information in the planning domain models. In contrast to our system, there are no ontologies describing the smart environment, the knowledge about expansion methods is distributed and collected from the present smart devices, and service composition is entirely performed at design time.

There are some similarities between our work and the work of Lundh et al. [6]. They address the issue of configuring network robot systems at run time using AI planning techniques, and the role of functional configurations in their work is comparable to the role played by workflows in ATRACO. Their configuration planner uses methods that describe alternative ways to combine functionalities (or other methods) for specific purposes. This technique is inspired by HTN planning. However, there are also differences with our work. First, their planning framework does not deal with causal links, and their probabilistic/possibilistic conditional action planner is based on a forward-chaining planner that searches in the space of world states, while the search engine of the PA explores the space of partial plans at different abstraction levels and addresses deterministic problems with complete information about world states. Second, with our framework, we can explicitly represent and cope with ordering constraints in plans and expansion methods. Third, their action planner is given some search control knowledge in the form of first-order linear temporal logic formulas as input, which is used to prune the search space.

Vuković, Kotsovinos, and Robinson proposed an approach to the composition of Web services based on the context information [19]. The architecture of their system is composed of different layers, and it includes a planning system in the abstract service composition layer. The planner is a forward-chaining planner. Like our system, theirs supports a dynamic on-the-fly adaptation of applications; i.e., it can recover from execution time failures of individual service instances. However, unlike the search engine of the PA, their planning system searches in the space of states and cannot deal with procedural knowledge. In addition, Web service instances are represented as actions in their planning domains, whereas abstract services are encoded as facts in the planning problems the PA addresses. In the same way as the action planner presented in [6], some search control knowledge is used to reduce the search effort. Finally, abstract execution plans are represented in BPEL4WS in their prototype, and the abstract services of these plans are bound to service instances at execution time.

The work of Paluska et al. focuses on automating high-level implementation decisions in pervasive applications [8]. Their system enables a model in which an application programmer can specify the behavior of an adaptive application as a set of open-ended decision points. Each of these decision points may be satisfied by a set of alternative, competing scripts. The set may be extended at run time without needing to modify or remove any existing scripts. Their approach is hierarchical, since the scripts may contain themselves decision points. In the same vein as our work, but without using AI planning techniques and ontologies, their system is able to bind resources at run time.

8.3 Artificial Intelligence Planning

In Artificial Intelligence (AI), the *classical planning problem* consists typically in a set of operators, a single initial world state, and a single goal world state. The instance of an operator is often called a task, and a world state is a set of positive literals. A given world state can evolve; i.e., we can go from this state to another state: A state changes when a task is executed, since every executed task has usually at least one effect on the state. The objective of AI planning is to select and order tasks that allow one to attain the goal state from the initial state. Tasks can be executed in a state, if and only if some preconditions hold in this state. Each task is thus associated with preconditions and effects. A *plan* consists of tasks and ordering and causal relationships between these tasks. In a book, Ghallab, Nau, and Traverso present a survey about AI planning [5].

We consider an illustrative example of such a classical planning problem. A person is living in a flat in the Ulmstreet, and there are services for tuning the ambient temperature and the luminance level in the flat. In the initial state, there are devices that can cool, provide the current ambient temperature and the luminance level, increase and decrease the luminance level in the flat. The person can move with means of transport, can install a radiator in the flat, and can buy a radiator in a hardware store called Buildhouse. In the goal state, we want the ambient temperature and the luminance level in the flat to be tuned according to the person preferences. An intuitive and pragmatic solution would be the following procedure: The person has to go to Buildhouse, buy a radiator there, go back to home, install the radiator, and then devices can start tuning the ambient temperature and the luminance in the flat. We can model this planning problem with five operators: a person moves from one location to another one; a person buys a radiator in a hardware store; a person installs a radiator at a location; devices tune ambient temperature at a location; devices tune luminance level at a location. The operators have preconditions; e.g., devices can tune the ambient temperature at a location, if (a) a person is present at this location; (b) the location is home; (c) there is a device that can heat; (d) there is a device that can cool; (e) there is a device that can sense the current ambient temperature. The operators have also some effects; e.g., when a radiator is installed at a location that is a flat, then this flat can be heated.

A plan that solves the planning problem described above consists of six tasks that are partially ordered (see the acyclic digraph in Fig. 8.1, where the nodes represent tasks, and the arrows depict ordering constraints): A person goes from Ulmstreet (their home) to Buildhouse (a hardware store); the person buys a radiator; the person goes from Buildhouse back to Ulmstreet; devices start tuning luminance level and at the same time the person installs a radiator; after this installation, devices (including the radiator) start tuning the ambient temperature in the flat in the Ulmstreet. In Figs. 8.1 and 8.2, the first task and the last task are dummy tasks that represent the initial state and the goal state of the planning problem, respectively.

A planning problem is a complex problem to solve, since it is highly combinatorial: A large number of tasks to select and a huge number of conflicts that appear between the tasks. The acyclic digraph on the left of Fig. 8.2 shows the causal relationships between the tasks: the nodes represent tasks, and the annotated arrows depict causal links. In ATRACO, an intelligent planner is encapsulated within the Planning Agent (PA) and used to solve planning problems. Section 8.3.1 introduces the basic ingredients of the planning framework on which the intelligent planner is based, and Sect. 8.3.2 gives a review about planning strategies that guide the planner towards solution plans. Schattenberg et al. [15, 16] give more technical details about the framework and the planning strategies, respectively.

8.3.1 A Formal Framework for Refinement Planning

This section provides the concepts for a generic AI planning approach: Planning by plan refinement. PANDA, an existent development platform, is based on a planning framework that integrates POCL planning and HTN planning [2]. The planning framework is hybrid, since it is composed of and combines elements coming from both POCL planning and HTN planning. The Planning Agent (PA) of ATRACO encapsulates a search engine which is a particular configuration of PANDA. The formal framework uses an action description language (ADL) like representation

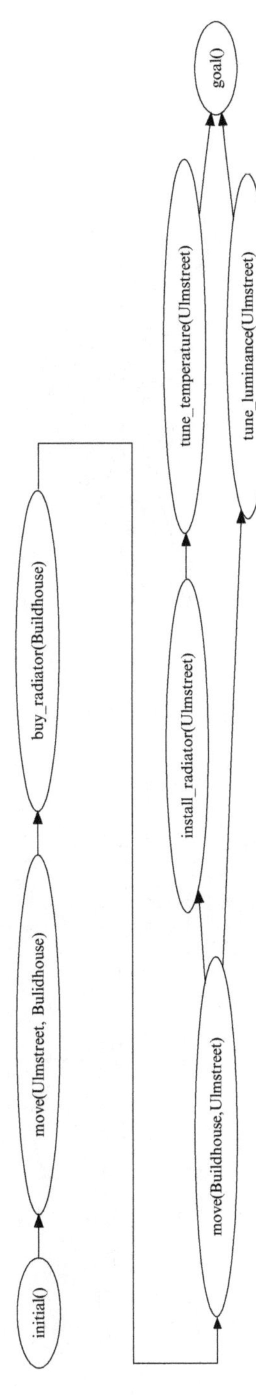

Fig. 8.1 Ordering structure of a plan

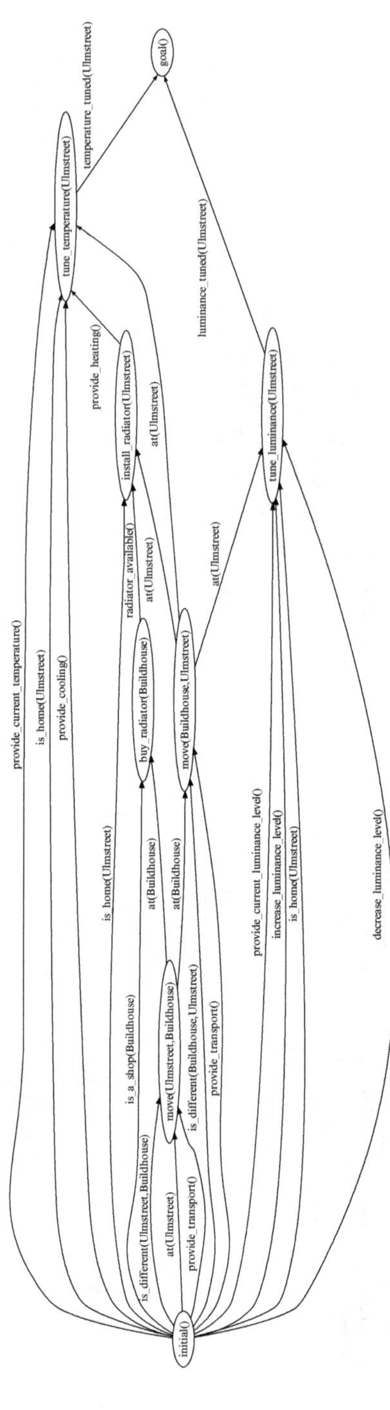

Fig. 8.2 Causal structure of a plan

of states and basic actions (*primitive tasks*). States and *preconditions* and *effects* of tasks are specified through formulae of a fragment of first-order logic. *Abstract tasks* can be refined by the so-called *expansion methods* which provide *task networks* (*partial plans*). The task networks describe how the corresponding abstract task can be solved. Partial plans may contain abstract as well as primitive tasks. With that, hierarchies of tasks and associated methods can be used to encode the various ways to accomplish an abstract task.

A domain model $D = (\mathtt{T}, \mathtt{EM})$ for hybrid planning consists of a set of task schemata $\mathtt{T}$ and a set of expansion methods $\mathtt{EM}$ for implementing the complex tasks in $\mathtt{T}$. Note that there are in general multiple methods provided for each complex task schema. A *task schema* $t(\overline{\tau}) = (\text{prec}(t(\overline{\tau})), \text{add}(t(\overline{\tau})), \text{del}(t(\overline{\tau})))$ specifies the *preconditions* and the *positive* and *negative effects* of the task. Preconditions and effects are sets of literals, $\overline{\tau} = \tau_1, \ldots, \tau_n$ are the task parameters. A ground instance of a task schema is called an *operation*. A *state* is a finite set of ground atoms, and an operation $t(\overline{c})$ is called *applicable* in a state s, if the positive literals of $\text{prec}(t(\overline{c}))$ are in s and the negative are not. The result of applying the operation in a state s is a state $s' = (s \cup \text{add}(t(\overline{c}))) \setminus \text{del}(t(\overline{c}))$.

In hybrid planning, abstract tasks like primitive tasks show preconditions and effects. The associated state transition semantics is based on axiomatic state refinements that relate task preconditions and effects across various abstraction levels. In contrast to that, in POCL planning, there are no abstract tasks and no expansion methods, but all tasks are described at the same level of abstraction. Though HTN planning allows us to describe tasks at different abstraction levels and decompose them with the help of expansion methods, it does not associate preconditions and effects with the tasks, which precludes any causal reasoning.

A *partial plan* or *task network* is a tuple $P = (\mathit{TE}, \prec, \mathit{VC}, \mathit{CL})$ with the following sets of *plan components*: TE is a set of *task expressions* (plan steps) $te = l : t(\overline{\tau})$ where l is unique label and $t(\overline{\tau})$ is a (partially) instantiated task. $\prec$ is a set of *ordering constraints* that impose a partial order on the plan steps in TE. VC is a set of *variable constraints*. They include equations and inequations of the form $v \dot{=} \tau$ and $v \dot{\neq} \tau$ which *codesignate* and *non-codesignate* variables occurring in TE with variables or constants. Moreover, VC contains *co-typing constraints* $v \dot{\in} Z$ and *non-cotyping constraints* $v \dot{\notin} Z$, where Z is a sort symbol, that restrict further codesignations. CL is a set of *causal links* and provides the usual means to establish and maintain causal relationships among the tasks in a partial plan. A causal link $cl = \langle te_i, \phi, te_j \rangle$ indicates, that formula ϕ expressed in first-order logic, which is an effect of task te_i, supports (a part of) the precondition of task te_j.

Task networks are also used as pre-defined *implementations* of complex tasks. An *expansion method* $em = (t(\overline{\tau}), (\mathit{TE}_{em}, \prec_{em}, \mathit{VC}_{em}, \mathit{CL}_{em}))$ relates such a complex task schema $t(\overline{\tau})$ to a task network.

A *planning problem* $\pi = (D, s_{\text{init}}, s_{\text{goal}}, P_{\text{init}})$ for hybrid planning consists of a domain model D, an initial world state s_{init}, a goal world state s_{goal}, and an initial task network P_{init}. The solution to a planning problem is obtained by transforming P_{init} stepwise into a partial plan P that meets the following solution criteria: (1) All steps in P are primitive tasks; (2) P is executable in s_{init} and generates a state s_{end}

such that $s_{\mathrm{goal}} \subseteq s_{\mathrm{end}}$ holds. Plan P is thereby called *executable* in a state s and *generates* a state s', if all ground linearizations of P, that means all linearizations of all ground instances of the task expressions in TE that are compatible with $\prec$ and VC, are executable in s and generate a state $s'' \supseteq s'$.

On the one hand, the POCL planning process consists in inserting tasks, ordering constraints, and causal links into partial plans until all preconditions are supported by formulae and all causal conflicts disappear. On the other hand, the HTN planning process endeavors to decompose all abstract tasks, until all tasks in the partial plan are primitive.

The XML code in Listing 8.1 defines formally our running planning problem in the planning language of PANDA. Note that there is no initial task network in this problem.

```
<?xml version="1.0" encoding="UTF-8"?>
<!DOCTYPE problem SYSTEM "problem.dtd">
<problem domainModel="TuneTemperatureDomainModel"
                name="TuneTemperatureProblem">
   <initialStateDescription>
      <fact name="provide_cooling"/>
      <fact name="increase_luminance_level"/>
      <fact name="decrease_luminance_level"/>
      <fact name="provide_current_luminance_level"/>
      <fact name="is_home">
         <constant name="Ulmstreet" sort="location"/>
      </fact>
      <fact name="provide_transport"/>
      <fact name="is_a_shop">
         <constant name="Buildhouse" sort="location"
                                     type="rigid"/>
      </fact>
       <fact name="is_different">
           <constant name="Buildhouse" sort="location"/>
           <constant name="Ulmstreet" sort="location"/>
       </fact>
       <fact name="is_different">
           <constant name="Ulmstreet" sort="location"/>
           <constant name="Buildhouse" sort="location"/>
       </fact>
      <fact name="at">
         <constant name="Ulmstreet" sort="location"/>
      </fact>
      <fact name="provide_current_temperature"/>
   </initialStateDescription>
   <goalStateDescription>
      <fact name="temperature_tuned">
         <constant name="Ulmstreet" sort="location"/>
      </fact>
      <fact name="luminance_tuned">
          <constant name="Ulmstreet" sort="location"/>
      </fact>
   </goalStateDescription>
</problem>
```

Listing 8.1 Illustrative planning problem

In this framework, the search space associated with a planning problem depends on several parameters such as the number of operators, the number of expansion methods, the number of preconditions and effects of each operator, the number of parameters of each operator, the number of predicates, and the number of objects.

Using the framework for the implementation of a search engine is particularly advantageous, since (a) it allows us to easily encode and efficiently deal with procedural knowledge given at different abstraction levels (HTN planning) and (b) it offers a great flexibility thanks to the capacity of reasoning about causal relations between tasks (POCL planning). With these features, we can model and solve a large variety of real-world planning problems.

8.3.2 Planning Strategies

Transforming partial plans into their refinements is done by using the so-called *plan modifications*. Given a partial plan $P = (TE, \prec, VC, CL)$ and a domain model D, a plan modification is defined as $\mathtt{m} = (E^{\oplus}, E^{\ominus})$, where $E^{\oplus}$ and $E^{\ominus}$ are disjoint sets of elementary additions and deletions of plan components over P and D, and $E^{\oplus} \cup E^{\ominus} \neq \emptyset$. Consequently, all elements of $E^{\ominus}$ are from TE, $\prec$, VC, or CL, respectively, while $E^{\oplus}$ consists of new plan components. This generic definition makes the changes that a modification imposes on a plan explicit. With this, the available options for a search strategy become comparable qualitatively and quantitatively. $\mathtt{P}$ denotes the set of all plans, while $\mathtt{M}$ constitutes the set of all plan modifications. The application of plan modifications is characterized by the generic plan transformation function $app : \mathtt{M} \times \mathtt{P} \rightarrow \mathtt{P}$ which takes a plan modification $\mathtt{m} = (E^{\oplus}, E^{\ominus})$ and a plan P, and returns a plan P' that is obtained from P by adding all components of $E^{\oplus}$ and removing those of $E^{\ominus}$.

$\mathtt{M}$ is grouped into modification classes $\mathtt{M}_{\mathtt{y}}$. As an example, the class $\mathtt{M}_{\mathtt{AddCL}}$ contains plan modifications $\mathtt{m} = (\{\langle te_i, \phi, te_j\rangle,\ v_1 \dot{=} \tau_1, \ldots, v_k \dot{=} \tau_k\}^{\oplus}, \{\}^{\ominus})$ for manipulating a given partial plan $P = (TE, \prec, VC, CL)$ by adding causal links. The plan steps te_i and te_j are in such a modification in TE, and the codesignations represent variable substitutions. They induce a VC'-compatible substitution σ' with $VC' = VC \cup \{v_1 \dot{=} \tau_1, \ldots, v_k \dot{=} \tau_k\}$ such that $\sigma'(\phi) \in \sigma'(\text{add}(te_i))$ for positive literals ϕ, $\sigma'(|\phi|) \in \sigma'(\text{del}(te_i))$ for negative literals, and $\sigma'(\phi) \in \sigma'(\text{prec}(te_j))$.

The complete collection of plan modifications for our hybrid planning framework presented above is introduced in [15]. This also covers the decomposition of abstract plan steps and the insertion of new plan steps, ordering constraints, and variable (in-) equations.

For a partial plan P that is a refinement of the initial task network of a given problem, but is not yet a solution, the so-called *flaws* are used to make the violations of the criteria defined above explicit. Flaws list those plan components that constitute the deficiencies of the partial plan. The set of all flaws is denoted by $\mathtt{F}$, and subsets $\mathtt{F}_{\mathtt{x}}$ represent classes of flaws. For example, the class $\mathtt{F}_{\mathtt{CausalThreat}}$ contains flaws $\mathtt{f} = \{\langle te_i, \phi, te_j\rangle, te_k\}$ describing causal threats; i.e., such flaws indicate that a task te_k is possibly being ordered between plan steps te_i and te_j, and there exists a variable substitution σ that is consistent with the equations and in-equations imposed by the variable constraints in *VC* such that $\sigma(\phi) \in \sigma(\text{del}(te_k))$ for positive literals ϕ or $\sigma(|\phi|) \in \sigma(\text{add}(te_k))$ for negative literals. This means that the presence of te_k in P as it stands will possibly corrupt the executability of at least some ground linearizations of P.

Flaw classes also cover the presence of abstract tasks in the plan, ordering and variable constraints inconsistencies, unsupported preconditions of actions, etc. The complete class definitions can be found in [15]. It can be shown that these flaw definitions are complete in the sense that for any given planning problem π and plan P that is not flawed, P is a solution to π.

Based on the formal notions of plan modifications and flaws, a generic algorithm and planning strategies can be defined. A strategy specifies *how* and *which* flaws in a partial plan are eliminated through appropriate plan modification steps.

A class of plan modifications $\mathtt{M_y} \subseteq \mathtt{M}$ is called *appropriate* for a class of flaws $\mathtt{F_x} \subseteq \mathtt{F}$, if and only if there exist partial plans P which contain flaws $\mathtt{f} \in \mathtt{F_x}$ and plan modifications $\mathtt{m} \in \mathtt{M_y}$, such that the refined plans $P' = app(\mathtt{m}, P)$ do not contain these flaws any longer.

It is easy to see that the plan modifications perform a strict refinement, that means, that a subsequent application of any modification instances cannot result in the same plan twice; the plan development is inherently acyclic. Given that, any flaw instance cannot be re-introduced once this has been eliminated. This qualifies the appropriateness relation as a valid strategic advice for the plan generation process and motivates its use as the trigger function for plan modifications: the α modification triggering function relates flaw classes with their potentially solving plan modification classes. As an example, causal threat flaws can be solved by expanding abstract actions which are involved in the threat (by overlapping task implementations), by promotion or demotion, or by separating variables through inequality constraints [2]: $\alpha(\mathtt{F_{CausalThreat}}) = \mathtt{M_{ExpTask}} \cup \mathtt{M_{AddOrdCstr}} \cup \mathtt{M_{AddVarCstr}}$. Please note that α states nothing about the relationship between the actual flaw and modification instances.

The modification triggering function allows for a simple plan generation process: (1) the flaws of the current plan are collected; if no flaw is detected, the plan is a solution; (2) suitable plan modifications are generated using the modification trigger; if for some flaws no plan modification can be found, the plan is discarded (dead-end); (3) selected plan modification is applied and generates further refinements of the plan; (4) the next candidate plan is selected and we proceed with (1).

Note that the tasks of the generated solution plans are partially ordered, which is desirable for realistic applications where activities are often to be performed in parallel.

The search space for refinement planning can be represented by a graph shown in Fig. 8.3. Large elliptic nodes represent partial plans. Small boxes inside nodes correspond to flaws, and arrows symbolize plan modifications. Triangles on the figure mean portions of the search space that are not explicitly shown. For example, partial plan P_2 is created by applying plan modification $\mathtt{m}_2$ to partial plan P_{init}

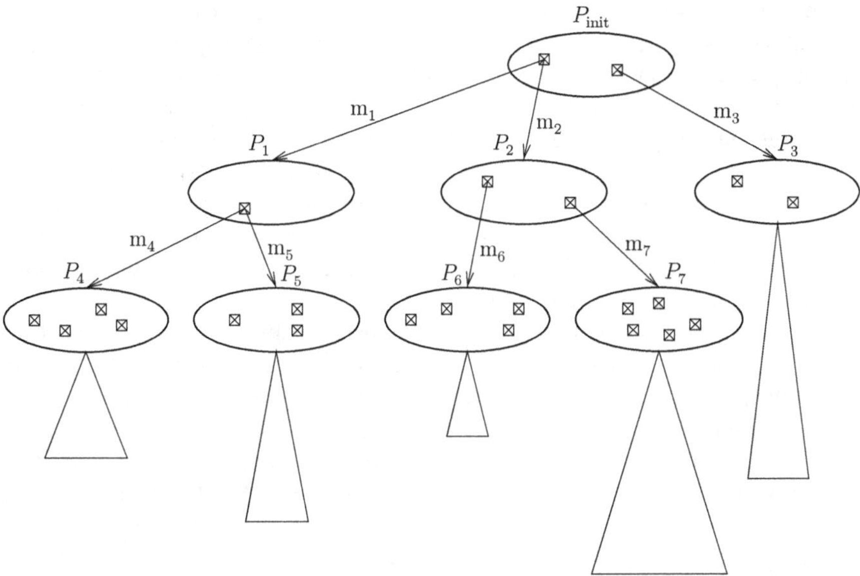

Fig. 8.3 Search space for refinement planning

(the initial task network). During the search for finding the plan presented in Figs. 8.1 and 8.2, the planning system explores 44 partial plans, including dead-ends.

The above mentioned triggering function completely separates the computation of flaws from the computation of modifications, and in turn both computations are independent from search-related considerations. The system architecture relies on this separation and exploits it in two ways: module invocation and interplay are specified through the α-trigger, while reasoning about search can be performed on the basis of flaws and modifications without taking their actual computation into account. Hence, we map flaw and modification classes directly onto groups of modules which are responsible for their computation.

A *detection module* x is a function that, given a partial plan P and a problem specification π, returns all flaws of type x that are present in the plan: $f_x^{\text{det}} : \mathtt{P} \times \Pi \rightarrow 2^{\mathtt{F}_\mathtt{x}}$ where $\mathtt{P}$ is the set of all plans, and Π is the set of all planning problems.

A *modification module* y is a function which computes all plan modifications of type y that are applicable to a given plan P and that are appropriate for given flaws $\mathtt{f}$ with respect to a given domain model D: $f_y^{\text{mod}} : \mathtt{P} \times \mathtt{F}_\mathtt{x} \times \mathtt{D} \rightarrow 2^{\mathtt{M}_\mathtt{y}}$ for $\mathtt{M}_\mathtt{y} \subseteq \alpha(\mathtt{F}_\mathtt{x})$ where $\mathtt{D}$ is the set of all domain models.

Please note that plan modifications carry a reference to the flaw instance they address; i.e., any plan modification is unambiguously linked with its triggering flaw.

While the plan deficiency detectors and the refinement generators provide the basic plan generation functionality, strategy functions can be designed for reasoning about which paths in the refinement space to pursue. To this end, we split up reasoning about search into two compartments. The first compartment is an option evaluation that is performed in the local view of the currently processed plan; it reasons about the detected flaws and proposed refinements in the current plan and assesses the modifications. The second component is responsible for the global view on the refinement space and evaluates the alternative search options.

We begin with the definition of a strategic function that selects all plan modifications that are considered to be worthwhile options, thereby determining the ordered set of successors for the current plan in the plan refinement space. In doing so, the following function also determines the branching behavior of the upcoming refinement-based planning algorithm.

Given a plan P, a set of flaws $\mathtt{f}$, and a set of plan modifications $\mathtt{m}$, a *modification-selection module* is a function $f^{\text{modSel}} : \mathtt{P} \times 2^{\mathtt{F}} \times 2^{\mathtt{M}} \to 2^{\mathtt{M} \times \mathtt{M}}$ that selects some (or all) of the plan modifications and returns them in a partial order for application to the passed plan.

Strategies discard a plan P, if any flaw remains unaddressed by the associated modification modules. That means, that we reject any plan P for any planning problem π, if for any f_x^{det} and $f_{y_1}^{\text{mod}}, \ldots, f_{y_n}^{\text{mod}}$ with $\mathtt{M}_{\mathtt{y_1}} \cup \ldots \cup \mathtt{M}_{\mathtt{y_n}} = \alpha(\mathtt{F_x})$ the following holds: $\bigcup_{1 \le i \le n} f_{y_i}^{\text{mod}}(P, f_x^{\text{det}}(P, \pi), D) = \emptyset$.

The second aspect of search control concerns the selection of those plans that are to be processed next by the detection and modification modules. These unevaluated partial plans, the leaves of the search tree, are usually called the *fringe*. In other words, concrete implementations of the following module are responsible for the general search schema, ranging from uninformed procedures such as depth-first, breadth-first, etc., to informed, heuristic schemata.

A *plan-selection module* is a function that returns a partial order of plans for a given sequence of plans. This function is described as $f^{\text{planSel}} : \mathtt{P}^* \to 2^{\mathtt{P} \times \mathtt{P}}$.

Based on the building blocks defined so far, we can now assemble a planning system that integrates both HTN planning and POCL planning. A software artifact that implements the generic refinement planning algorithm (Algorithm 1) is making the flaw detection and modification generating modules operational by stepwise collecting plan deficiencies, collecting appropriate modifications, selecting worthwhile modifications, and finally selecting the next plan in the fringe of the search tree. Please note that the algorithm is formulated independently from the deployed modules, since the options to address existing flaws by appropriate plan modifications are defined via α. The body of the algorithm is basically divided into four sections:

- Termination (2–3): If no more plans in the fringe are due to examination, that means, that the search space is exhausted and does not contain any solution; the algorithm terminates.
- Flaw detection (5–9): The results of all deployed detection functions $\mathfrak{Det}$ are collected. If no deficiency can be spotted, the current plan is considered to be a solution to the given planning problem.
- Plan modifications generation (10–18): The applicable plan modification steps are accumulated per flaw class and per class instance from the set of available

Algorithm 1 The generic refinement planning algorithm

Require: Sets of modules $\mathfrak{Det} = \{f_1^{\text{det}}, \ldots, f_d^{\text{det}}\}$ and $\mathfrak{Mod} = \{f_1^{\text{mod}}, \ldots, f_m^{\text{mod}}\}$
Require: Selection modules f^{modSel} and f^{planSel}

plan$(P_1 \ldots P_n, \pi)$: $\{P_1$ is the plan that is worked on$\}$
if $n = 0$ **then**
3: **return** failure
$P \leftarrow P_1$; Fringe $\leftarrow P_2 \ldots P_n$
$F \leftarrow \emptyset$
6: **for all** $f_x^{\text{det}} \in \mathfrak{Det}$ **do** {Flaw detection}
$F \leftarrow F \cup f_x^{\text{det}}(P, \pi)$
if $F = \emptyset$ **then**
9: **return** P
$M \leftarrow \emptyset$
for $x = 1$ to d **do** {Modification generation}
12: $F_x = F \cap \mathtt{F}_x$ {Process flaws class-wise as returned by corresponding f_x^{det}}
for all $\mathtt{f} \in F_x$ **do**
for all $f_y^{\text{mod}} \in \mathfrak{Mod}$ with $\mathtt{M}_y \subseteq \alpha(\mathtt{F}_x)$ **do**
15: $M \leftarrow M \cup f_y^{\text{mod}}(P, \mathtt{f}, D)$
if f was un-addressed **then**
$P_{\text{next}} \leftarrow f^{\text{planSel}}(\text{Fringe})$
18: **return** plan$(P_{\text{next}} \circ (\text{Fringe} \setminus P_{\text{next}}), \pi)$
for all m in *linearize*$(f^{\text{modSel}}(P, F, M))$ **do** {Strategic choices}
Fringe $\leftarrow$ *app*$(\mathtt{m}, P) \circ$ Fringe
21: $P_{\text{next}} \leftarrow$ *first*(*linearize*$(f^{\text{planSel}}(\text{Fringe})))$
return plan$(P_{\text{next}} \circ (\text{Fringe} \setminus P_{\text{next}}), \pi)$

plan modification generators $\mathfrak{Mod}$ according to the α-defined assignments. If any flaw is found unaddressed by its associated plan modification generation functions, the current plan is discarded and the algorithm is called recursively with a newly selected current candidate plan.

- Strategy (19–22): All plan modifications that pass the strategic plan modification selection function are applied to the current plan and thereby constitute the set of its refinements, that is, the set of its successor plans. This fringe extension is established by the strategic decisions in $f^{\text{modSel}}$and inserted at the beginning of the fringe. The plan selection function f^{planSel}finally chooses the next current plan and the procedure is called recursively.

The algorithm uses a function *linearize* to compute linear sequences of plans, respectively plan modifications that are consistent with the partial orders obtained from the appropriate strategic selection function.

The hybrid planning framework introduced in the previous section and the generic refinement planning algorithm (Algorithm 1) are very adequate to building planning systems for real-world applications in IEs, since: (a) we can encode formally the procedural knowledge about everyday activities of people thanks to abstract tasks and expansion methods (owing to HTN planning); and (b) the algorithm offers the means necessary to resolve causal conflicts between tasks that share resources of the IE (thanks to POCL planning). In addition, the integration of POCL and HTN planning techniques allows for the causal reasoning at different abstraction levels in the task hierarchy due to decomposition axioms, which is very powerful.

A large number of search strategies can be realized in the proposed refinement-planning framework by sequencing the respective selection modules and using the returned partially ordered sets of modifications, respectively plans, to modulate preceding decisions: If the primary strategy does not prefer one option over the other, the secondary strategy is followed and so on, until finally a random preference is assumed.

Some strategies are *unflexible* in the sense that they represent a fixed preference schema on the flaw type they want to get eliminated and then select appropriate modification methods. A traditional form of modification selection is either to prefer or to disfavor categorically specific classes of plan modifications; e.g., we may prefer the decomposition of tasks to task insertions.

With our refinement planning framework, it is also possible to design modification-selection strategies that are capable of operating on a more general level than unflexible strategies by exploiting flaw/modification information: They are neither flaw-dependent, as they do not primarily rely on a flaw type preference schema, nor modification-dependent, since they do not have to be biased in favor of specific modification types. A representative is the modification-selection strategy Least Committing First (LCF) which selects those modifications that address flaws for which the smallest number of alternative plan modifications has been found:

$$\begin{aligned}
&\mathtt{m}_i < \mathtt{m}_j \in f_{\mathrm{LCF}}^{\mathrm{modSel}}(P, \{\mathtt{f}_1, \ldots, \mathtt{f}_m\}, \{\mathtt{m}_1, \ldots, \mathtt{m}_n\}) \\
&\quad \text{if } 1 \le x_i, x_j \le m,\ 1 \le i, j \le n,\ \mathtt{m}_i \in \mathit{modFor}(\mathtt{f}_{x_i}, P),\ \mathtt{m}_j \in \mathit{modFor}(\mathtt{f}_{x_j}, P) \\
&\qquad \text{and } |\mathit{modFor}(\mathtt{f}_{x_i}, P)| < |\mathit{modFor}(\mathtt{f}_{x_j}, P)|.
\end{aligned}$$

It can easily be seen that this is a *flexible* strategy, since it does not depend on the actual types of issued flaws and modifications: It just compares answer set sizes in order to keep the branching in the search space low. In [16], more details about flexible strategies are presented.

In the next section, we will explain (a) how the planning techniques offered by the formal framework can participate in the composition of services in IE, and (b) how we have shaped a search engine for the PA, based on this formal framework, that is particularly suited for solving planning problems during the composition of services.

8.4 Planning in Intelligent Environments

In ATRACO, each Planning Agent (PA) encapsulates an AI planning system. This system is a search engine for hybrid planning that relies on the formal framework presented in Sect. 8.3.1. This search engine is actually a specific configuration of the existent development platform PANDA. For specifying a PANDA planning problem, we need two sources: the planning domain and the planning problem.

The planning domain contains the various operators and the alternatives (by means of expansion methods) to implement each abstract operator. Furthermore, predicates and types of objects of the application domain are represented.

The planning problem describes the problem instance to be solved. The description is made up of three parts: The initial world state, the specification of the goal world state, and the initial task network. In addition, we declare the objects of the problem at hand; e.g., the abstract services that are available in the room where a particular user is.

Searching for solutions to a planning problem is complex, since the search space associated with such a problem may be huge (e.g., see Fig. 8.3). In order to cope with this issue, we need planning strategies that guide the search towards solutions.

In ATRACO, we address the deployment of ASs using a service-oriented architecture approach in which resources in the AmI environment provide independent, loosely coupled services, see Sect. 1.3. A new PA is created, each time a new AS is instantiated; when an AS is dissolved, the associated PA is destructed. The PA is part of the advanced service composition mechanism of ATRACO.

In the literature, many research efforts addressing the Web service composition problem via AI planning have been reported (see Sect. 8.2). In terms of Web services, the initial world state and the goal world state are specified in the requirement of Web services requesters. The set of operators correspond to the set of available Web services, unlike our approach where abstract services are represented by facts. For example, McDermott [9] presented a Web service composition method based on PDDL and introduced a type of knowledge called value of an action that persists and is not treated as a truth literal.

In our approach, each user goal corresponds to a particular planning domain. From the abstract services offered currently by the AmI environment and known by Ontology Manager (OM), we generate the planning problem π to be solved by the PA. The planning problem depends on the user goal to be attained.

The OM is responsible for the creation of the Sphere Ontology which will include references to all the IE resources relevant to the AS that have been discovered.

The PA is in charge of finding a solution plan P to π. This plan describes abstract tasks that require abstract services in order to support the user goals. From the set of applied plan refinements that have led to P, the PA generates a workflow expressed in the ATRACO-BPEL language.

ATRACO-BPEL abstract service workflows comprise a set of activities and abstract services associated with these activities. These activities are structured with special constructs and correspond to the task of the plan generated by the PA. Actually, an abstract service workflow represents the abstract process model of an AS.

A flow diagram for the PA is shown in Fig. 8.4. The PA communicates with the Sphere Manager (SM) and OM for the creation of abstract service workflows.

The SM is in charge of (a) binding services provided by IE resources to the abstract services of the abstract service workflows, and then (b) executing and

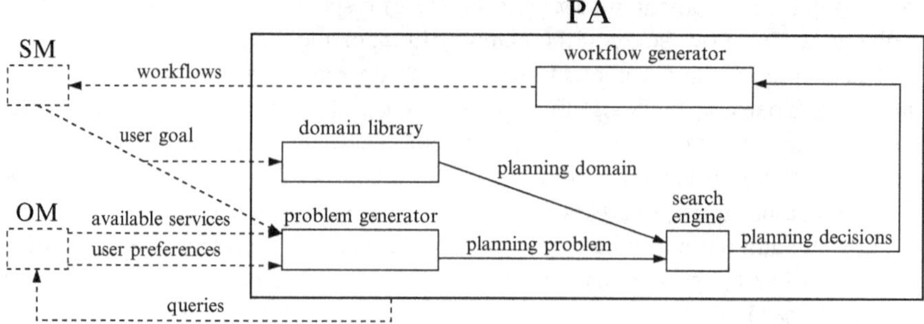

Fig. 8.4 Flow diagram of the Planning Agent

controlling the execution of the executable service workflows that result from the binding. During the execution of these workflows, the SM can handle execution errors and rebind abstract services if necessary. If the rebinding fails, the SM calls the PA again for replanning.

The life cycle of the PA is described as follows:

1. The SM instantiates a new PA for user goal *goal*.
2. The PA receives *goal* from the SM. The planning domain that matches *goal* is selected, and a new planning problem $\pi = (D, s_{\text{init}}, s_{\text{goal}}, P_{\text{init}})$ is then defined dynamically; i.e., s_{init}, s_{goal}, and P_{init} are defined.
3. A backtracking search starts in the space of partial plans in order to find a plan P that solves π.
4. From the set of plan modifications that have been applied during search and that have led to the solution plan, a workflow W, expressed in the ATRACO-BPEL language, is generated by the PA.
5. The PA transmits W to the SM.
6. The SM makes dynamic service binding for W (see Sect. 1.4.2) and executes W:

   - During the execution of W, if a concrete service that is bound to an abstract service of W is no longer available, SM searches for a new concrete service to be bound to this abstract service. If no concrete service is found, the binding fails and replanning takes place:

     a. a new planning problem π' is defined by querying OM (π' differs from π either in the initial state or in the goal state);
     b. we proceed with step 3 by replacing π with π'.

   - When the execution of W is finished, the corresponding PA is destructed.

The initial generation of W in step 3 relies on the generic refinement planning algorithm (see Algorithm 1) and supports the creation of ASs. The replanning procedure described in step 7 provides the structural adaptation of ASs.

8.4.1 Generation of Planning Problems

The search engine of the PA is based on the formal framework for refinement-based planning presented in Sect. 8.3.1. In this framework, states and *preconditions* and *effects* of tasks are specified through formulae of a fragment of first-order logic. A world state is thus represented by a list of facts. Contrary to other approaches presented in the literature (e.g., [9, 17]), we do not create planning domains dynamically, but instead the PA creates planning problems dynamically: (1) Depending on the user goal to be achieved, the currently available abstract services are modeled either by the facts of the initial world state or by the facts of the goal world state (e.g., `ambientLightLevel(service0)` in the domain "entertainment"); (2) if the initial world state s_{init} is not empty, then the user goal is modeled by tasks of the initial task network P_{init}. For determining s_{init} and s_{goal} of the planning problem at hand, the PA queries the OM to get the list of available abstract services that are relevant to the current planning domain; e.g., the PA asks the OM whether there are some abstract services available in the IE for sensing and controlling the ambient temperature in the room where a particular user is. Section 3.3 introduces ontologies and ontology managers.

In the planning formalism presented in Sect. 8.3.1, an ATRACO planning problem is described as follows:

- the user goal indicates the planning domain to be selected and is possibly modeled by the tasks of the initial task network P_{init};
- the currently available abstract services that can support the current user's goals are modeled by the facts of the world state s_{init} or s_{goal}.

We have designed a planning domain for AS "prepare an unexpected dinner," and an excerpt of which is included in Appendix.

8.4.2 Backtracking Search for Solution Plans

Some of the ATRACO planning domains we have designed contain a large number of expansion methods, the initial task network contains some tasks, and the goal state of each ATRACO planning problem π is then empty ($s_{\text{goal}} = \emptyset$). In this context, the insertion of tasks during backtracking search is not a necessity: in practice, we can find concrete plans without applying plan modifications of class $\mathtt{M}_{\mathtt{InsTask}}$. The configuration of PANDA reduces quite a bit the size of the search space to be explored. In addition, search is not going to be trapped into partial plans that contain arbitrary long sequences of reversible tasks. Such reversible tasks and causal links are inserted recursively in POCL planning in order to support open preconditions, but the insertion creates cycles. Avoiding the creation of these cycles during search makes the backtracking search more efficient.

For addressing the planning problems, we use the following unflexible plan-selection strategy:

$$P_i < P_j \in f^{\text{planSel}}_{\text{MorePlanSteps}}(P_1, \ldots, P_n) \quad \text{if } |TE_i| > |TE_j|$$

that prefers partial plans with more plan steps. We apply this plan-selection strategy, since each solution plan should contain as many tasks (i.e., services) as possible to support a user's goals.

The PA uses a flexible modification-selection strategy to solve ATRACO planning problems: LCF (see Sect. 8.3.2).

In theory, searching for solutions to a planning problem is complex, since the search space associated with such a problem may be huge. The complexity of the backtracking search done by the PA is limited in practice for ATRACO, since (1) the planning domains and the planning problems are designed at a high level: the number of abstract services is small; i.e, the size of world states is small, (2) the number of preconditions and effects per tasks is small, and (3) the number of causal interactions between the task networks of different abstract tasks is small. For our practical application, the size of search space is small enough to be explored in a few seconds using a Pentium 4 processor 3 GHz. For example, for a planning domain with 5 primitive tasks, 3 abstract tasks, 8 expansion methods, 8 predicates, and 2 preconditions and 2 effects per task, a corresponding planning problem consisting of one single abstract task and 6 objects is solved after about 8s, after having explored 154 partial plans including dead-ends. Note, however, that this planning domain involves a large number of causal interactions between tasks.

Moreover, we can still expect a significant speedup of the planning process by integrating some inference mechanisms. For example, there is not yet any special provision to deal with duplicate partial plans and cycles in the search space such as the approach presented in [18]. In addition, there is a recent paper that presents a landmark technique for HTN planning [4]. This technique filers out some parts of a planning domain that are irrelevant for the planning problem at hand.

8.4.3 Generation of Workflows

In ATRACO, the SM manages and executes workflows with services. A workflow consists of abstract and primitive tasks that are partially ordered. We have decided to use a variant of the language BPEL (also known as WS-BPEL, Web Service Business Process Execution Language) for the description of the ATRACO workflows. This XML language is called ATRACO-BPEL. The PA is in charge of generating workflows expressed in ATRACO-BPEL at design time, and then the PA transmits these workflows directly to the SM (see step 5 of the life cycle of the PA above).

There are three kinds of basic BPEL activities called respectively <bpel:invoke>, <bpel:receive>, and <bpel:reply>, and they are associated with primitive tasks of the planning domain. The structured activities of the BPEL-like language, such as <bpel:while> and <bpel:repeatUntil> used to express the repeated execution of activities are associated with expansion methods of the planning domain. The structured activities of type <bpel:pick> are also associated with expansion methods.

The BPEL constructs <bpel:partnerLink> and <ATRACO:role> used to express the abstract services required and provided by each basic activity are directly inserted into a workflow depending on what primitive tasks are present in the corresponding solution plan. These constructs are used to associate activities with abstract services.

The BPEL construct <bpel:sequence> used to describe precedence constraints between workflow activities corresponds to an abstract task and the expansion method that has been applied to decompose it and that describes a network of tasks that are totally ordered with ordering constraints. The construct <bpel:link> is used to synchronize workflow activities that belong to different structured activities. This construct also corresponds to any ordering constraint in the solution plan that has been explicitly added during the planning process; e.g., the insertion of such an ordering constraint is sometimes necessary to remove causal threats.

The BPEL construct <bpel:flow> used to describe workflow activities that execute in parallel corresponds to an abstract task and the expansion method that has been applied to decompose it and that describes a network of tasks that are not ordered.

For the generation of abstract service workflows in our BPEL-like language, the workflow generator of the PA looks for and analyze the set of plan modifications $\mathtt{M}_{\mathtt{success}}$ that have led to solution plans and that belong to two classes: $\mathtt{M}_{\mathtt{ExpTask}}$ (decomposition of abstract tasks) and $\mathtt{M}_{\mathtt{AddOrdCstr}}$ (addition of ordering constraints). The formal definition of these plan modification classes is given in [15]. Each plan modification of $\mathtt{M}_{\mathtt{ExpTask}}$ is interpreted as a BPEL structured activity such as <bpel:flow> composed possibly of other structure activities and basic activities (<bpel:invoke>, <bpel:receive>, and <bpel:reply>). The analysis of the plan modifications of $\mathtt{M}_{\mathtt{AddOrdCstr}}$ leads to the identification of <bpel:sequence> or <bpel:link> constructs. Since we have decided to deactivate the modification module $f^{mod}_{\mathtt{InsTask}}$ of PANDA in the PA, there are no plan modifications of class $\mathtt{M}_{\mathtt{InsTask}}$ in $\mathtt{M}_{\mathtt{success}}$.

The search space explored during the planning process is represented by a tree. The cost of generating workflows after the planning process is linear in the number of successful plan modifications $\mathtt{M}_{\mathtt{success}}$, since all the information we need is stored inside one single path: the cost depends on the depth of the tree (i.e., the number of applied plan modifications) from the initial task network P_{init} to the solution plan.

The XML code in Listing 8.2 defines a simple ATRACO-BPEL workflow with abstract services. The workflow consists of three activities that are not ordered (i.e., they are to be executed in parallel): musicControl, lightsControl, and photoViewerControl.

```
<bpel:process name="entertainment"
              targetNamespace="http://eclipse.org/bpel/sample"
  xmlns:tns="http://eclipse.org/bpel/sample"
  xmlns:bpel="http://docs.oasis-open.org/wsbpel/2.0/
                                            process/executable"
  xmlns:ATRACO="http://daisy.cti.gr/ATRACO_BPEL">
  <!--  ============================================  -->
  <!--  PARTNER LINKS                                  -->
  <!--  List of services participating in this BPEL    -->
  <!--  process                                        -->
  <!--  ============================================  -->
  <bpel:partnerLinks>
    <bpel:partnerLink name="ListenMusic"
          partnerLinkType="Trigger"
          partnerRole="ATRACO:ListenMusic"/>
    <bpel:partnerLink name="SetupLight"
          partnerLinkType="Continuous"
          myRole="ATRACO:LightControl"
          partnerRole="ATRACO:TriggerLight"/>
    <bpel:partnerLink name="ShowPhotos"
          partnerLinkType="Trigger"
          partnerRole="ATRACO:ShowPhotos"/>
  </bpel:partnerLinks>
  <!--  ============================================  -->
  <!--  ORCHESTRATION LOGIC                            -->
  <!--  Set of activities coordinating the flow of     -->
  <!--  messages across the                            -->
  <!--  services integrated within this business process  -->
  <!--  ============================================  -->
  <bpel:sequence name="main">
    <bpel:flow name="Flow">
      <bpel:reply name="musicControl"
                  partnerLink="ListenMusic"
                  operation="music"/>
      <bpel:invoke name="lightsControl"
                   partnerLink="SetupLight"
                   operation="light"/>
      <bpel:reply name="photoViewerControl"
                  partnerLink="ShowPhotos"
                  operation="photo"/>
    </bpel:flow>
  </bpel:sequence>
  <!--  ============================================  -->
  <!--  ROLES                                          -->
  <!--  ============================================  -->
  <ATRACO:roles>
    <ATRACO:role name="ListenMusic" type="output"
                 agentMonitored="no"
                 interactionType="direct">
      <ATRACO:device type="Music"/>
    </ATRACO:role>
    <ATRACO:role name="ShowPhotos" type="output"
                 agentMonitored="no"
```

```
                     interactionType="direct">
          <ATRACO:device type="ShowPhoto"/>
        </ATRACO:role>
        <ATRACO:role name="LightControl" type="input"
                     agentMonitored="no" interactionType="no">
          <ATRACO:service type="AmbientLightLevel"
                          triggerValue="On" resetValue="Off"
                          quantity="1" specialRules=""/>
        </ATRACO:role>
        <ATRACO:role name="TriggerLight" type="output"
                     agentMonitored="yes"
                     interactionType="direct">
          <ATRACO:device type="OutputLightLevel"/>
        </ATRACO:role>
      </ATRACO:roles>
    </bpel:process>
```

Listing 8.2 Illustrative workflow with abstract services

8.5 ATRACO Contribution

The Planning Agent (PA) proposes to the Sphere Manager (SM) the service of generating workflows with abstract services that describe what basic activities are to be executed and in which order they are to be executed for achieving user goals. The PA uses some services offered by the Ontology Manager (OM) in order to define the initial world state of the planning problem at hand.

The SM is in charge of (a) binding services provided by IE resources to the abstract services of the abstract service workflows, and then (b) executing and controlling the execution of the executable service workflows that result from the binding. The OM creates the Sphere Ontology which will include references to all the IE resources relevant to the Activity Sphere that have been discovered. More details about SM and OM are given in Sect. 1.3.3.

In the prototype system, the PA integrates some components of PANDA, and also includes a generator of ATRACO-BPEL workflows. The workflows are stored in files to be parsed and used by SM, which creates and executes workflows with concrete services.

8.6 Conclusion

Sphere Adaptation (SA) is a major aspect of adaptation and evolution with AmI environments. In the ATRACO project, we address SA in the context of a distributed, ontology-based framework with a service-oriented architecture approach. The objective is to support users in their home for various everyday activities.

In this chapter, we elaborated the concept of SA within ATRACO. We described how Artificial Intelligence planning techniques contribute to SA for the realization and the structural adaptation of activity spheres (ASs).

An existing formal framework for refinement-based hybrid planning that integrates HTN planning and partial-order causal link planning was presented. In addition, we detailed an existent development platform for hybrid planning, PANDA, which is based on the formal framework. We explained that the large flexibility, the high reasoning capacities, and the important expressive power offered by this formal framework and the development platform are well suited to participating in the composition of services in Intelligent Environments. We introduced the Planning Agent (PA) that encapsulates a search engine based on PANDA and which takes part in the advanced service composition of ATRACO and is responsible for the generation of workflows with abstract services at design time. The workflows are part of ASs and consist of the basic activities to be executed in order to achieve user goals in different and changing environments. At run time, the activities are bound to concrete services dynamically by the Sphere Manager (SM). This approach for the composition of services is indeed powerful, since it can cope with complex and large-scale environments. The adaptations of services and devices are indeed possible due to the dynamic binding of services at run time. The flexibility offered by abstract service workflows prevents the running system from replanning each time a service appears or disappears in the AmI environment.

This chapter also explained how we configure and adapt PANDA to shape the search engine of the PA: particular modules and planning strategies are selected in order to deal efficiently with ATRACO planning problems. The problems are tractable and solved in a couple of seconds at design time, since the number of abstract services is limited and there are few causal interactions between tasks. Finally, we detailed how the PA generates workflows with abstract services in the ATRACO-BPEL language in cooperation with SM and the Ontology Manager, based on particular information included in the ontology of the execution environment.

8.7 Further Readings

A journal paper of Ramos, Augusto, and Shapiro introduces some general ideas about exploiting Artificial Intelligence (AI) for addressing the issues of AmI [11].

In a book, Ghallab, Nau, and Traverso present a survey about AI planning [5]. They explain a broad range of planning approaches. In particular, a book chapter is dedicated to HTN planning.

The PhD thesis of Schattenberg is focused on hybrid planning and scheduling, and it describes in detail the PANDA experimental platform [14].

The most prominent journals that focus on AI are Artificial Intelligence (AI Journal) and Journal of Artificial Intelligence Research (JAIR). The International Joint Conference on Artificial Intelligence (IJCAI), the National Conference on Artificial

Intelligence (AAAI), and the European Conference on Artificial Intelligence (ECAI) are renowned general AI conferences. The International Conference on Artificial Intelligence Planning and Scheduling (AIPS) and the International Conference on Automated Planning and Scheduling (ICAPS) are devoted specifically to AI planning and scheduling.

Appendix

This appendix consists of an excerpt of the planning domain for the activity sphere "prepare an unexpected dinner."

```
<?xml version="1.0" encoding="UTF-8"?>
<!DOCTYPE domainModel SYSTEM "model.dtd">
<domainModel name="Kitchen_threads">
   <sortDefinition name="hotplate" type="concrete"/>
   <sortDefinition name="oven" type="concrete"/>
   <sortDefinition name="microwave" type="concrete"/>
   <sortDefinition name="device" type="abstract">
      <documentation>Locations with additional functionality
          and only suitable for a number of containers.</
          documentation>
      <subSortStatement subsort="hotplate"/>
      <subSortStatement subsort="oven"/>
      <subSortStatement subsort="microwave"/>
   </sortDefinition>

   <!-- ... -->

   <sortDefinition name="resource" type="abstract">
      <documentation>Parent sort to all objects of the
          kitchen subjected to planning.</documentation>
      <subSortStatement subsort="food"/>
      <subSortStatement subsort="equipment"/>
   </sortDefinition>


   <!-- ... -->

   <relationDeclaration name="has_context" type="flexible">
      <sortExpression name="resource"/>
      <sortExpression name="context"/>
   </relationDeclaration>

   <!-- ... -->

   <decompositionAxiom name="clean_or_no_context_axiom">
      <varDeclaration name="coca_res" sort="resource"/>
      <varDeclaration name="coca_cxt" sort="context"/>
      <leftHandSide>
         <atomic name="clean_or_no_context">
```

```
         <variable name="coca_res"/>
         <variable name="coca_cxt"/>
      </atomic>
   </leftHandSide>
   <rightHandSide>
      <atomic name="clean">
         <variable name="coca_res"/>
      </atomic>
      <not>
         <atomic name="clean_or_no_context">
            <variable name="coca_res"/>
            <variable name="coca_cxt"/>
         </atomic>
      </not>
   </rightHandSide>
</decompositionAxiom>

<!-- ... -->

<taskDeclaration name="move_container" type="primitive">
   <documentation>Move a container from one non_storage
       location to another.</documentation>
   <varDeclaration name="move_container_obj" sort="
       container"/>
   <varDeclaration name="move_container_from" sort="
       non_storage"/>
   <varDeclaration name="move_container_to" sort="
       non_storage"/>
   <and>
      <atomic name="free">
         <variable name="move_container_to"/>
      </atomic>
      <atomic name="ready">
         <variable name="move_container_obj"/>
      </atomic>
      <atomic name="at">
         <variable name="move_container_obj"/>
         <variable name="move_container_from"/>
      </atomic>
   </and>
   <and>
      <atomic name="free">
         <variable name="move_container_from"/>
      </atomic>
      <atomic name="at">
         <variable name="move_container_obj"/>
         <variable name="move_container_to"/>
      </atomic>
      <not>
         <atomic name="at">
            <variable name="move_container_obj"/>
            <variable name="move_container_from"/>
         </atomic>
      </not>
```

```
            <not>
               <atomic name="free">
                  <variable name="move_container_to"/>
               </atomic>
            </not>
         </and>
      </taskDeclaration>

      <!-- ... -->

      <taskDeclaration name="procedure_fry" type="complex">
         <varDeclaration name="p_fry_food" sort="food"/>
         <varDeclaration name="p_fry_container" sort="
             fry_in_here"/>
         <varDeclaration name="p_fry_tool" sort="tool"/>
         <varDeclaration name="p_fry_cxt" sort="context"/>
         <and>
            <atomic name="ready">
               <variable name="p_fry_container"/>
            </atomic>
            <atomic name="ready">
               <variable name="p_fry_tool"/>
            </atomic>
         </and>
         <and>
            <atomic name="has_context">
               <variable name="p_fry_container"/>
               <variable name="p_fry_cxt"/>
            </atomic>
            <atomic name="has_context">
               <variable name="p_fry_food"/>
               <variable name="p_fry_cxt"/>
            </atomic>
            <atomic name="has_context">
               <variable name="p_fry_tool"/>
               <variable name="p_fry_cxt"/>
            </atomic>
         </and>
      </taskDeclaration>

      <!-- ... -->

      <methodDeclaration name="method_procedure_fry" taskRef="
          procedure_fry">
         <varDeclaration name="method_procedure_fry.p_fry_food"
             sort="food"/>
         <varDeclaration name="method_procedure_fry.
             p_fry_container" sort="fry_in_here"/>
         <varDeclaration name="method_procedure_fry.p_fry_tool"
             sort="tool"/>
         <varDeclaration name="method_procedure_fry.p_fry_cxt"
             sort="context"/>
         <taskNode name="mpfmc_into_pan" taskRef="move_food">
```

```
      <varDeclaration name="mpfmc_into_pan.move_food_obj"
          sort="movable"/>
      <varDeclaration name="mpfmc_into_pan.move_food_from"
           sort="container"/>
      <varDeclaration name="mpfmc_into_pan.move_food_to"
          sort="container"/>
   </taskNode>
   <taskNode name="mpfmc_context_container" taskRef="
       set_context">
      <varDeclaration name="mpfmc_context_container.
          scxt_obj" sort="resource"/>
      <varDeclaration name="mpfmc_context_container.
          scxt_cxt" sort="context"/>
   </taskNode>
   <taskNode name="mpfmc_context_food" taskRef="
       set_context">
      <varDeclaration name="mpfmc_context_food.scxt_obj"
          sort="resource"/>
      <varDeclaration name="mpfmc_context_food.scxt_cxt"
          sort="context"/>
   </taskNode>
   <taskNode name="mpfmc_context_tool" taskRef="
       set_context">
      <varDeclaration name="mpfmc_context_tool.scxt_obj"
          sort="resource"/>
      <varDeclaration name="mpfmc_context_tool.scxt_cxt"
          sort="context"/>
   </taskNode>
   <taskNode name="mpfmc_flip" taskRef="use_tool">
      <varDeclaration name="mpfmc_flip.use_tool_tool" sort
          ="tool"/>
      <varDeclaration name="mpfmc_flip.use_tool_on" sort="
          container"/>
   </taskNode>
   <taskNode name="mpfmc_onto_hotplate" taskRef="
       move_container">
      <varDeclaration name="mpfmc_onto_hotplate.
          move_container_obj" sort="container"/>
      <varDeclaration name="mpfmc_onto_hotplate.
          move_container_from" sort="non_storage"/>
      <varDeclaration name="mpfmc_onto_hotplate.
          move_container_to" sort="non_storage"/>
   </taskNode>
   <orderingConstraint predecessor="mpfmc_context_tool"
       successor="mpfmc_flip"/>
   <orderingConstraint predecessor="
       mpfmc_context_container" successor="mpfmc_into_pan
       "/>
   <orderingConstraint predecessor="mpfmc_context_food"
       successor="mpfmc_into_pan"/>
   <orderingConstraint predecessor="mpfmc_into_pan"
       successor="mpfmc_flip"/>
   <orderingConstraint predecessor="mpfmc_onto_hotplate"
       successor="mpfmc_into_pan"/>
```

```
        <valueRestriction type="eq" variable="method_procedure_fry.p_fry_container">
          <variable name="mpfmc_into_pan.move_food_to"/>
        </valueRestriction>
        <valueRestriction type="eq" variable="mpfmc_context_container.scxt_obj">
          <variable name="method_procedure_fry.p_fry_container"/>
        </valueRestriction>
        <valueRestriction type="eq" variable="method_procedure_fry.p_fry_cxt">
          <variable name="mpfmc_context_container.scxt_cxt"/>
        </valueRestriction>
        <valueRestriction type="eq" variable="method_procedure_fry.p_fry_food">
          <variable name="mpfmc_into_pan.move_food_obj"/>
        </valueRestriction>
        <valueRestriction type="eq" variable="mpfmc_context_container.scxt_obj">
          <variable name="method_procedure_fry.p_fry_container"/>
        </valueRestriction>
        <valueRestriction type="eq" variable="mpfmc_context_food.scxt_cxt">
          <variable name="method_procedure_fry.p_fry_cxt"/>
        </valueRestriction>
        <valueRestriction type="eq" variable="mpfmc_context_food.scxt_obj">
          <variable name="method_procedure_fry.p_fry_food"/>
        </valueRestriction>
        <valueRestriction type="eq" variable="method_procedure_fry.p_fry_cxt">
          <variable name="mpfmc_context_tool.scxt_cxt"/>
        </valueRestriction>
        <valueRestriction type="eq" variable="method_procedure_fry.p_fry_tool">
          <variable name="mpfmc_flip.use_tool_tool"/>
        </valueRestriction>
    </methodDeclaration>

    <!-- ... -->

</domainModel>
```

Listing 8.3 Excerpt of the planning domain for AS “Prepare an unexpected dinner”

References

1. Amigoni, F., Gatti, N., Pinciroli, C., Roveri, M.: What planner for ambient intelligence applications? IEEE Trans. Syst. Man Cybern. Part A Syst. Hum. **35**(1), 7–21 (2005)
2. Biundo, S., Schattenberg, B.: From abstract crisis to concrete relief—a preliminary report on combining state abstraction and HTN planning. In: Proceedings of the 6th European Conference on Planning (ECP’01), pp. 157–168. Toledo, Spain (2001). Preprint

3. El Falou, M., Bouzid, M., Mouaddib, A.I., Vidal, T.: A distributed planning approach for web services composition. In: Proceedings of the 2010 IEEE International Conference on Web Services (ICWS), pp. 337–344. IEEE Computer Society, Miami (2010)
4. Elkawkagy, M., Schattenberg, B., Biundo, S.: Landmarks in hierarchical planning. In: Coelho, H., Studer, R., Wooldridge, M. (eds.) ECAI. Frontiers in Artificial Intelligence and Applications, vol. 215, pp. 229–234. IOS Press, Amsterdam (2010)
5. Ghallab, M., Nau, D.S., Traverso, P.: Automated Planning: Theory and Practice. Morgan Kaufmann, Los Altos (2004)
6. Lundh, R., Karlsson, L., Saffiotti, A.: Autonomous functional configuration of a network robot system. Robot. Auton. Syst. **56**(10), 819–830 (2008)
7. Marquardt, F., Uhrmacher, A.M.: Evaluating AI planning for service composition in smart environments. In: Wiberg, M., Zaslavsky, A.B. (eds.) Proceedings of the 7th International Conference on Mobile and Ubiquitous Multimedia (MUM 2008), pp. 48–55. ACM, Umeå (2008)
8. Mazzola Paluska, J., Pham, H., Saif, U., Chau, G., Terman, C., Ward, S.: Structured decomposition of adaptive applications. In: PerCom, pp. 1–10. IEEE Computer Society, Salt Lake City (2008)
9. McDermott, D.V.: Estimated-regression planning for interactions with web services. In: Ghallab, M., Hertzberg, J., Traverso, P. (eds.) Proceedings of the Sixth International Conference on Artificial Intelligence Planning and Scheduling (AIPS), pp. 204–211. AAAI, Toulouse (2002)
10. Pistore, M., Traverso, P., Bertoli, P.: Automated composition of web services by planning in asynchronous domains. In: Biundo, S., Myers, K.L., Rajan, K. (eds.) Proceedings of the Fifteenth International Conference on Automated Planning and Scheduling (ICAPS), pp. 2–11. AAAI, Monterey (2005)
11. Ramos, C., Augusto, J.C., Shapiro, D.: Ambient intelligence–the next step for artificial intelligence. IEEE Intell. Syst. **23**(2), 15–18 (2008)
12. Ranganathan, A., Campbell, R.H.: Autonomic pervasive computing based on planning. In: ICAC, pp. 80–87. IEEE Computer Society, Salt Lake City (2004)
13. R-Moreno, M.D., Borrajo, D., Cesta, A., Oddi, A.: Integrating planning and scheduling in workflow domains. Expert Syst. Appl. **33**(2), 389–406 (2007)
14. Schattenberg, B.: Hybrid planning and scheduling. Ph.D. thesis, University of Ulm, Institute of Artificial Intelligence, Ulm (2009). URN: urn:nbn:de:bsz:289-vts-68953
15. Schattenberg, B., Weigl, A., Biundo, S.: Hybrid planning using flexible strategies. In: Furbach, U. (ed.) Proceedings of the 28th German Conference on Artificial Intelligence (KI). Lecture Notes in Artificial Intelligence, vol. 3698, pp. 258–272. Springer, Koblenz (2005)
16. Schattenberg, B., Bidot, J., Biundo, S.: On the construction and evaluation of flexible plan-refinement strategies. In: Hertzberg, J., Beetz, M., Englert, R. (eds.) Proceedings of the 30th German Conference on Artificial Intelligence (KI). Lecture Notes in Artificial Intelligence, vol. 4667, pp. 367–381. Springer, Osnabrück (2007)
17. Sirin, E., Parsia, B., Vu, D., Hendler, J., Nau, D.: HTN planning for web service composition using SHOP2. Web Semant. Sci. Serv. Agents World Wide Web **1**(4), 377–396 (2004)
18. Smith, D.E., Peot, M.A.: Suspending recursion in causal-link planning. In: Drabble, B. (ed.) Proceedings of the Third International Conference on Artificial Intelligence Planning and Scheduling (AIPS), pp. 182–190. AAAI, Edinburgh (1996)
19. Vuković, M., Kotsovinos, E., Robinson, P.: An architecture for rapid, on-demand, service composition. Serv. Oriented Comput. Appl. **1**, 197–212 (2007). http://dx.doi.org/10.1007/s11761-007-0016-x
20. Zhao, H., Doshi, P.: Haley: A hierarchical framework for logical composition of web services. In: Proceedings of the 2007 IEEE International Conference on Web Services (ICWS), pp. 312–319. IEEE Computer Society, Salt Lake City (2007)

Chapter 9
From Scenarios to "Free-Play": Evaluating the User's Experience of Ambient Technologies in the Home

Joy van Helvert and Christian Wagner

Abstract The ambient intelligence (AmI) vision of interactive home environments with embedded technologies that learn from our behaviour and provide services in anticipation of our needs has been with us since the 1990s. While the technical knowledge and capability to realise the physical aspects of the vision is now within our grasp, user involvement in developing and refining the concepts underlying this new intimate relationship between humans and their technologies appears so far to have been limited. This may, in part, be due to the very nature of the research and innovation process, in that technical competence is often far ahead of the potential users ability to envisage how such services could be usefully and affordably incorporated in their everyday lives. It is understandable therefore that user involvement in the early stages of development may be seen as hindering innovation. Instead, one approach has been to capture the user perspective in visionary scenarios, often written by the researchers themselves. These are useful for elaborating the vision and driving further technical innovation, however, they often assume an established relationship between system and user and thus avoid more mundane issues such as how the user might practically incorporate AmI technologies in his or her everyday life today or in the near future. If we are to make the transition from future vision to present reality we must at some point move away from the visionary scenario and engage with users in the process of evolving our existing home environments to incorporate practical and grounded AmI solutions. This chapter looks first at the notion of user experience and options for the location of AmI evaluation studies in general. It moves on to describe some of the key features of the Adaptive and Trusted Ambient Ecologies (ATRACO) concept and prototype from the user perspective and proceeds to discuss related research. The second part of the chapter describes a preliminary evaluation of

---

J. van Helvert
University of Essex, Colchester CO4 3SQ, UK
e-mail: jvanhe@essex.ac.uk

C. Wagner (✉)
The Computational Intelligence Centre, School of Computer Science and Electronic Engineering, University of Essex, Colchester CO4 3SQ, UK
e-mail: chwagn@essex.ac.uk

© Springer International Publishing Switzerland 2016

S. Ultes et al. (eds.), *Next Generation Intelligent Environments*,
DOI 10.1007/978-3-319-23452-6_9

ATRACO followed by a description of the design for the participant oriented final study that draws on Dervin's Sense-Making approach. The authors conclude that the user experience evaluation has significantly contributed to the development of the ATRACO concepts in a way that ensures they are relevant to everyday users.

9.1 Introduction

Our home environments have changed in the last 100 years. Take, for example, an average British Victorian semi-detached home; while its character and solidity remain unchanged, the patterns of everyday life conducted within it have been revolutionised by wave upon wave of technological innovation. If the house could talk it would tell, for example, of the advent of domestic electricity bringing light and home appliances, in turn liberating women from many domestic chores. It would tell of the transformation of its fire places as sources of heat to central pieces of indoor decoration, and the availability of hot water from the tap leading to the conversion of a bedroom into an internal bathroom and the incorporation of the scullery (separate room for washing dishes and cutlery) into the kitchen. It would tell of the arrival of television and its influence on the time and attention of household members and its effect on the fabric of the family life. It would note the appearance of the telephone, connecting friends and relatives instantly from across the neighbourhood to across the world; and towards the end of the 1990s, it would tell of the rise of the internet and wireless networks, leading to computers in children's bedrooms and the conversion of the loft space to allow its occupants to work from home. Right now the house is witnessing a revolution in personal relationships through on-line social networking and mobile connectivity, and new patterns of living resulting from the convergence of devices, for example, the streaming of media from the computer to the living room TV. Given this continual process of technological transformation, the unavoidable and exciting question is: What inventions, what technology and what stories are in the pipeline for the next 20 years? How will the dynamic relationship between space, lifestyle and technology continue to develop and importantly, how can we steer technological innovation towards maximum benefit for the user?

Researchers in ambient intelligence (AmI) make it their aim to investigate and explore the future potential of technology in and around the home by envisaging spaces that will be sensitive and responsive to human presence and behaviour through the use of networked sensors and microchips embedded in the fabric of the built environment. Based on detailed private profiles we will carry with us, it is predicted these spaces will endeavour to know us personally, they will learn from our patterns of activity, our likes and dislikes, our aims and desires, and seek to support our needs [2, 17, 23].

Their predictions and visions stem from futuristic narrative scenarios, originating in 1990s, that were generated by technologists who could see the potential of embedded networked devices within the home. Their influence has been far-reaching, galvanising research across disciplines such as "electrical engineering, computer science, industrial design, user interfaces and the cognitive sciences" [1].

Narrative scenarios are recognised as a valuable tool and a necessary part of the research process. Based on storytelling, they have the "power to create in our minds an image of a world…so that we almost feel we are there…[They] bring with them a wealth of context, mostly unwritten, from our shared culture…[and provide] a frame of reference in which the reader can evaluate the story as a whole, and make sense of its individual statements…Stories provide an internal logic – of a sequence of events in time; of causality – which is valuable to engineering as it permits, indeed encourages validation" [3]. Scenarios are used extensively in design and engineering to communicate new ideas, stimulate further research, guide user evaluations and communicate with the public.

Used as a visionary tool in the field of AmI, they are also intended to align resources, innovation focus and commercial interest in order to bring the dream into reality. Therefore, it is appropriate to ask why, despite the technology being within our grasp, do the scenarios for AmI in the home remain just out of reach [2, 16, 17]?

One reason may be that the scenarios written by technical visionaries project a landscape of future daily life that ordinary people do not identify with and find difficult to evaluate. Many scenarios assume an established relationship between the system and user and avoid mundane issues such as how the user might practically begin to incorporate the technologies into his or her current daily life routines. Also, in contrast to the straightforward life enhancement offered, for example, by the transition from gas to electric lighting in the 1890s (where some home owners were so eager that they converted their houses long before the power supply was available in their area [11]), the apparent benefits of ambient technologies in the home are multiple, subtle and complex, at times addressing needs that we may have not yet fully articulated to ourselves. Thus, scenarios that paint pictures of frequently all-knowing systems, intervening in every aspect of daily life, can engender impressions of powerlessness and subjugation. In short, it seems there is a disconnection between the way potential users envisage their future home life and the technologists vision; and while potential users are not captivated by a particular technological advance, private investment, research and product development in that sector remains minimal. This, in turn, can inhibit public interest and prevent the kind of self-fuelling adoption we have seen with social networking.

Interestingly Alvin Toffler [20] identified this type of disconnection at a societal level and termed it "Future Shock". He explains the phenomenon in relation to the more widely used term "Culture Shock":

> Culture shock is the effect that immersion in a strange culture has on the unprepared visitor…It is what happens when the familiar psychological cues that help an individual to function in a society are suddenly withdrawn and replaced by new ones that are strange or incomprehensible…Future shock is a time phenomenon, a product of the greatly accelerated rate of change in society. It arises from the superimposition of a new culture on an old one. It is culture shock in one's own society. But its impact is far worse…most people are grotesquely unprepared to cope with it.

If, as Toffler suggests, the man or woman in the street is ill prepared to cope with the new world as envisaged in the AmI scenarios, then perhaps, there is a point at which we need to abandon the predictions and visions and acknowledge users as stakeholders that are directly engaged in the co-creation and evaluation of these new highly personalised environments. After all, as Rodden and Benford [18] point out, the new smart home will evolve from our existing homes and the process will necessarily involve ordinary people. If this is the case, we need new approaches to evaluation that engage users and incorporate understanding of their existing living patterns.

The concern of this chapter is to illustrate how the design of a user-centred evaluation of the Adaptive and Trusted Ambient Ecologies (ATRACO) prototype attempts to address difficulties with scripted scenario-based approaches that manage and confine user response, by adopting an iterative process of co-creation that acknowledges existing patterns of living and integrates the participant's voice alongside that of the researcher and/or technologist.

Specifically, this chapter looks first at the notion of user experience and, intrinsically related, at the options for the setting/location of AmI evaluation studies in general. It moves on to describe some of the key features of ATRACO from the user perspective in the context of user experience and proceeds to discuss related research. The second part of the chapter describes the preliminary evaluation of ATRACO and its limitations followed by a review of the resulting, modified design for the final, participant oriented study. Conclusions are drawn to close the chapter.

9.2 User Experience (UX) and Study Settings

Prototypes and demonstrators[1] are the logical outputs from the technical AmI research process. They provide not only the opportunity to test the technical feasibility of the concepts, designs, interoperation of components and usability of the interface, but also the actual user experience: how the user feels about the system, it's acceptability, it's ease of use and the level of comfort or satisfaction it induces [19]. User experience (UX) has become a strong theme in academic approaches to design and evaluation; it goes beyond the efficient accomplishment of a single task with a single system to consider multiple tasks and/or systems in a broader flow of interaction emphasising the users' perceptions of fulfilment [9, 12, 14, 19, 21]. It is particularly relevant in the context of AmI system evaluation where the number of possible choices presented to the individual user have increased "to a level that no longer allows evaluation of each individual option" [1]. Add this to the high levels of personalisation and adaptability/learning capabilities of AmI systems and it is clear that traditional structured scenario or use case driven approaches that script or predetermine the user's path through the system are no longer adequate.

[1]Prototypes and demonstrators in terms of AmI implementations such as intelligent homes or ambient intelligent subsystems such as intelligent kitchens, intelligent offices, etc.

Law et al. [13] suggest that as yet there is no consensus within the academic community on the nature and scope of UX , although "most. . . agree that it is dynamic, context-dependent and subjective". Hassenzahl in [9] defines experience as:

> . . . an episode, a chunk of time that one went through – with sights and sounds, feelings and thoughts, motives and actions; they are closely knitted together, stored in memory, labelled, relived and communicated to others. An experience is a story, emerging from the dialogue of a person with his or her world through action.

A "user" experience might be described similarly but the dialogue is between the person and a system or device designed to serve a particular set of user needs. The story that emerges is multi-dimensional, dependent not only on the usability and efficiency of the system but also a range of contextual factors such as past experience and attitudes towards technology, cultural and personal beliefs/values and even prevailing mood on the day and time of evaluation. It is difficult to reduce the story to its constituent parts without losing essential information related to their interconnectedness [9].

The importance of context in UX evaluation means the setting of the study is likely to have an impact on the outcome. With regard to AmI in the home, naturalistic household settings allow participants to align their experience with their own everyday life patterns and thus give valuable accounts of what would or would not work for them if they were living with the system on a daily basis. This could mean using the technology in their own homes or alternatively in a simulated home environment such as a "Smart Space" or "Living Lab".

Genuine in-home field studies can be costly and problematic where systems are designed to be embedded in the fabric of the building. Researcher observations in the home can also raise privacy issues and multiple instances of a system located in different participants' homes can generate additional variations in the data. Alternatively, the Living Lab or Smart Space provides a single controlled environment with embedded technologies such as networking and sensors that looks and feels like a home. Such environments are research friendly while the familiarity of the surroundings allows participants to relax and interact with the system in a relatively natural way. The two phases of prototype evaluation discussed here, take place in the setting of a Living Lab.

9.3 The User Experience Within ATRACO

In order to further discuss approaches to evaluation is it necessary to provide some further detail about ATRACO and how the user might experience it. As a Future and Emerging Technology project, the focus of the research has been to develop an underlying technical framework for the development of a symbiotic relationship between the user and her/his home devices and services. It can be described as an ambient ecology consisting of people, context-aware artefacts and digital commodities (e.g. services and content) that can be grouped together in what is referred to as an Activity Sphere (AS), where each AS supports a particular type of user activity; for example, cooking, relaxing, watching TV, etc. The components within each AS are related with each other, and the purpose of a specific AS is to learn and adapt to support a user's activity in a meaningful way; from simple co-operation to "smart", or anticipatory behaviour.

From the perspective of the user, the ambient ecology is the space surrounding and including her or him. It encompasses digital services and physical objects (TV, computer, washing machine, music centre, home security, etc.) as well as devices such as sensors (temperature, light, location, etc.), touch screens and displays. Within a specific AS, all members of the AS are interrelated via the medium of ATRACO which interprets the conditions and activities in the space, and provides appropriate services to the user according to their needs. As a simple example, when the user wants to relax, ATRACO can pool all available relaxation options as part of "relaxation AS" (TV, music, games...), put them on standby and activate them according to the users preferences (e.g. favourite TV station, favourite relaxing music, most played game and so on). The user can initiate any of the options via an integrated interface that recognises several modalities of interaction such as physical controls (e.g. dimmers, switches), voice and touch screen commands. Also, once an option is selected, ATRACO makes appropriate changes to the environment to support the activity, such as adjusting the lighting or opening/closing the curtains. Importantly, however, functionality within ATRACO is not scripted; in other words, while functionality is made available, it is the user who decides what aspects to engage with in the context of her/his current activity. For example, the lights or curtains are only adjusted as part of a specific AS if ATRACO has previously learnt the user's preference to adopt this specific state (e.g. closed curtain during relaxation).

As such, the AS is a key concept of service delivery, it groups devices and services dynamically according to the type of activity the user wants to perform. "Relax" is one example of an AS, encompassing the appliances and services described above. "Work" is another example; it could encompass a completely different set of appliances and services or it could share some of the same. A sphere can be associated with a particular space or geographical location; for example, relaxation might be centred around the couch or TV area of the living space, but it can also be adapted to other locations. When a sphere is mobile, it adapts its services to the appliances/devices available in the particular space. For example, the bedroom might not be equipped with a TV but may provide a radio which can be incorporated into the "Relax" AS when executed in the bedroom. Similarly, adaptation occurs if a device in the ecology fails: ATRACO aims to continue to support the user activity by transferring the roles adopted by a particular device to another device with compatible attributes. For example, if the user is listening to a particular radio program when the Hi-Fi fails, the system will locate the nearest available device able to emit sound, such as the TV, switch it on and reroute the music to it so that the user can continue the entertainment experience.

It is clear that the underlying framework of ATRACO would potentially allow for a complex matrix of possibilities of interaction between the user, the services and the devices that populate her/his space. At present this vision is in the process of being partially realised in prototype form; some aspects will however remain conceptual. Any evaluation will be required to address the acceptability of this abstract view as well as the constituent parts embodied in the prototype. While there are a wide range of assessment approaches under the heading of UX , it is necessary to consider what is relevant to the particular ATRACO context.

9.4 Related Work: Assessing the UX of AmI

Considering approaches that have been used to asses AmI systems, several have focused on evaluating UX in terms of quality metrics. For example, Wang et al. [22] consider user group experience in an AmI or Smart Museum setting and aim to control context by classifying groups according to the relationships between the individuals within them (homogeneous groups are assumed to have equal relationships, heterogeneous groups are assumed to have unequal/hierarchical relationships and loosely coupled groups are assumed to be strangers/no relationship). They propose a framework for the evaluation of group experiences using metrics such as user rating (on a scale from one to ten) and duration of user attention in relation to group classifications. They claim the framework can also be deployed in the evaluation of home settings such as the "optimisation of family TV viewing". While this is a useful approach for systems that perform with a tightly defined set of services for users in a group, it can be argued that it reduces real-world complexity to an extent that cannot adequately represent the continually evolving and multi-dimensional user-system relationship as it exists in a real-world setting and as is partially incorporated by ATRACO.

In a contrasting approach, Mourouzis et al. [15] combine contextual and psychological perspectives in a heuristic framework for evaluating UX in terms of the extent to which an interactive product is user-oriented. They claim their approach is suitable for diverse user groups such as those with different cultural backgrounds or with different levels of physical ability. Their framework focuses on three groups of metrics:

1. The characteristics of the user (gender, physical and cognitive abilities, language, culture, etc.).
2. The context of system use (tasks, social and environmental conditions).
3. The user's behavioural situation (perceived usefulness and perceived ease of use).

In addition, it incorporates the notion of progressive levels of usage from discovering the product, through to exploring it at a high level to using it in depth for specialised purposes. Although comprehensive in its recognition of many of salient aspects of user experience , including a focus on both the internal state of the participant and the context of use, the researchers' choice of metrics/questions still limits and controls the participant's response, i.e. the full scope of the participant's perspective is lost. However, the concept of exploring how the user progresses or becomes familiar with the system is highly relevant to the evaluation of AmI environments and in particular to the learning and anticipatory elements of ATRACO. It shifts the focus of inquiry from the participant's state after completing the experience to what happens in the process of moving through it.

Hole and Williams [10] are also concerned with what happens throughout the process of user–system interaction. They use an Emotion Sampling Device (ESD) which can be accessed by mobile phone or PDA to gather the participants' emotional responses each time they hit a positive or negative event during the period of interaction with the system. Participants register their feelings by answering a set of questions which are then analysed to identify the emotions experienced. According to the authors, Emotion Sampling aims to provide insight into the "hidden reasons for users' responses"; their "love/hate/tolerate" moments which the authors claim can provide useful indications for improving specific aspects of a product or service. This approach has a significant degree of participant orientation in that there are no restrictions on the way the system is used and the participant is empowered with identifying her/his own moments of emotional upheaval or salience. However, in the context of the aims of this chapter, it can be argued that the approach is limited as it is the researchers who define and categorise the actual emotion, not the participant. Additionally, the approach does not consider user characteristics or context of use. It would therefore be used most appropriately in conjunction with other techniques.

Ethnography is a research strategy used predominantly in Sociology and Anthropology that has more recently been applied to aspects of user inquiry in both academic and commercial technology research. It is concerned with a holistic understanding of the way groups of people live; their material, cultural and spiritual practices. Koskela and Vaananen-Vainio-Mattila [12] use an ethnographic approach to design and evaluate three prototype interfaces (for laptop, remote control and mobile phone) to provide access to a range of standard smart devices (automated curtains, status aware plant pots, etc.). The process starts with a highly user-oriented requirements elicitation phase involving contextual inquiry, home interviews that identify people's living patterns and focus groups within which participants are allowed to choose options for the development of the user interfaces (UIs). Outcomes from the focus groups are also used to shape the evaluation study. The smart devices with their prototype UIs are installed in an ordinary apartment. Following this, two participants, selected for their neutral approach to new technology, are asked to live in the space and use the interfaces as part of their normal patterns of living. The authors study the participants' device usage over a 6 month period by monitoring their interactions with the UIs and conducting contextual interviews as well as participatory walkthroughs. The study shows that incremental smart additions to existing home devices are welcomed by the participants; an outcome supporting the argument that progress towards the AmI home will be evolutionary. It also reveals additional requirements particularly with respect to tailoring UIs to different types of activity patterns. Overall, this study demonstrates a high degree of user-orientation, including elements of co-creation and the careful preparation of the prototypes to fit into everyday life patterns, leading to a focussed and productive evaluation. However, the number of participants was small and the 6 months duration of the experiments, while highly valuable, would be difficult to reproduce as a general approach as significant resources are required to sustain long-term studies. Nevertheless, we feel that the inquiry into existing living patterns is of particular relevance to ATRACO as the concept of Activity Spheres is intended to support natural behaviour in the home.

These examples illustrate some of the diversity of approaches to UX evaluation. Of course, each study depends on numerous local conditions such as the stated aims of the study (more requirements—usability scores), the time and resources available, whether it is a research exercise or refining of a near market product, etc. While none of the approaches detailed above are wholly suitable to the evaluation of ATRACO, they provide insights to help shape a user-oriented study within the ATRACO context. Specifically, the inclusion of user background/characteristics, existing living pattern inquiry, a focus on what happens throughout the experience as opposed to end point satisfaction, and empowering the user to identify her/his own moments of emotional upheaval or salience were identified as highly relevant.

9.5 Own Approach to User Evaluation

ATRACO is an EU funded project focussed predominantly on developing the underlying standards and frameworks to deliver an intelligent home environment based on the concept of the ASs (as described above). The components of the prototype are being developed in the various European partner institutions and are subsequently integrated and installed in the University of Essex's Living Lab, referred to as the "iSpace". From the outset of the project, prototype development and evaluation has been considered as an iterative and bi-directional process. At the time of writing, a preliminary prototype with limited functionality has been developed, deployed and evaluated and the outcomes are being used to shape both the final prototype and the design of the corresponding evaluation study which is scheduled to take place in early 2011. In this part of the chapter we will describe the iSpace evaluation environment, the context of the ATRACO evaluation, and proceed to give details of the preliminary evaluation and its outcomes. Following on from this, we will present the outline of the final study.

9.5.1 The iSpace

Simulated living spaces or "Living Labs" are becoming more commonplace around the globe and the research carried out within them ranges from the longer term visionary experimentation to the refining of commercial products prior to launch [7]. Fowler et al. define a Living Lab as

Fig. 9.1 The iSpace at the University of Essex, view of the main living room area

> . . . an environment that is designed to support innovation through co-creation and evaluation of products and services being used in realistic but familiar contexts.

The University of Essex iSpace (see Figs. 9.1 and 9.2) is a simulated home environment comprising a two bedroom apartment with specially designed walls and ceilings able to conceal the extensive networking infrastructure. There are numerous sensors and networked artefacts throughout, including location tracking systems and pressure sensitive furniture. The core space is an open-plan kitchen, sitting and dining area which generates a relaxed modern home ambience. It includes appliances such as two large plasma TV's, a music system, light sensors, remotely operated curtains and a networked picture frame. The main bedroom is also equipped with a touch screen enabled flat screen, and the smaller bedroom doubles as an office containing a desk and computer alongside a fold-away bed.

Importantly, the iSpace provides ATRACO with an appropriate test-bed for the integration of the prototypes and their subsequent user evaluations. It enables participants to interact with and experience ATRACO in a relaxed and natural setting that they can begin to imagine being their own home. This in turn helps enable and stimulate co-creative interview responses and focuses group discussions to provide a rich data resource for user experience evaluation.

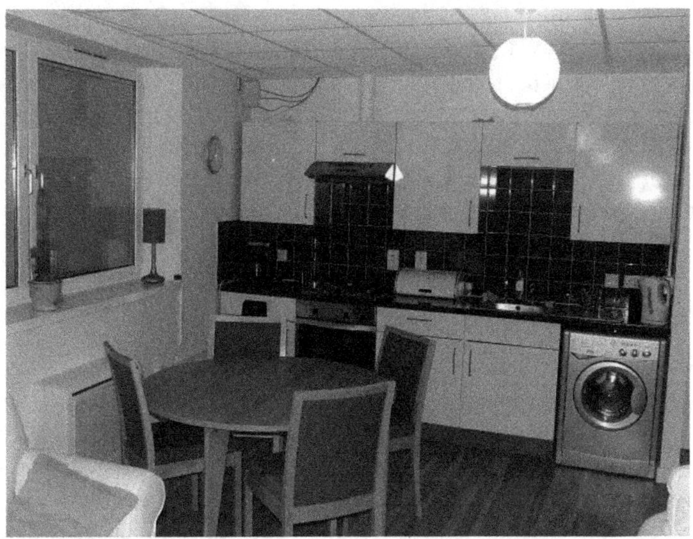

Fig. 9.2 The iSpace at the University of Essex, view of the open-plan kitchen area

9.5.2 The ATRACO Evaluation Context

As each partner institution in the ATRACO project is responsible for developing a different component of the prototype system, it was important to ensure the broad design of both rounds of evaluation were produced collaboratively involving all the relevant stakeholders. At the beginning of each iteration, the components had not yet been integrated and given that ATRACO offers multiple possibilities for supporting daily living, there was a certain amount of flexibility in the nature and scope of the prototype design. This meant that the development of both prototypes and the activities required to evaluate them were/are interdependent to a certain degree, with the broad design for each emerging through informal rounds of collaboration between the partners.

In the lifecycle of a commercial product, user evaluation would generally be conducted once the core technology was robust and the user interface reasonably polished, allowing participants to get a feel for a device or service that may be incremental to technologies they are already familiar with and could envisage owning or subscribing to. In contrast, as a research project, ATRACO's aims are longer term and more focused on the exploration of visionary concepts, i.e. less focussed on an immediate product. This presents a challenge for evaluation design as participants with limited technical expertise might not be familiar with the concepts or technologies deployed and could potentially find it difficult to imagine how they would be relevant to their lives. Furthermore, terms of reference for the project meant that prototype development was focussed primarily on the underlying technical and conceptual frameworks allowing interoperability between devices—the user interface was only a secondary concern. Participants would therefore experience ATRACO through an unrefined front end, again making it potentially more difficult to envisage its benefits. Finally, for various reasons and in particular time constraints, a longitudinal study involving participants living in the iSpace was/is not an option for either the preliminary or final evaluation. This meant that it was necessary to find alternative approaches to evaluate the learning and user experience aspects of ATRACO.

9.5.3 Preliminary Evaluation

As a future and emerging technologies research project with no pre-existing user base, the user requirements for ATRACO were driven by a number of visionary narrative scenarios compiled by members of the project during its initial phase. In the first round of evaluation, these scenarios were simplified into five vignettes (short interaction scenes) illustrating a range of possibilities for the deployment of ATRACO in the home. Due to its early stage of development, the prototype was only able to support the specific interactions defined in the scenarios. Figure 9.3 shows the layout of the iSpace and the devices available to ATRACO.

Each vignette prescribed either a single interaction or a flow of interactions between the participant and the system. In this first stage of the evaluation, individual participants were introduced to the iSpace and asked to complete the five vignettes imagining that the iSpace was their home. A short interview was conducted after each vignette in which the participant was asked questions about

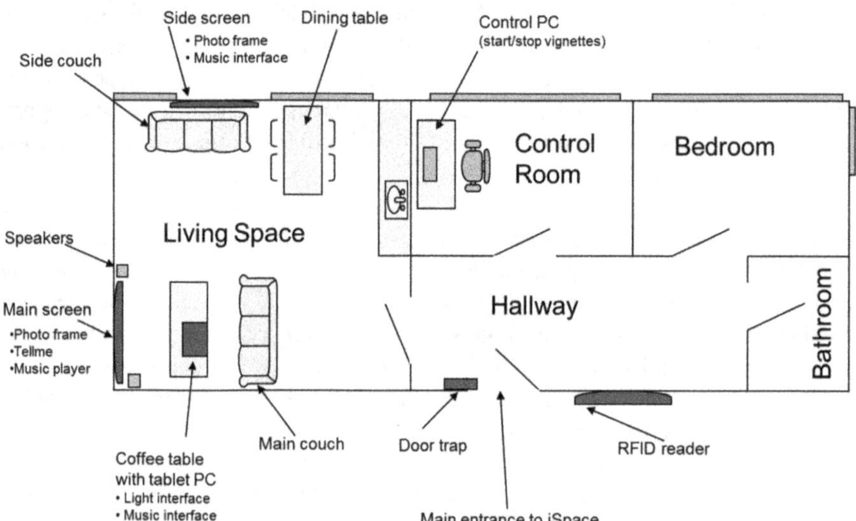

Fig. 9.3 The iSpace layout for the preliminary study

the appropriateness of the system response in relation to specific interactions within the vignette. This was followed by some broader questions about their overall experience. The approach attempted to build a representative image of what the participant experienced throughout each vignette as well as their overall satisfaction. The specific vignettes investigated were the following:

- **Arriving home after work:** The participant starts outside the iSpace. She/he holds an RFID tag to the reader, the door unlocks and the participant enters the hallway and then walks into the living space. The voice interaction system says "please adjust the lights" as the participant walks towards the main couch. The lights are off (there is sufficient light from the hallway to see but the light level is not pleasant for everyday use). The participant sits on the main couch and adjusts the lights using a touch screen interface on a tablet PC located on the coffee table. The system asks "would you like to view photos?" If the participant replies with a "yes", a rolling photo display appears on the main screen. The system proceeds to ask "would you like some music?" On a positive response, the music starts to play. Vignette ends.
- **Guest and privacy:** The participant sits on the sofa looking at holiday photographs on the TV. The tablet PC on the coffee table displays a touch screen interface allowing the participant to control the photo viewing using "next" and "previous" options. Some of the photographs have been marked as "public", i.e. anyone can see them, others are "private" and as such marked only for the "consumption" of the owner. The researcher is outside the iSpace (playing the part of an arriving guest). After a couple of minutes the researcher places her/his RFID tag against the reader outside the front door. The system announces to the participant that "X has arrived" (where X is the researchers name) and asks "would you like her/him to enter?" If the participant responds by saying "yes", the door opens and the researcher enters the iSpace—as she/he does so, the picture set is modified to display only "public" pictures; the private pictures are removed from the slideshow. Vignette ends.
- **Photo frame adaptation:** The participant sits on the sofa viewing holiday photos displayed in the photo frame on the main screen. The tablet PC on the coffee table displays a touch screen interface allowing the participant to control the photo viewing using "next" and "previous" options. The researcher is in the second bedroom/office of the iSpace. After a few minutes the researcher presses a button that simulates a failure of the photo frame device on the main screen. The system autonomously adapts to the failure by moving the photo display to the side screen. Vignette ends.
- **Follow me:** The touch screen music interface is displayed on the side screen. Holding a tag, the participant walks from the doorway to the side screen and starts playing some music by touching the appropriate symbol(s) on the user interface. She/he then move across the room to sit on the sofa in front of the main screen. After a few seconds the interface autonomously moves from the side screen to the tablet PC on the coffee table in front of the participant. Vignette ends.

- **Light adaptation:** The lights are on (to the level specified by the user during the first vignette) and the lighting control interface is visible on the tablet PC on the coffee table. The participant has been asked to readjust the lights if there is any change in light level (requirement). The participant sits on the sofa with a magazine. She/he adjusts the lights so they are comfortable for reading. The researcher is in the control room/office and after a few minutes presses a button to simulate a partial failure of the lights (i.e. some bulbs fail). The participant re-adjusts the lights to a comfortable level and then continues reading. Vignette ends.

Nine participants took part (in all five vignettes) and in order to provide at least a minimal insight into their background, the qualitative interview data was supplemented by a short questionnaire. It gathered demographic data, such as age, gender, employment and accommodation status, and asked participants to gauge their level of enthusiasm for new technologies by rating themselves out of ten in three areas: personal technologies, work technologies and home technologies. The questionnaire also asked them to tick boxes indicating which technologies they currently owned and which they were aiming to acquire in the next 12 months.

In the second stage of the preliminary evaluation, six of the same participants attended a focus group together in which they were presented with a futuristic scenario based on potential ATRACO capabilities. They discussed various aspects of the scenario in relation to their own visions of future home living.

The interview and focus group data were analysed in terms of themes, resonances and recurrences with the context provided by the questionnaire and demographic data. Standard ethical practice in relation to human research participants was followed throughout.

9.5.4 Preliminary Evaluation Outcomes and Reflections

The initial study found that the majority of participants were positive about the notion of living with some aspects of AmI in the home, although two participants rejected it altogether. Key concerns were:

- Maintaining control over the environment.
- The changeable nature of human moods in relation to an electronic presence.
- The erosion of everyday life skills and domestic cultural norms and rituals.
- The risk of data loss and unauthorised access to the system.

Overall, the study produced valuable data to help shape the final prototype and evaluation approach, including highlighting the importance of ASs and communicating how they are formed and operate as a central concept of ATRACO.

However, reflecting on the study design it was clear that with only minimal and in particular rather scripted exposure to the prototype, participants found it difficult to grasp some of the concepts. It also became clear that embodied in the design of the scenarios and vignettes were many assumptions about the user's acceptance of the vision. The constrained and scripted path of interactions meant that participants found it hard to relate their experience to their own patterns of living, particularly with respect to the learning and adaptation aspects of ATRACO. Also, a lack of any illustration of how they might establish and personalise such a system for their own home gave rise to feelings of being controlled by the technology.

9.6 Design of the Final Evaluation Study

Building on the findings of the preliminary study, a key feature of the final prototype was identified to be that it should support "free-play"; in other words, the unconstrained use of the system within a particular AS. This means the evaluation design can move away from scenarios and vignettes, allowing participants to explore freely the full range of functionality within the sphere according to their own instinctive patterns of thought and action. It is hoped that this will enable them to more closely relate the functionality of ATRACO with their current patterns of activity within the home and therefore overcome feelings of alienation. The prototype will also include a user interface that will allow individual participants to set up and personalise the system according to their own requirements, including for example, their personal choice of music and photos or choice of male or female speech interface, etc. Again, it is hoped that this will promote a feeling of connection and ownership with the participants.

The format of the final study will be AS focussed and iterative, with participants returning for up to four separate sessions to "play" with different ASs or to experience the same AS with multiple users. This is intended to support participants in moving from a superficial to a deeper understanding of ATRACO (and its underlying concepts), and in particular, to allow them over time to experience the "smart" or learning and anticipatory aspects of the system. Demographic details and metrics relating to participant's disposition towards technology will be collected in the same way as in the earlier study.

9.6.1 Approach to Data Collection

The move away from the structured "walkthrough" approach taken in the preliminary evaluation study and allowing the participant uninterrupted "free-play" with the prototype raises the question of how to gain insight into what moments of understanding, struggle, anxiety, pleasure, etc. emerge throughout this much more interactive experience. In this context, we refer to Dervin's Sense-Making [6] which is a methodology drawn from the field of communications and grounded in the paradigm of phenomenology. It provides a framework for investigation based on moments when the making of "sense" is interrupted, made or remade in communication. It departs from other approaches by theorising humans as being able to reflect on their own experiences in a structured way that emphasises the voice of the participant alongside that of the researcher.

In developing the ontological and epistemological assumptions of Sense-Making, Dervin draws primarily on the work of Richard F. Carter [4, 5] and his assumptions about the discontinuity of reality, "as well as ideas suggested by Giddens [8]". In this framework, human experience is seen as being pervaded by "gaps". These exist across time, between entities (human, system or otherwise) and between spaces, as Dervin [6] explains:

> This discontinuity condition exists between reality and human sensors, between human sensors and the mind, between mind and tongue, between tongue and message created, between message created and channel [mode of communication], between human at time one and human at time two, between human one at time one and human two at time one, between human and culture, between human and institution, between institution and institution between nation and nation and so on. Discontinuity is an assumed constant of nature generally and the human condition specifically.

As depicted in Fig. 9.4, the human subject in Sense-Making is seen as moving through time-space, possessing an innate need to "bridge" any gaps that are encountered (here, "bridging" is the act of communicating, either through internal dialogue or external interaction). The subject is also seen as being "situated in

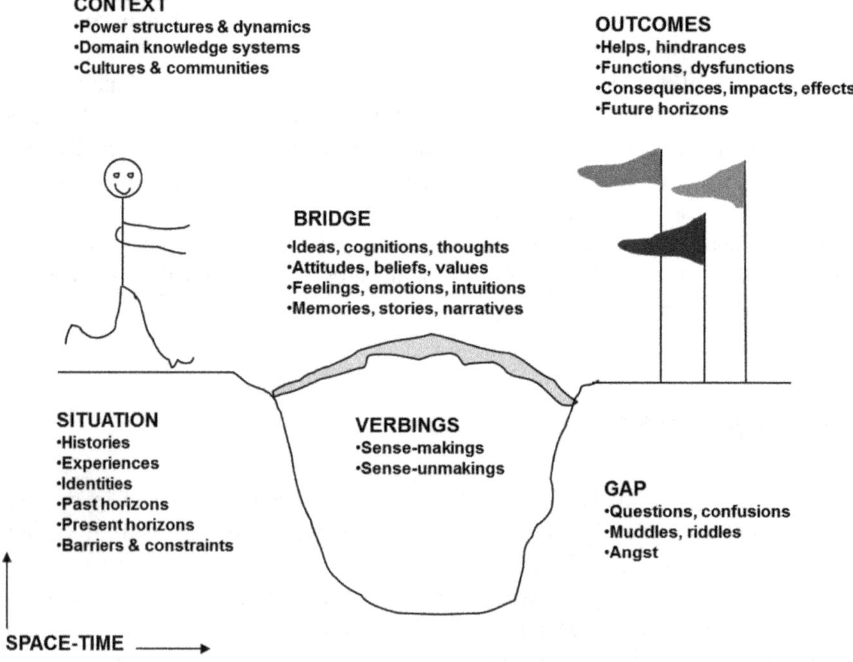

Fig. 9.4 Sense-Making Metaphor (reproduced with permission from Dervin et al. [6])

cultural/historical moments in time-space and that culture, history, and institutions define much of the world within which... [she/he] lives" [6]. At the same time, the actor is assumed to construct her/his own personal sense of her/his relationship to such phenomena drawing on and interpreting her/his existing knowledge. Gaps then, are not rigidly or objectively defined, they are essentially personal moments of struggle, angst or uncertainty; moments where sense cannot immediately be made; moments of reaching out for clues from past experience, from current context or from future expectations, dreams or aspirations.

From a Sense-Making perspective, gathering differing accounts of gaps faced and gaps bridged in relation to the same phenomenon may reveal insight, not into the phenomenon itself, but into the processes, patterns, themes and recursivities relating to the "experience" of interaction with the phenomena.

Sense-Making, which we deem highly relevant to the UX evaluation requirements of ATRACO, is implemented through a specific approach to interview which as such, in the context of ATRACO we propose to conduct as follows:

- Prior to each interaction session "training" or talking to the participant about being conscious/mindful of her/his own "gaps" as they move through the experience.
- The free-play session then takes place and is recorded on video.
- The researcher and the participant then replay the video together. The participant is asked to provide a commentary explaining her/his actions and to stop the video when a gap moment is reached. At each gap moment, the researcher encourages the participant to elaborate in detail, describing the nature of the gap and connecting it to her/his past experience (e.g. existing patterns of living and interaction with personal technologies), future aspirations, attitudes, values, etc. The participants are also asked to suggest ideal solutions or preferences in relation to aspects of the system that they do not feel comfortable with.
- The interview concludes with a set of more general questions about the AS concept and the overall experience.

It is envisaged that this approach will result in rich data that include detailed descriptions of the participants' responses as they interact with the system allied to contextual understanding. Subsequent analysis should provide insight into the practical, socially situated usage of each AS, its acceptability and a range of potential user needs that can help inform further refinement of the system and/or its concepts.

Possible drawbacks to the approach may include difficulty in getting participants to think and talk in terms of gap moments. This depends largely on the prior effort of the researcher to communicate the concept effectively. Similarly, the success of the interview is dependent on the skill of the interviewer in allowing participants to elaborate on their personal context while encouraging them to remaining focussed on the task of evaluation.

9.7 The ATRACO Contribution

As part of ATRACO, the user experience evaluation has been an essential tool, not simply as a posterior measure of the quality of the designed system and proposed concepts but more importantly, as an active part of the iterative process of developing and shaping the concepts and subsequent prototypes which were implemented to expose, visualise and evaluate ATRACO. As such, while the initially designed scenarios gave structure to the early stages of the project and particular its implementation, the gradual move towards a more "free" and responsive environment where the user is free to explore her/his augmented environment as part of specific ASs directly guided the implementation and design of the ATRACO components and subsequently their integration towards the final prototype. At the time of writing the integration of this final prototype is in its final stages and it is expected to provide a very rich environment to evaluate the ATRACO concepts, the underlying technologies as well as the user experience.

Finally, it should be noted that the direct involvement of real (lay) users as part of the user evaluation has given the ATRACO project the unique opportunity to ensure the development of concepts and components which are directly useful and acceptable to end-users, a criterion which should facilitate future real life adoption of some of the concepts explored.

9.8 Conclusions

Emerging technologies such as AmI are challenging traditional requirements elicitation and evaluation practices. For example, as discussed earlier, the complexity and personalisation of many AmI systems make it impossible to evaluate every possibility (of interaction, system state,...) with the user. This has shifted the emphasis from "optimum performance" as an objective measure, towards the more subjective notion of user experience or user satisfaction. In addition, the advent of Smart Spaces or Living Labs provides a new setting, combing aspects of the controlled clinical environment of the laboratory and the naturalistic but costly and unpredictable field trial.

The development of the ATRACO prototypes and their installation in the iSpace at the University of Essex, UK, has created a valuable opportunity to explore a new user-centred, cross disciplinary approach to evaluation in a simulated home environment. As part of ATRACO, the user experience evaluation has significantly contributed towards guiding the development of the ATRACO concepts (e.g. ASs and their application) and prototypes. It has allowed the project to not only provide advancements in technological, conceptual and visionary terms, but also to validate them by relying on real users, thus ensuring that the specific advancements are relevant to everyday users. This approach in turn is expected to become common practice in AmI research and drive the increasingly prominent real-world adoption of concepts and technologies developed in a research context.

Most importantly, the user experience evaluation studies conducted within ATRACO have highlighted the shortcomings of existing post-implementation, scenario-based user evaluations and have clarified the need for a significantly more interactive approach to user requirements elicitations, design and implementation as well as user experience evaluation. As an initial means of addressing this, the final evaluation as part of ATRACO will be based around the use of Dervin's Sense-Making, an approach borrowed from field of Communications which is, adapted to UX, a completely novel approach as far as the authors are aware. In the near future, the final evaluation will be conducted and the authors will present the details in a forthcoming journal article.

In summary, while visionary scenarios have their place in inspiring innovation and rallying resources, it appears the time is now right to look in the other direction and engage users, such as the occupiers of our Victorian semi-detached home, in a process of identifying the next step towards the AmI home. As Aarts [1] suggests "starting from the other side" is quite difficult because it is hard to obtain validated end-user insights that reveal unmet needs leading to successful introduction of new solutions. Therefore we need more insights into the nature of human behaviour. It is hoped the use of a Sense-Making approach to evaluate ATRACO will contribute to our understanding of how to achieve this. Finally, compared to the 1890s and the introduction of domestic electricity, the incredible social connectivity of our age may mean that once that steps are identified and the benefits articulated, people could very quickly move towards adoption—and then the house will have new stories to tell.

9.9 Further Readings

The reader might be interested in Brenda Dervin and Lois Foreman-Wernet with Eric Lauterbach (Eds.) (2003) Sense-Making Methodology Reader: Selected writings of Brenda Dervin. Creskill NJ: Hampton Press, and Marc Hassenzahl (2010) Experience Design: Technology for All the Right Reasons. San Rafael CA: Morgan and Claypool. Furthermore Malcolm McCullough (2004) Digital Ground: Architecture, Pervasive Computing, and Environmental Knowing. Cambridge, MA: MIT Press and Sharlene Nagy Hesse-Biber (in press 2011). The Handbook of Emergent Technologies in Social Research. New York: Oxford University Press provide more in-depth information.

References

1. Aarts, E.: Ambient intelligence: basic elements and insights. Inf. Technol. **50**, 7–12 (2008)
2. Aarts, E., Marzano, S. (eds.): Cultural Issues in Ambient Intelligence. 101 Publishers, Rotterdam (2003)

3. Alexander, I., Maiden, N. (eds.): Introduction: Scenarios in Systems Development. Wiley, Chichester (2004)
4. Carter, R.: Discontinuity and communication. In: Proceedings of the Conference on Communication Theory East and West. East-West Centre, San Francisco (1980)
5. Carter, R.: What does gap imply? In: Proceedings of the Annual Conference of the International Communication Association, San Francisco (1989)
6. Dervin, B., Foreman-Wernet, L., Lauterbach, E.: Sense-Making Methodology Reader: Selected Writings of Brenda Dervin. Hampton Press, Cresskill (2003)
7. Fowler, C., O'Neil, L., van Helvert, J.: Living Laboratories: Social Research Applications and Evaluation. Oxford University Press, New York (2011)
8. Giddens, A.: The Constitution of Society: Outline of the Theory of Structuration. Polity Press, Cambridge (1984)
9. Hassenzahl, M.: Experience Design: Technology for All the Right Reasons. Morgan and Claypool, New York (2010)
10. Hole, L., Williams, O.: Gaining insight into the user experience. Internet draft (2007). www.olliewilliams.co.uk/research/paper-2.pdf
11. Institution of Engineering and Technology: Lighting the Home. On-line exhibition (2010). http://www.theiet.org/about/libarc/archives/exhibition/domestic/lighting.cfm
12. Koskela, T., Vaananen-Vainio-Mattila, K.: Evolution towards smart home environments: empirical evaluation of three user interfaces. Ubiquit. Comput. **8**, 234–240 (2004)
13. Law, E.L.C., Roto, V., Hassenzahl, M., Veerman, A., Kort, J.: Understanding, scoping and defining user experience: a survey approach. In: Proceedings of the CHI 2009, Boston, pp. 719–728 (2009)
14. McCullough, M.: Digital Ground: Architecture, Pervasive Computing and Environmental Knowing. MIT Press, Cambridge (2004)
15. Mourouzis, A., Antona, M., Boutsakis, E., Stephanidis, C.: A user-orientation evaluation framework: assessing accessibility throughout the user experience lifecycle. In: Proceedings of the International Conference on Computers Helping People with Special Needs, Linz, pp. 412–428 (2006)
16. Mukherjee, S., Aarts, E., Doyle, T.: Special issue on ambient intelligence. Inf. Syst. Front. **11**, 1–5 (2009)
17. Riva, G.: The Psychology of Ambient Intelligence. IOS Press, Amsterdam (2005)
18. Rodden, T., Benford, S.: The evolution of buildings and the implications for the design of ubiquitous domestic environments. In: Proceedings of the SIGHCI 2003 Conference on Human Factors in Computing Systems, Fort Lauderdale, pp. 9–16 (2003)
19. Rogers, Y., Preece, J.: Interaction Design: Beyond Human-Computer Interaction. Wiley, Chichester (2007)
20. Toffler, A.: Future Shock. Bantam Books, New York (1970)
21. Vaananen-Vainio-Mattila, K., Roto, V., Hassenzahl, M.: Towards practical user experience evaluation methods. In: Proceedings of the 5th COST294-MAUSE Open Workshop on Meaningful Measures: Valid Useful User Experience Measurement (VUUM 2008), Reykjavik, pp. 9–16 (2008)
22. Wang, Z., Zhou, X., Yu, Z., Wang, H., Ni, H.: Quantitative evaluation of group user experience in smart spaces. Cybern. Syst. Int. J. **41**, 105–122 (2010)
23. Wright, S., Steventon, A.: Intelligent Spaces: The Vision, the Opportunities and the Barriers. Springer, London (2006)

Appendix: List of Reviewers

Juan Carlos Augusto University of Ulster, Jordanstown, UK

Abdelhamid Bouchachia Alpen-Adria-Universität, Klagenfurt, Austria

Amedeo Cesta Italian National Research Council, Rome, Italy

Hakan Duman BT Research and Technology, Ipswich, UK

Jérôme Euzenat Institut National de Recherche en Informatique et en Automatique & Laboratoire d'informatique de Grenoble, Grenoble, France

Damianos Gavalas University of the Aegean, Mytilene, Greece

David K. Hunter University of Essex, Colchester, UK

Christophe Jacquet Supélec, Gif-sur-Yvette, France

Frank Kargl University of Twente, Enschede, The Netherlands

Christophe Kolski Université de Valenciennes et du Hainaut-Cambrésis, Valenciennes, France

Andreas Komninos Glasgow Caledonian University, Glasgow, UK

Dimitrios Koutsomitropoulos University of Patras, Patras, Greece

Effie Lai-Chong Law University of Leicester, Leicester, UK

Antonio López University of Oviedo, Gijón, Spain

Ramón López-Cózar University of Granada, Granada, Spain

Elias Manolakos University of Athens, Athens, Greece

Michael McTear University of Ulster, Antrim, NIR, UK

Alessandro Saffiotti University of Örebro, Örebro, Sweden

Michel Sall Trialog, Paris, France

© Springer International Publishing Switzerland 2016

S. Ultes et al. (eds.), *Next Generation Intelligent Environments*,
DOI 10.1007/978-3-319-23452-6

Florian Schaub Ulm University, Ulm, Germany

Vera Stavroulaki University of Piraeus, Piraeus, Greece

Eran Toch Carnegie Mellon University, Pittsburgh, USA

Lorenzino Vaccari European Commission – Joint Research Center, Ispra, Italy

Victor Zamudio Instituto Tecnologico de León, León, Mexico

CPSIA information can be obtained
at www.ICGtesting.com
Printed in the USA
LVHW081101300620
659213LV00003BA/288

9 783319 234519